T0302175

BIOTRANSFORMATION AND METABOLITE ELUCIDATION OF XENOBIOTICS

BIOTRANSFORMATION AND METABOLITE ELUCIDATION OF XENOBIOTICS

Edited by

Ala F. Nassar

A JOHN WILEY & SONS, INC., PUBLICATION

Published by John Wiley & Sons, Inc., Hoboken, New Jersey
Published simultaneously in Canada

Library of Congress Cataloging-in-Publication Data:

Biotransformation and metabolite elucidation of xenobiotics / edited by Ala F. Nassar.
p. ; cm.
Includes bibliographical references and index.
ISBN 978-0-470-50478-9 (cloth)
1. Xenobiotics–Metabolism. 2. Metabolites–Identification. 3. Biotransformation (Metabolism) I. Nassar, Ala F.
[DNLM: 1. Xenobiotics–metabolism. 2. Xenobiotics–pharmacology. 3. Biotransformation–physiology. 4. Drug Discovery. QU 120 B6166 2010]
QP529.B583 2010
615.9–dc22

2010016629

Printed in Singapore

10 9 8 7 6 5 4 3 2 1

To Sam, Ray (Sam 2), Ruth, and JW

CONTENTS

PREFACE

The first seeds for this work were sown in 1994 at the University of Connecticut, when I was a graduate student, trying to analyze metabolites using LC-MS/MS and CE-MS/MS. I could not find a book describing good, modern strategies, techniques, and tools for metabolite identification. After 15 years in the field, working on a variety of drug metabolism studies, I still cannot find a good reference book to guide the drug metabolism scientist. Metabolite identification and characterization have progressed over the years to the point where they now play a major role in drug discovery and development. Clearly, there is a growing need for improving and expanding the education of scientists involved in investigations in these areas. The need for continuing education in the rapidly expanding and dynamic area of metabolite characterization studies is not being met at the university level, and it is often left to lead researchers and managers to teach this critical subject in a piecemeal fashion on the job in the pharmaceutical industries. While this book is written primarily for those scientists involved in designing, developing, and advancing metabolite characterization for drug discovery and development, the information should also be helpful to students, researchers, educators, and professionals in a number of other related fields.

The goal of the work presented herein is to provide a systematic approach for the education of researchers and students at the university level and to improve their knowledge of how and when to perform metabolite identification and characterization. Here we have made an attempt to bring together the most recent FDA guidelines, current tools for metabolite identification, and an outstanding introduction on human biotransformation by authors who not only are great scientists in the field but are also passionate about their work. The material presented here is intended to guide students, as well as teachers and professionals, in fields such as pharmacy, pharmacology, biochemistry, and analytical chemistry. Another group who would benefit from this book includes physicians, along with scientists and managers in the pharmaceutical industry, researchers in universities and R&D settings, and environmental and agrochemical industries. Any one of these people would benefit from a better understanding of the latest tools, methods, and regulatory requirements for performing metabolite characterization in a safe, effective, and efficient manner.

This textbook consists of eight chapters, starting with an excellent discussion on human biotransformation coauthored by one of the leaders in the field, Dr. Andrew Parkinson. Chapters 2 and 3 present a good background on the tools for metabolite identification, followed by Chapter 4, which discusses some strategies for using these tools to improve drug design, providing important examples of how to manipulate the relationship between metabolic stability, PK properties, and toxicity. Chapters 5 and 6 present recent and relevant case studies illustrating novel rearrangements of the DNA alkylating agent laromustine by collision-induced dissociation, along with identification of the *in vitro* metabolite/decomposition products of laromustine. Chapter 7 provides a well-written discussion of strategies for the detection of reactive intermediates in drug discovery and development. Chapter 8 presents a wonderful overview of the most recent FDA guidelines on human metabolites in safety testing (MIST) and regulatory agency needs, along with their impact on drug metabolism research in the pharmaceutical industry.

With great pleasure and respect, I extend my sincere thanks to my hardworking scientific colleagues for their sacrifices of family time, timely responses, excellent contributions, and consistent cooperation. Also, I would like to thank the publishers for making this book available at reduced cost to our scientific colleagues in developing countries to enhance their knowledge in the area of metabolite characterization and to help them in furthering their careers in this very important area of research.

ALA F. NASSAR

CONTRIBUTORS

MARK P. GRILLO, PhD, Pharmacokinetics and Drug Metabolism, Amgen Inc., South San Francisco, CA

TIFFINI N. HENSLEY, PhD, XenoTech LLC, Lenexa, KS

TUKIET T. LAM, Director, FT-ICR MS Resource/PhD, Yale University/WM Keck Foundation Biotechnology Resource Laboratory, New Haven, CT

YONGMEI LI, PhD, Boehringer Ingelheim Pharmaceuticals, Inc., Danbury, CT

GREG J. LOEWEN, PhD, XenoTech LLC, Lenexa, KS

ALA F. NASSAR, PhD, Chemistry Department, Brandeis University, Waltham, MA

BRIAN W. OGILVIE, PhD, XenoTech LLC, Lenexa, KS

BRANDY PARIS, PhD, XenoTech LLC, Lenexa, KS

ANDREW PARKINSON, PhD, XenoTech LLC, Lenexa, KS

J. GREG SLATTER, PhD, Pharmacokinetics and Drug Metabolism, Amgen Inc., Seattle, WA

CHAPTER 1

HUMAN BIOTRANSFORMATION

ANDREW PARKINSON, BRIAN W. OGILVIE, BRANDY L. PARIS,
TIFFINI N. HENSLEY, and GREG J. LOEWEN
XenoTech LLC, Lenexa, KS

1.1 INTRODUCTION

Biotransformation is the enzyme-catalyzed process of chemically modifying drugs and other xenobiotics (foreign chemicals) to increase their water solubility in order to facilitate their elimination from the body. For most orally active drugs, biotransformation is the process that alters the physiochemical properties of a drug away from those properties (namely lipophilicity) that favor the initial absorption of the drug from the gastrointestinal tract as well as its reabsorption from the kidney and gut to those properties (namely hydrophilicity) that favor elimination in the urine and feces. Human biotransformation is of interest for several reasons. First, for many drugs, the rate of biotransformation in the small intestine and liver determines the amount of drug that reaches the systemic circulation. Second, the rate of biotransformation in the liver (and occasionally some extrahepatic tissues) determines, in part, the dose of drug required to achieve therapeutically effective drug levels (which is also influenced by the affinity with which the drug binds to its therapeutic target) and the dosing interval (which determines whether the desired therapeutic goal can be achieved with once-a-day dosing, which for most drugs is highly desirable). Third, the metabolites formed by drug biotransformation must be considered in terms of their therapeutic properties and adverse effects. Fourth, the particular drug-metabolizing enzyme or enzymes involved in the metabolism of a drug can determine the drug's victim potential (i.e., its susceptibility to pharmacogenetic variation in enzyme expression and its interaction with other agents that alter the pharmacokinetics of the drug by inhibiting or inducing that particular drug-metabolizing enzyme or enzymes). The disposition of many drugs is also determined by (or influenced by) their transmembrane

Biotransformation and Metabolite Elucidation of Xenobiotics, Edited by Ala F. Nassar

transport by various uptake or efflux transporters. Like drug-metabolizing enzymes, drug transporters can influence the uptake, distribution, and elimination of drugs, but they do not chemically modify drugs by converting them to metabolites. This chapter focuses mainly on human drug biotransformation with a final section on drug transport.

1.2 BIOTRANSFORMATION VERSUS TRANSPORT

According to the Biopharmaceutical Classification System (BCS), drugs can be categorized into four classes based on their aqueous solubility and permeability (based on the extent of oral drug absorption; Wu and Benet, 2005). As shown in Fig. 1.1, highly permeable drugs (drugs with high oral absorption) belong to Classes 1 and 2, which are distinguished on the basis of their aqueous solubility (which is high for Class 1 drugs and low for Class 2). Class 3 and 4 drugs both have low permeability (poor oral absorption) but with different aqueous solubility (high for Class 3, low for Class 4). As a general rule, biotransformation represents the predominant route of elimination of Class 1 and 2 drugs (i.e., the highly permeable drugs that show high oral absorption). Drugs in Classes 3 and 4 tend to be eliminated unchanged. Whereas metabolism plays an important role in the disposition of Class 1 and 2 drugs, transporters (especially uptake transporters) play an important role in the disposition of Class 3 and 4 drugs. Efflux transporters tend to play an important role in the disposition of drugs with low aqueous solubility (Classes 2 and 4).

Because they are well absorbed from the gastrointestinal tract, BCS Class 1 and 2 drugs contain a large number of orally active drugs. Their high permeability is largely due to their lipophilicity. This same property prevents their

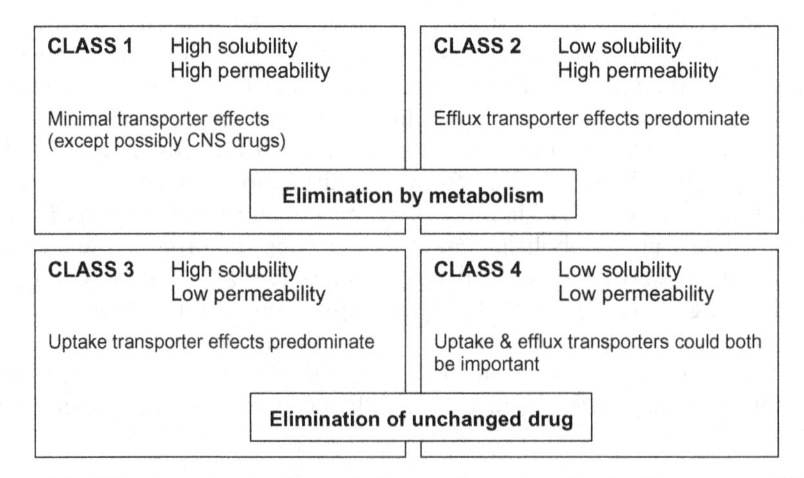

Figure 1.1 The general role of metabolism and transport in the disposition of drugs according to their Biopharmaceutical Classification (BCS).

elimination in urine and feces because, even if eliminated, the unchanged drugs can readily be reabsorbed from the kidney and intestine. For this reason, biotransformation is the predominant route of elimination of lipophilic, high permeable drugs (those in BCS Classes 1 and 2) because their conversion to water-soluble metabolites permits their excretion in urine and feces. As a general rule, biotransformation to water-soluble metabolites is required to eliminate drugs with a $\log D_{7.4} \geq 1$. For highly lipophilic drugs, two or more biotransformation reactions may be required to achieve urinary or biliary excretion.

1.3 PHARMACOKINETIC VARIATION AND THERAPEUTIC INDEX

In a clinical setting, there is always some interindividual variation in systemic exposure to an oral drug, which is usually measured as plasma C_{max} (the maximum concentration of drug in plasma) or AUC (the area under the plasma concentration vs. time curve). This pharmacokinetic variability has three major causes: variation in the extent of absorption (perhaps due to food effects or differences in the extent of first-pass metabolism in the intestine and liver); variation in the rate of systemic clearance (due largely to differences in rates of hepatic uptake and metabolism), and variation in rates of excretion (due to impaired renal or biliary elimination as a result of impaired kidney and liver function, respectively). The variation in systemic exposure due to variation in rates of drug metabolism can arise for both genetic and environmental reasons. Genetically determined differences in the expression of a drug-metabolizing enzyme, such as CYP2D6, give rise to four phenotypes designated as UM, EM, IM, and PM, which respectively stand for ultra-rapid metabolizers (individuals who express multiple copies of the gene encoding the drug-metabolizing enzyme), extensive metabolizers (individuals who express at least one fully functional allele), intermediate metabolizers (individuals who express one or two partially functional alleles), and poor metabolizers (individuals who express no functional alleles and are completely deficient in the affected drug-metabolizing enzyme) (Zineh et al., 2004). Even for enzymes that show little pharmacogenetic variation (such as CYP3A4), considerable variation in enzyme expression can be caused by environmental factors including medication. In the case of drug-metabolizing enzymes like CYP3A4, UMs can be created pharmacologically by enzyme-inducing drugs whereas PMs can be created pharmacologically by inhibitory drugs. This topic is discussed later in the section on Drugs as Victims and Perpetrators. Whatever its cause, the impact of interindividual differences in systemic exposure to an oral drug depends on the magnitude of the pharmacokinetic differences relative to the drug's therapeutic index, as illustrated in Fig. 1.2. When a drug has a large therapeutic index, it is possible that no dosage adjustment is required either to achieve therapeutic efficacy in UMs or to prevent adverse effects in PMs, assuming that the therapeutic and adverse effects are both mediated by

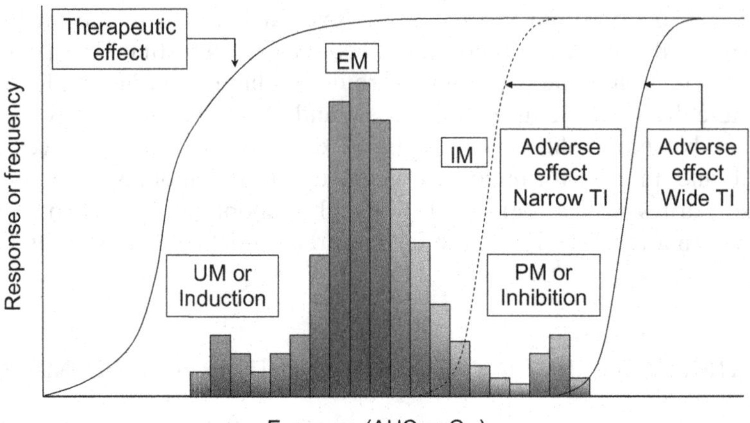

Figure 1.2 The impact of genetic polymorphisms (such as the UM, EM, IM, and PM phenotypes of CYP2D6) and the corresponding impact of enzyme induction or inhibition of a drug-metabolizing enzyme like CYP3A4 on systemic exposure to a drug whose rate of clearance is largely determined by its rate of biotransformation.

the parent drug (which is often but not always the case, as discussed below in the section on Parent Drug versus Metabolite). When a drug has a narrow therapeutic index, the standard dosage may need to be increased in UMs (to achieve a therapeutic effect) or decreased in PMs (to prevent an adverse effect). The U.S. Food and Drug Administration (FDA) defines a narrow therapeutic range as either "less than a 2-fold difference in median lethal dose (LD_{50}) and median effective dose (ED_{50}) values" or "less than a 2-fold difference in the minimum toxic concentrations and minimum effective concentrations in the blood, and safe and effective use of the drug products require careful titration and patient monitoring." An ideal drug is not necessarily a drug that shows no interindividual differences in systemic exposure but rather one with a therapeutic index that encompasses interindividual differences in systemic exposure such that everyone derives therapeutic benefit from the same dose and no dosage adjustment is required for UMs or PMs.

1.4 PARENT DRUG VERSUS METABOLITE

The metabolism of a drug will terminate its pharmacological action and lessen its toxicity when the therapeutic and adverse effects are due to the parent drug. Conversely, the metabolism of a drug will increase its therapeutic and adverse effects when one or more metabolites mediate the pharmacological and toxicological effects. This principle is illustrated in Fig. 1.3 for drugs that are metabolized by CYP2D6, a genetically polymorphic drug-metabolizing enzyme. When the parent drug is pharmacologically active (as in the case of

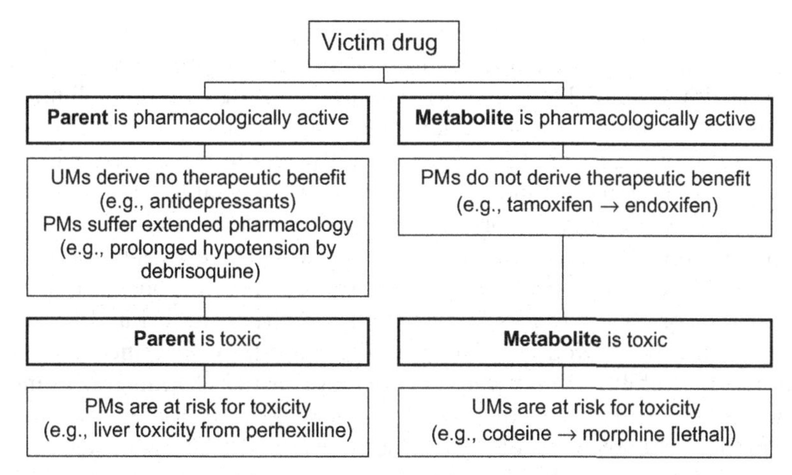

Figure 1.3 The therapeutic and safety consequences of ultra-rapid metabolism (UM) and poor metabolism (PM) of a drug as a function of whether the pharmacological or adverse effects of the drug are mediated by the parent drug or its metabolite(s). The examples shown are drugs metabolized by the genetically polymorphic enzyme CYP2D6.

the antihypertensive drug debrisoquine and various antidepressants), UMs are at risk of deriving no therapeutic benefit whereas PMs are at risk of exaggerated or prolonged pharmacological effects. The converse occurs when the pharmacological effects of a drug are mediated by a metabolite; in this case, PMs are at risk from deriving little or no therapeutic benefit, which is why, for example, CYP2D6 PMs are at increased risk for breast cancer recurrence following tamoxifen adjuvant therapy because these individuals cannot convert tamoxifen to endoxifen, which is 30- to 100-fold more potent than tamoxifen in suppressing estrogen-dependent cell proliferation (Goetz et al., 2005).

When an adverse effect is due to the parent drug (such as the hepatotoxic effects of the antihypertensive drug perhexilline), PMs are at increased risk. Conversely, when an adverse effect is due to a metabolite, UMs are at increased risk. For example, in the case of codeine, which is converted by CYPD6 to morphine (an effective analgesic that causes respiratory depression), CYP2D6 UMs are at increased risk of morphine toxicity. The death of an elderly man administered a recommended dose of codeine and the death of a baby who was breast-fed by a woman on codeine have been attributed to the CYP2D6 UM genotype that causes rapid conversion of codeine to morphine (which causes respiratory depression) (Gasche et al., 2004; Koren et al., 2006).

1.5 DRUGS AS VICTIMS AND PERPETRATORS

From a drug interaction and pharmacogenetic perspective, drugs can be evaluated for their victim and perpetrator potential. *Victims* are those drugs whose clearance is predominantly determined by a single route of elimination, such as metabolism by a single cytochrome P450 (CYP) enzyme. Such drugs have a high victim potential because a diminution or loss of that elimination pathway, either due to a genetic deficiency in the relevant CYP enzyme or due to its inhibition by another, concomitantly administered drug, will result in a large decrease in clearance and a correspondingly large increase in exposure to the victim drug (e.g., plasma AUC). *Perpetrators* are those drugs (or other environmental factors) that inhibit or induce the enzyme that is otherwise responsible for clearing a victim drug. Genetic polymorphisms that result in the partial or complete loss of enzyme activity (i.e., the IM and PM genotypes) can also be viewed as perpetrators because they have the same effect as an enzyme inhibitor: they cause a decrease in the clearance of—and an increase in exposure to—victim drugs. Likewise, genetic polymorphisms that result in the overexpression of enzyme activity (i.e., the UM genotype) can be viewed as perpetrators because they have the same effect as an enzyme inducer: they cause an increase in the clearance of—and a decrease in exposure to—victim drugs. Several drugs whose elimination is largely determined by their metabolism by CYP2C9, CYP2C19, or CYP2D6, three genetically polymorphic enzymes, are victim drugs because their clearance is diminished in PMs, that is, individuals who are genetically deficient in one of these enzymes. Drugs whose disposition is dependent on uptake or efflux by a transporter or on metabolism by a drug-metabolizing enzyme other than CYP can also be considered from the victim/perpetrator perspective. From a drug interaction perspective, *victim* drugs are also known as *objects*, whereas *perpetrators* are also known as *precipitants*.

Reaction phenotyping is the process of identifying which enzyme or enzymes are largely responsible for metabolizing a drug or drug candidate. When biotransformation is known or suspected to play a significant role in the clearance of a drug candidate (which applies to most drugs in BCS Classes 1 and 2, which are well-absorbed drugs with a $\log D_{7.4} \geq 1.0$), then an *in vitro* reaction phenotyping or enzyme mapping study is required prior to approval by the FDA and other regulatory agencies (Tucker et al., 2001; Bjornsson et al., 2003a,b; Williams et al., 2003; US FDA, 2006; Huang et al., 2008). Reaction phenotyping allows an assessment of the victim potential of a drug candidate. The FDA also requires drug candidates be evaluated for their potential to inhibit the major CYP enzymes involved in drug metabolism which can be evaluated *in vitro* with human liver microsomes or recombinant human CYP enzymes (US FDA, 2006; Huang et al., 2008; Grimm et al., 2009). This allows an assessment of the perpetrator potential of the drug candidate. Drugs can also cause pharmacokinetic drug interactions by inducing CYP and other drug-metabolizing enzymes and/or drug transporters, which can be evaluated *in vitro* with cultured human hepatocytes (US FDA, 2006; Huang et al., 2008; Chu et al., 2009).

Terfenadine, cisapride, astemizole, and cerivastatin are all victim drugs, so much so that they have all been withdrawn from the market or, in the case of cisapride, made available only with severe restrictions. The first three are all victim drugs because they are extensively metabolized by CYP3A4 (US FDA, 2006; Ogilvie et al., 2008). Inhibition of CYP3A4 by various antimycotic drugs, such as ketoconazole, and antibiotic drugs, such as erythromycin, decrease the clearance of terfenadine, cisapride, and astemizole, and increase their plasma concentrations to levels that, in some individuals, cause ventricular arrhythmias (QT prolongation and torsade de pointes) which can result in fatal outcomes (Backman et al., 2002; Ogilvie et al., 2008). Cerivastatin is extensively metabolized by CYP2C8. Its hepatic uptake by OATP1B1 and CYP2C8-mediated metabolism are both inhibited by gemfibrozil (actually by gemfibrozil glucuronide), and the combination of cerivastatin (Baycol®) and gemfibrozil (Lopid®) was associated with a high incidence of fatal, cerivastatin-induced rhabdomyolysis, which prompted the worldwide withdrawal of cerivastatin (Ozdemir et al., 2000; Shitara et al., 2004, Ogilvie et al., 2006, 2008; http://www.emea.europa.eu/pdfs/%20human/referral/Cerivastatin/081102en.pdf).

Posicor® (mibefradil) is the only drug withdrawn from the U.S. market largely because of its perpetrator potential (Huang et al., 2008). This calcium channel blocker not only caused *extensive* inhibition of CYP3A4, but it also caused *prolonged* inhibition of the enzyme by virtue of being a metabolism-dependent inhibitor of CYP3A4. By inactivating CYP3A4 in an irreversible manner, such that restoration of normal CYP3A4 activity required the synthesis of new enzyme, mibefradil inhibited CYP3A4 long after treatment with the drug was discontinued.

Victim potential can be quantified on the basis of fractional metabolism according to the following equation:

$$\text{Fold increase in exposure} = \frac{\text{AUC}_\text{i} \text{ or } \text{AUC}_\text{PM}}{\text{AUC}_\text{ui} \text{ or } \text{AUC}_\text{EM}} = \frac{1}{1-f_m} \qquad \text{(Eq. 1.1)}$$

where AUC_i (AUC inhibited) and AUC_ui (AUC uninhibited) are the plasma AUC values of the victim drug in the presence and absence of inhibitor, respectively, and where AUC_PM and AUC_EM are plasma AUC values in genetically determined PMs and EMs, respectively. The relationship between fractional metabolism (f_m) by a single enzyme and the fold increase in drug exposure that results from the loss of that enzyme is shown in Fig. 1.4, which shows that the relationship is not a linear one. (The same principle applies to a drug whose elimination is determined by a transporter, although in this case the term f_e is used instead of f_m). Loss of an enzyme that accounts for 50% of a drug's clearance ($f_\text{m} = 0.5$) results in a twofold increase in AUC, whereas it results in a 10-fold increase in AUC when the affected enzyme accounts for 90% of a drug's clearance ($f_\text{m} = 0.9$). In the case of oral drugs that undergo significant pre-systemic clearance (i.e., first-pass metabolism in the intestine and/or liver), the impact of enzyme inhibition (or the PM genotype) can be twofold: it can increase AUC of the victim drug by (1) decreasing pre-systemic clearance (which increases oral bioavailability) and (2) decreasing systemic clearance. In the case of drugs administered intravenously, enzyme inhibition

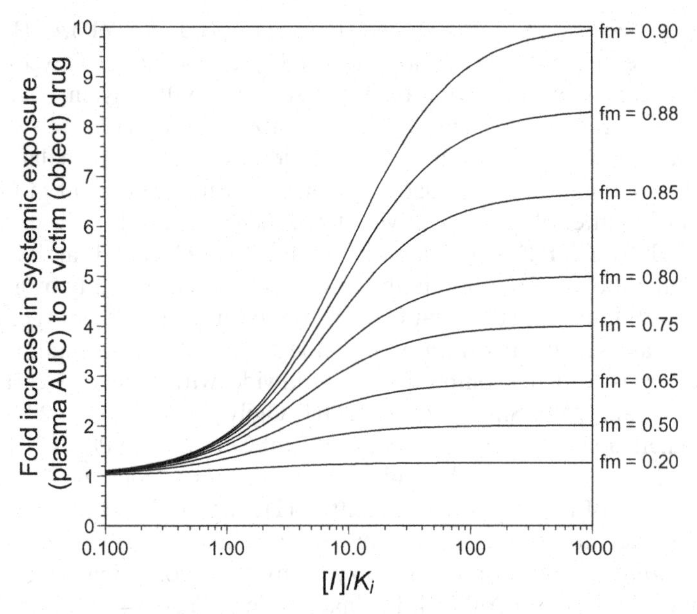

Figure 1.4 The effect of fractional metabolism of a drug by cytochrome P450, $f_{m(\mathrm{CYP})}$, on the theoretical increase in systemic exposure to a victim drug with increasing enzyme inhibition by a perpetrator drug (as reflected by increasing $[I]/K_i$ values).

increases AUC only by decreasing systemic clearance. Consequently, the magnitude of the increase in AUC for certain drugs depends on their route of administration, as illustrated by the interaction between ketoconazole (perpetrator) and midazolam (victim): ketoconazole, a CYP3A4 inhibitor, increases the AUC of midazolam 3- to 5-fold when midazolam is administered intravenously, but it causes a 10- to 16-fold increase when midazolam is administered orally (Tsunoda et al., 1999). The same is true when the perpetrator is an enzyme inducer. For example, the CYP3A4 inducer rifampin decreases the AUC of midazolam by a factor of 9.7 when midazolam is administered orally, but it decreases the AUC by a factor of only 2.2 when midazolam is administered intravenously (Gorski et al., 2003). The difference caused by route of administration is more dramatic for the inductive effect of rifampin on the AUC of nifedipine (12-fold for oral vs. 1.4-fold for intravenous; Holtbecker et al., 1996) and even more dramatic for *S*-verapamil (30-fold for oral vs. 1.3-fold for intravenous; Fromm et al., 1996).

When an enzyme accounts for only 20% of a drug's clearance ($f_m = 0.2$), complete loss of the enzyme activity causes only a 25% increase in AUC, which is normally considered to be bioequivalent (the so-called bioequivalence goalposts range from 80% to 125%, meaning that AUC values within this range can be considered equivalent and, therefore, acceptable). Therefore, an "unacceptable" increase in AUC requires an f_m of greater than 0.2. Actually, the FDA urges the characterization of all elimination pathways that account for 25% or more of a drug's clearance (i.e., $f_m \geq 0.25$).

Fractional metabolism by an enzyme determines the magnitude of the increase in drug exposure in individuals lacking the enzyme, but it does not determine its pharmacological or toxicological consequences. These are a function of therapeutic index, which is a measure of the difference between the levels of drug associated with the desired therapeutic effect and the levels of drug associated with adverse events. For drugs with a large therapeutic index, a high degree of clearance by a polymorphically expressed enzyme is not necessarily an obstacle to regulatory approval. For example, dextromethorphan is extensively metabolized by CYP2D6. Its fractional metabolism is estimated to be 0.93 to 0.98 such that the AUC of dextromethorphan in CYP2D6 PMs or in EMs administered quinidine is about 27–48 times greater than that in EMs (Gorski et al., 2004; Pope et al., 2004). Dextromethorphan has a large therapeutic index; hence, despite this large increase in exposure in CYP2D6 PMs, dextromethorphan is an ingredient in a large number of over-the-counter (OTC) medications. Strattera® (atomoxetine) is another example of a drug whose clearance is largely determined by CYP2D6 ($f_{m(CYP2D6)} \approx 0.9$). Its AUC in CYP2D6 PMs is about 10 times that of EMs, but Strattera has a sufficiently large therapeutic index that it was approved by the FDA in 2002.

Genetic polymorphisms give rise to four basic phenotypes based on the combination of allelic variants that encode a fully functional enzyme (the wild type or *1 allele designated "+"), a partially active enzyme (designated "*"), or an inactive enzyme (designated "–"). These four basic phenotypes are (1) EMs, individuals who have at least one functional allele (+/+, +/*, or +/–), (2) PMs, individuals who have no functional alleles (–/–), (3) IMs, individuals who have two partially functional alleles or one partially functional and one nonfunctional allele (*/* or */–), and (4) UMs, individuals who, through gene duplication, have multiple copies of the functional gene ([+/+]n). This traditional classification scheme has been revised recently on the basis of an activity score, which assigns to each allelic variant a functional activity value from 1 (for the wild type or *1 allele) to 0 (for any completely nonfunctional allele), as reviewed by (Zineh et al., 2004). The basis of the activity score, as it applies to CYP2D6, is illustrated in Table 1.1.

A wide range of activity is also observed for many of the CYP enzymes that have a very low incidence of genetic polymorphisms. This variation arises because CYP inhibitors can produce the equivalent of the PM phenotype whereas CYP inducers can produce the equivalent of the UM phenotype. For example, CYP3A4 shows a low incidence of functionally significant genetic polymorphisms, but PMs can be produced pharmacologically with inhibitors, such as ketoconazole, erythromycin, and mibefradil, whereas UMs can be produced pharmacologically with inducers such as rifampin, St. John's wort, and the enzyme-inducing antiepileptic drugs (EIAEDs), such as phenobarbital, phenytoin, carbamazepine, and felbamate (Parkinson and Ogilvie, 2008).

TABLE 1.1 The Relationship between Genotype and Phenotype for a Polymorphically Expressed Enzyme with Active (wt), Partially active (*x*), and Inactive (*y*) Alleles

Genotype	Alleles	Conventional Phenotype[a]	Activity Score[b]	Activity Score Phenotype[c]
Duplication of active alleles (n = 2 or more)	(wt/wt)n	UM	$2 \times n$	UM
Two fully active wild type (wt) alleles	wt/wt	EM	$1 + 1 = 2$	High EM
One fully active + one partially active allele	wt/*x	EM	$1 + 0.5 = 1.5$	Medium EM
One fully active + one inactive allele	wt/*y	EM	$1 + 0 = 1$	Low EM
Two partially active alleles	*x/*x	EM or IM	$0.5 + 0.5 = 1$	Low EM
One partially active + one inactive allele	*x/*y	IM	$0.5 + 0 = 0.5$	IM
Two inactive alleles	*y/*y	PM	0 + 0 = zero	PM

[a]The phenotypes are ultra-rapid metabolizer (UM), extensive metabolizer (EM), intermediate metabolizer (IM), and poor metabolizer (PM), based on the particular combination of alleles that are fully active (wt), partially active (*x), or inactive (*y).

[b]In the case of CYP2D6 (Zineh et al., 2004), various activity scores have been determined experimentally as follows:

Activity score = 1.0 for each *1 (wt), *2, *35, and *41 [2988G].
Activity score = 0.75 for each *9, *29, *45, and *46.
Activity score = 0.5 for each *10, *17, *41 [2988A].
Activity score = zero for each *3, *4, *5, *6, *7, *8, *11, *12, *15, *36, *40, *42.

[c]Activity scores are classified as follows (from Zineh et al., 2004):

UM activity score = >2.0 (e.g., [*1/*1]n where n is 2 or more gene duplications).
High EM activity score = 1.75 to 2.0 (e.g., *1/*1).
Medium EM activity score = 1.5 (e.g., *1/*17, or *9/*9).
Low EM activity score = 1.0 to 1.25 (e.g., *1/*4, *17/*17, or *9/*17).
IM activity score = 0.5 to 0.75 (e.g., *4/*9 or *4/*17).
PM activity score = zero (e.g., *4/*4).

The variation in drug exposure that results from genetic polymorphisms or CYP inhibition/induction, and the importance of therapeutic index to drug safety was discussed previously (in the section on Pharmacokinetic Variation and Therapeutic Index) and is illustrated in Fig. 1.2. Drugs with a narrow therapeutic index are candidates for therapeutic drug monitoring, as in the case of the anticoagulant warfarin. The adverse effect of warfarin, namely excessive anticoagulation that can result in fatal hemorrhaging, is an extension of its pharmacological effect (inhibition of the synthesis of vitamin K-dependent clotting factors, which results in a prolongation of partial thromboplastin time or PTT). Warfarin is a racemic drug (a mixture of *R*- and *S*-enantiomers). The *S*-enantiomer, whose disposition is largely determined by CYP2C9, is four times more pharmacologically active than the *R*-enantiomer, whose disposition is largely determined by enzymes other than CYP2C9. The anticoagulant effect of warfarin (measured as an increase in prothrombin time based on the international normalized ratio [INR]) is monitored during initial treatment with a low dose of warfarin, during dose escalation and then periodically during maintenance dosing to select the appropriate dose of anticoagulant on an individual-by-individual basis. Genotyping analysis provides useful information on dose selection: individuals who are homozygous or heterozygous for certain allelic variants (the *3 and, to a lesser extent, the *2 allele) are CYP2C9 PMs or IMs and, as such, require less warfarin compared with an EM (an individual who is homozygous for the *1 [or wild-type] allele) (Majerus and Douglas, 2006). CYP2C9 genotype is not the only factor that influences dosing with warfarin. Genetic polymorphisms in the therapeutic target, vitamin K epoxide reductase (VKDR, gene symbol: VKORC1), can impact warfarin dosing because different levels of warfarin are required to inhibit the variants of VDKR.

To achieve the same degree of anticoagulation, the dose of warfarin must be decreased during concomitant therapy with a CYP2C9 inhibitor and, conversely, it must be increased during concomitant therapy with a CYP2C9 inducer. This is why drug candidates that are identified *in vitro* as CYP2C9 inhibitors or inducers may need to be examined for their ability to cause significant interactions with warfarin in the clinic. Warfarin is a victim drug; its disposition is heavily reliant on CYP2C9 activity. It has been identified by the FDA as a *CYP2C9 substrate with a narrow therapeutic range*. Other substrates, inhibitors, and inducers that are recognized by the FDA as sensitive *in vivo* substrates, potent inhibitors, or efficacious inducers are discussed later in the section on Drug Metabolism from an Enzyme Perspective: Cytochrome P450.

When a potential victim drug is identified by reaction phenotyping, or when a potential perpetrator is identified by assessing CYP inhibition or induction, the FDA's information on clinically acceptable CYP probe substrates and effective CYP inhibitors and inducers can point to the type of clinical drug interaction study that would test *in vivo* the veracity and clinical significance of *in vitro* data on reaction phenotyping, CYP inhibition, and CYP induction. For example, in the case of a drug candidate that is identified as a CYP2C9 substrate *in vitro* (i.e., a possible CYP2C9 victim drug), clinical studies might be carried out to assess whether the drug's disposition is affected by the same CYP2C9 genetic polymorphisms and the same CYP2C9 inhibitors/inducers that are known to influence the disposition of warfarin. Conversely, in the case of a drug candidate that is identified as a potent inhibitor (or efficacious inducer) of CYP2C9 (i.e., a CYP2C9 perpetrator), clinical studies might be carried out to assess whether the drug alters the disposition of warfarin (and other drugs whose clearance is dependent on CYP2C9).

1.6 PERPETRATOR–PERPETRATOR–VICTIM INTERACTIONS

As summarized in the preceding section (and as illustrated in Fig. 1.4 and represented by Eq. 1.1), the impact of a perpetrator drug on the disposition of a victim drug depends on the extent to which the perpetrator inhibits the enzyme (i.e., whether the perpetrator partially or completely inhibits the enzyme) and on the extent to which that same enzyme contributes to the elimination of the victim drug. For example, if two enzymes, such as CYP3A4 and CYP2D6, contributed equally (50% each) to the clearance of a victim drug, and if the clearance of the victim drug depended entirely on biotransformation by these two enzymes (such that $f_{mCYP3A4} = 0.5$ and $f_{mCYPD6} = 0.5$), then it is evident from Equation 1.1 that complete inhibition of CYP3A4 would result in a twofold increase in exposure to the victim drug (as measured by the increase in plasma AUC). (The twofold increase is apparent from Eq. 1.1 because the fold increase in plasma AUC $= 1 \div (1 - 0.5) = 2$.) Likewise, complete inhibition of CYP2D6 (or a genetic deficiency of CYP2D6) would also increase the AUC of the victim drug by twofold. A "worst-case" scenario can be imagined, in which multiple perpetrators simultaneously block different metabolic pathways of a victim drug. Such a scenario can result in a "maximum exposure," which could occur in PMs when there is concurrent inhibition of one or more non-polymorphic pathways (Collins et al., 2006).

In the preceding example, complete inhibition of *both* CYP3A4 and CYP2D6 (or complete inhibition of CYP3A4 in a CYP2D6 PM) would not simply result in a fourfold increase in the AUC of the victim drug but would result in a far greater increase in AUC. In fact, if the clearance of the victim drug were *entirely* dependent on metabolism by only CYP3A4 and CYP2D6 (with no residual clearance), then complete loss of both enzyme activities would theoretically cause an infinite increase in victim drug AUC. (The infinite increase is apparent from Equation 1.1 because the fold increase in plasma AUC $= 1 \div (1 - [0.5 + 0.5]) = 1 \div 0 = \infty$.) This extreme situation has not been observed clinically, but there are clinical cases where two perpetrators exert a much greater impact on exposure to a victim drug than would be predicted from the sum or product of the individual perpetrator effects. An example of a perpetrator–perpetrator–victim interaction is illustrated in Fig. 1.5 for the interaction between voriconazole (the victim drug) and ritonavir (a perpetrator drug) in CYP2C19 PMs (in whom the genetic deficiency of CYP2C19 represents the second "perpetrator") (Mikus et al., 2006). Voriconazole is metabolized by CYP3A4 and CYP2C19. Systemic exposure to voriconazole (based on plasma AUC) increases modestly ($\leq$3-fold) when only CYP2C19 activity is lost (as evident in Fig. 1.5 from a comparison of CYP2C19 PMs vs. EMs) or when only CYP3A4 activity is lost (as evident in Fig. 1.5 from the effects of administering the CYP3A4 inhibitor ritonavir to CYP2C19 EMs). However, when both enzyme activities are lost (due to administration of ritonavir to CYP2C19 PMs), there is a dramatic increase (>25-fold) in the plasma AUC of voriconazole. In theory, such "maximum

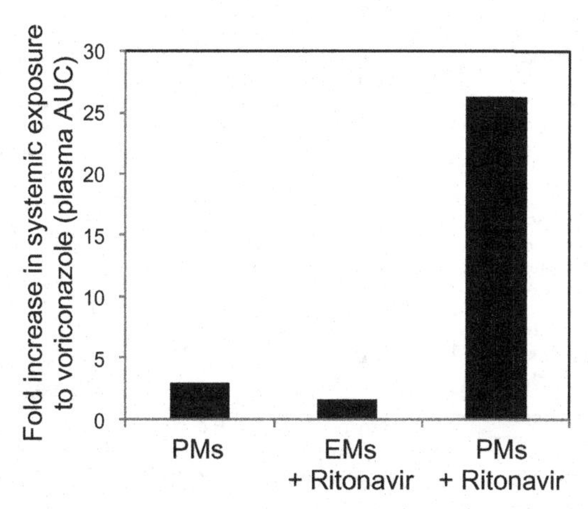

Figure 1.5 An example of a perpetrator–perpetrator–victim drug interaction. Voriconazole is metabolized by CYP3A4 and the genetically polymorphic enzyme CYP2C19. Systemic exposure to voriconazole (based on plasma AUC) increases modestly (≤3-fold) when only CYP2C19 activity is lost (compare CYP2C19 poor metabolizers [PMs] vs. extensive metabolizers [EMs]) or when only CYP3A4 activity is lost (due to administration of the CYP3A4 inhibitor ritonavir to CYP2C19 EMs). However, when both enzyme activities are lost (due to administration of ritonavir to CYP2C19 PMs), there is a dramatic increase (>25 fold) in the plasma AUC of voriconazole.

exposure" could also be achieved in CYP2C19 EMs by a drug that completely inhibits both CYP2C19 and CYP3A4, for which reason a drug that inhibits multiple pathways of drug clearance (which could include both metabolism- and transport-mediated routes of elimination) carries more potential perpetrator liability than a drug that inhibits only a single enzyme or transporter.

1.7 DRUG METABOLISM FROM A CHEMICAL PERSPECTIVE

It is beyond the scope of this chapter to discuss the impact of all chemical groups on the disposition of drugs, but the importance of chemical structure on drug disposition and metabolism cannot be overemphasized. The importance of chemical structure can be appreciated from the relatively large amount of information that can be gathered simply by knowing whether a drug is a lipophilic amine or carboxylic acid, as illustrated in Figs. 1.6 and 1.7.

1.7.1 Acids and Bases

As shown in Fig. 1.6, amine-containing drugs ($R\text{–}NH_2$) tend to be positively charged in the acidic environment of the stomach ($R\text{–}NH_3^+$) but are neutral (or a mixture of neutral and charged species) in the higher pH of the small intestine. Because the neutral species readily crosses biological membranes,

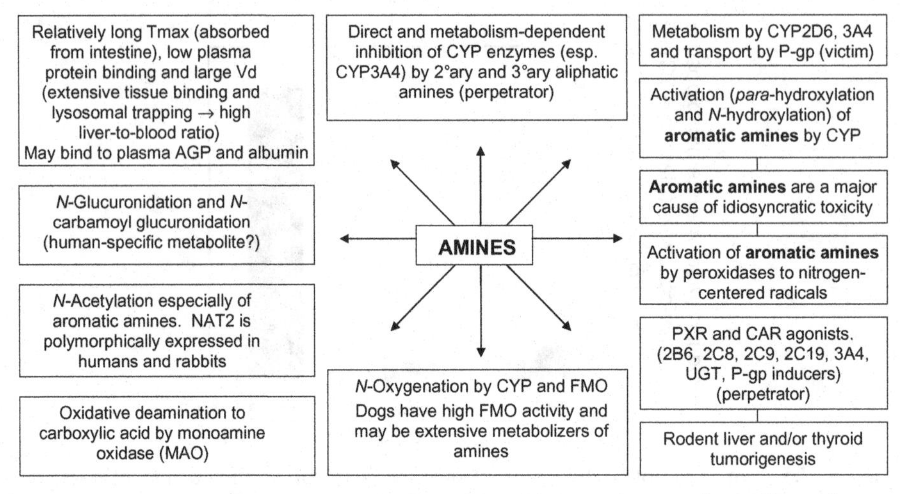

Figure 1.6 General features of the disposition of drugs that are lipophilic amines.

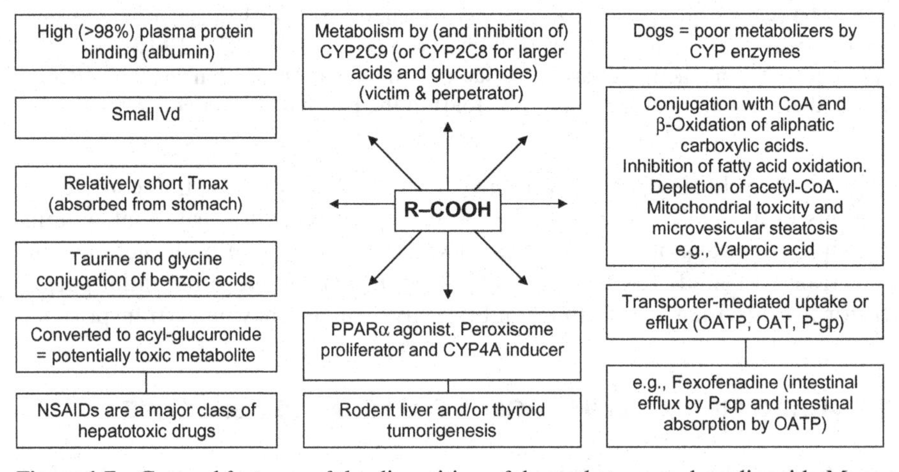

Figure 1.7 General features of the disposition of drugs that are carboxylic acids. Many of the features of carboxylic acid-containing drugs also apply to sulfonamide- and tetrazole-containing drugs.

lipophilic amines tend to be absorbed from the small intestine and, consequently, they tend to have a relatively long T_{max} (the time to reach C_{max}). They readily enter tissues from the blood (pH 7.2–7.4) where they bind to negatively charged lipids and proteins. Consequently, amine-containing drugs tend to have a large volume of distribution (Vd). In contrast, as shown in Fig. 1.7, carboxylic acid-containing drugs tend to be neutral in the acidic environment of the stomach (R–COOH) but are negatively charged ($R–COO^-$) in the higher pH of the small intestine. Because the neutral species readily crosses biological membranes, carboxylic acid-containing drugs tend to be absorbed from the stomach and, consequently, they tend to have a relatively short T_{max}. The absorption of acidic drugs from the intestine (as occurs with fexofenadine) is often mediated by an intestinal uptake transporter like organic-anion-transporting polypeptide (OATP). Carboxylic acid-containing drugs tend to bind extensively (>98%) to plasma proteins (e.g., albumin); consequently, they tend to have a small Vd.

In contrast to acid drugs, basic drugs tend not to be highly bound to plasma proteins. If they are, binding tends to be α1-acid glycoprotein (AGP) (and in contrast to acidic drugs, which are generally highly bound to albumin). If the pKa of the amine group is ≥8, an amine-containing drug may likely be subject to hepatic lysosomal trapping, meaning it will enter the lysosome as a neutral species but then become positively charged and, hence, trapped in the low pH environment (pH 4–5) of the lysosome. Assuming the pH of the lysosome is 4.8 and the pH of the cytosol is 7.4, the theoretical accumulation ratio due to the pH difference alone is 400 for a monobasic drug with a pKa ≥ 9 and 160,000 for a dibasic drug with similarly high pKa values (MacIntyre and Cutler, 1988). These accumulation ratios assume that the lysosomes and liver are two separate compartments of equal volume, whereas in reality, the volume of lysosomes are estimated to be about 0.7% of hepatic volume (MacIntyre and Cutler, 1988). Taking both lysosomal volume and accumulation ratios to account, lysosomal trapping effectively increases the distribution volume of the liver approximately threefold for a monobasic drug with a high pKa ($0.007 \times 400 \approx 3$-fold) and increases it approximately 1,000-fold for a dibasic drug ($0.007 \times 160{,}000 \approx 1{,}000$-fold). Monobasic and especially dibasic drugs that have relatively high lipophilicity (logD > 1.5) and basicity (pKa > 8) will exhibit lysosomal trapping in the liver (and other tissues), which accounts for the high liver-to-blood ratios observed with lipophilic amines such as propranolol, imipramine, quinidine, fluvoxamine, fluoxetine, dextromethorphan, chloroquine, and lignocaine (MacIntyre and Cutler, 1988; Hallifax and Houston, 2007).

From a victim potential, many amine-containing drugs are substrates for the polymorphically expressed enzyme CYP2D6, and lipophilic amines with a high molecular weight tend to be substrates for CYP3A4, an enzyme present in liver and small intestine (where it plays an important role in first-pass metabolism) and an enzyme that is inhibited or induced by a large number of drugs. In contrast, carboxylic acid-containing drugs tend to be substrates for the polymorphically expressed enzyme CYP2C9. Larger acidic drugs (or small acidic drugs that are converted to relatively large acyl glucuronides) may be substrates for CYP2C8.

From a perpetrator potential, several secondary and tertiary amines (such as erythromycin, troleandomycin, clarithromycin, and diltiazem) are metabolism-dependent (irreversible) inhibitors of CYP3A4, whereas several acidic drugs cause clinically significant inhibition of CYP2C9 (which can result in adverse interactions with warfarin, a widely used drug with a narrow therapeutic index). In this regard, it is worth mentioning that, like carboxylic acid-containing drugs, drugs with other acidic groups (a sulfonamide or tetrazoles, for example) also tend to bind extensively to plasma protein and cause clinically significant inhibition of CYP2C9.

Several amine-containing drugs are CYP inducers by virtue of their ability to activate the pregnane X receptor (PXR) and/or the constitutive androstane receptor (CAR), the nuclear receptors that function as ligand-activated transcription factors of CYP2B, 2C, and 3A enzymes. In rodents, prolonged activation of PXR/CAR (by various antiepileptic drugs and antihistamines, for example) may lead to liver and/or thyroid tumor formation by epigenetic mechanisms that are considered clinically irrelevant (based on the lack of liver or thyroid tumor formation in humans administered the antiepileptic drugs phenobarbital or phenytoin for 35+ years; Curran and DeGroot, 1991). In contrast to amines, carboxylic acid-containing drugs rarely activate PXR/CAR and, hence, rarely cause clinically significant induction of CYP2B, 2C, and 3A enzymes. In rodents, however, carboxylic acid-containing drugs may activate the peroxisome proliferator activated receptor-alpha (PPARα), which leads to a proliferation of hepatic peroxisomes and induction of microsomal CYP4A enzymes. Prolonged treatment of rodents with peroxisome proliferators (PPARα agonists), like prolonged treatment of rodents with PXR/CAR agonists, may be associated with liver and/or thyroid tumor formation by epigenetic mechanisms. Induction of CYP4A enzymes and hepatic peroxisome proliferation are considered responses to acid-containing drugs that occur in rodents but not in humans or other primates.

With respect to oxidative metabolism, dogs tend to be EMs of amine-containing drugs (often due to their high levels of flavin monooxygenase or FMO) but tend to be PMs of acid-containing drugs (at least with respect to metabolism by CYP and FMO).

Amine-containing drugs (especially those containing an aromatic amine) can be converted to reactive metabolites by CYP, FMO, aldehyde oxidase (AO) and peroxidases. Furthermore, as a class of drugs, aromatic amines are a major cause of idiosyncratic toxicity. (Formation of reactive metabolites is discussed later in the section on Reactive Metabolites.) Depending on their structure, amine-containing drugs can be metabolized extensively by several oxidative enzymes, such as CYP, FMO, and monoamine oxidase (MAO), and they can be conjugated directly by *N*-acetyltransferase (NAT) and uridine diphosphate (UDP)-glucuronosyltransferase. Acetylation of aromatic amines by NAT-2 may result in polymorphic drug metabolism in humans and rabbits. Formation of *N*-glucuronides or *N*-carbamoyl-glucuronides may result in the formation of human-specific or human-abundant metabolites because rodents tend to catalyze low rates of *N*-glucuronidation and *N*-carbamoyl-glucuronidation.

Acid-containing drugs can also be directly conjugated with glucuronic acid, and the resulting acyl glucuronide may be sufficiently reactive to cause liver toxicity. This topic is discussed later (see section on Reactive Metabolites).

Nonsteroid anti-inflammatory drugs (NSAIDs) are acid-containing drugs that, as a class, are a major cause of liver toxicity, for which reason the development of many NSAIDs has been halted during clinical development. Acid-containing drugs can also be directly conjugated with glycine or taurine (as can bile acids). The acid moiety of carboxylic drugs can *not* be metabolized by CYP, although enzymes like CYP2C9 and CYP2C8 can metabolize sites distal to the acid moiety. Like fatty acids, certain aliphatic carboxylic acids (those of the type $R\text{–}(CH_2CH_2)_n\text{–}COOH$) can undergo mitochondrial β-oxidation resulting in one or more cycles of 2-carbon chain-shortening. This represents an example of drug-metabolism by an *endobiotic*-metabolizing enzyme. Metabolism of carboxylic acid-containing drugs by β-oxidation can lead to mitochondrial toxicity, as seen with the anticonvulsant valproic acid.

The preceding comparison of the disposition of acidic and basic drugs serves to illustrate that several pharmacokinetic properties can be predicted (or at least anticipated) based on the fundamental structure of a drug. Needless to say, the foregoing deals with general properties, and there are undoubtedly exceptions to each of the generalities listed above. Despite these exceptions, the foregoing generalities can offer some guidance to the design and prioritization of *in vitro* studies and perhaps even clinical studies. Considerably more guidance can be obtained by examining the structure of a drug (or drug candidate) to predict which enzymes may and may not participate in its metabolism. This is particularly important in terms of selecting an appropriate test system to study the *in vitro* metabolism of a drug candidate to evaluate its metabolic stability, its victim and perpetrator potential, and its metabolic profile (and the potential for the formation of human-specific or human-abundant metabolites). This is the topic of the next section.

1.7.2 Metabolism of Functional Groups

The chemical structure of a drug (or drug candidate) determines which enzyme or enzymes are involved its metabolism. In general, orally active drugs that are well absorbed from the gut are lipophilic compounds ($logD_{7.4} > 1$) that require biotransformation (chemical modification) to facilitate their urinary and/or biliary elimination. As shown in Fig. 1.8, the oxidative, reductive, and hydrolytic drug-metabolizing enzymes that introduce or expose a functional group are classified as Phase 1 enzymes (such as CYP). The drug-metabolizing enzymes that conjugate the metabolite with a water-soluble molecule (such as glucuronic acid or sulfonic acid) are classified as Phase 2 enzymes. However, some drugs are conjugated directly such that Phase 2 metabolism occurs in the absence of Phase 1 metabolism, as illustrated in Fig. 1.9 for acetaminophen, which undergoes direct metabolism by both Phase 1 and Phase 2 enzymes. Furthermore, some drug conjugates are metabolized by CYP, such that Phase 2 metabolism *precedes* Phase 1 metabolism. This is frequently observed with CYP2C8, which preferentially metabolizes large acid drugs, including the glucuronide conjugates of 17β-estradiol, diclofenac, and gemfibrozil (Kumar

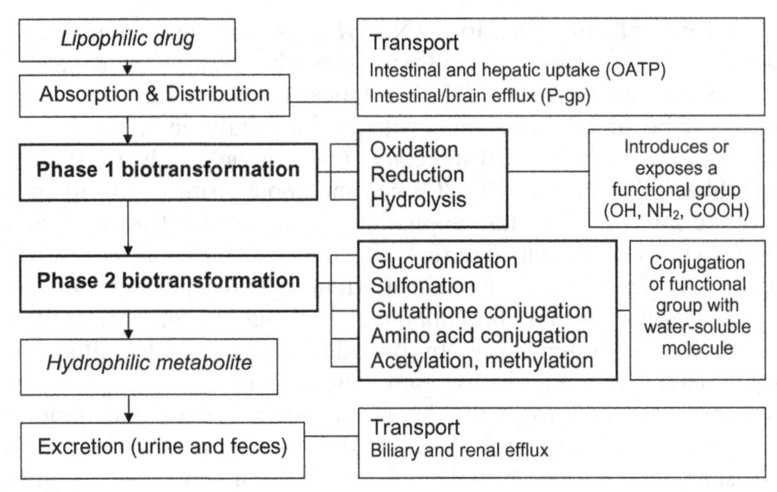

Figure 1.8 A summary of the major pathways of drug biotransformation and their overall role in drug disposition. In general, orally active drugs that are well absorbed from the gut are lipophilic compounds ($\log D_{7.4} > 1$) that require biotransformation (chemical modification) to facilitate their urinary and/or biliary elimination. The oxidative, reductive, and hydrolytic drug-metabolizing enzymes that introduce or expose a functional group are often classified as Phase 1 enzymes (such as cytochrome P450). The drug-metabolizing enzymes that conjugate the metabolite with a water-soluble molecule (such as glucuronic acid or sulfonic acid) are often classified as Phase 2 enzymes. However, some drugs are conjugated directly such that Phase 2 metabolism occurs in the absence of Phase 1 metabolism, and some drug conjugates are metabolized by cytochrome P450, such that Phase 2 metabolism precedes Phase 1 metabolism. Conjugates formed by Phase 2 metabolism are transported out of the liver by canalicular or sinusoidal transporters for elimination in bile or urine, respectively. A more complete list of drug-metabolizing enzymes is shown in Table 1.3.

et al., 2002; Ogilvie et al., 2006; Parkinson and Ogilvie, 2008). Conjugates formed by Phase 2 metabolism are transported out of the liver by canalicular or sinusoidal transporters for elimination in bile or urine, respectively.

Table 1.2 shows a variety of chemical groups (but by no means all the chemical groups present in drugs) together with the enzymes that are typically involved in their metabolism. From Table 1.2 it is apparent that CYP enzymes can metabolize a very wide variety of functional groups, which accounts for the prominent role of CYP enzymes in the metabolism of those drugs in BCS Classes 1 and 2. However, drugs with such functional groups as an ester, amide, or lactone may undergo hydrolysis, whereas UDP-glucuronosyltransferase (UGT), sulfotransferase (SULT), and other conjugating enzymes may directly conjugate those drugs containing such functional groups as a primary, secondary, or tertiary amine ($R\text{–}NH_2$, $R_1R_2\text{–}NH$, and $R_1R_2R_3\text{–}N$), an aliphatic hydroxyl group (alcohol; R–OH), an aromatic hydroxyl group (phenol; R–OH), a free sulfhydryl group (R–SH), or a carboxylic acid (R–COOH).

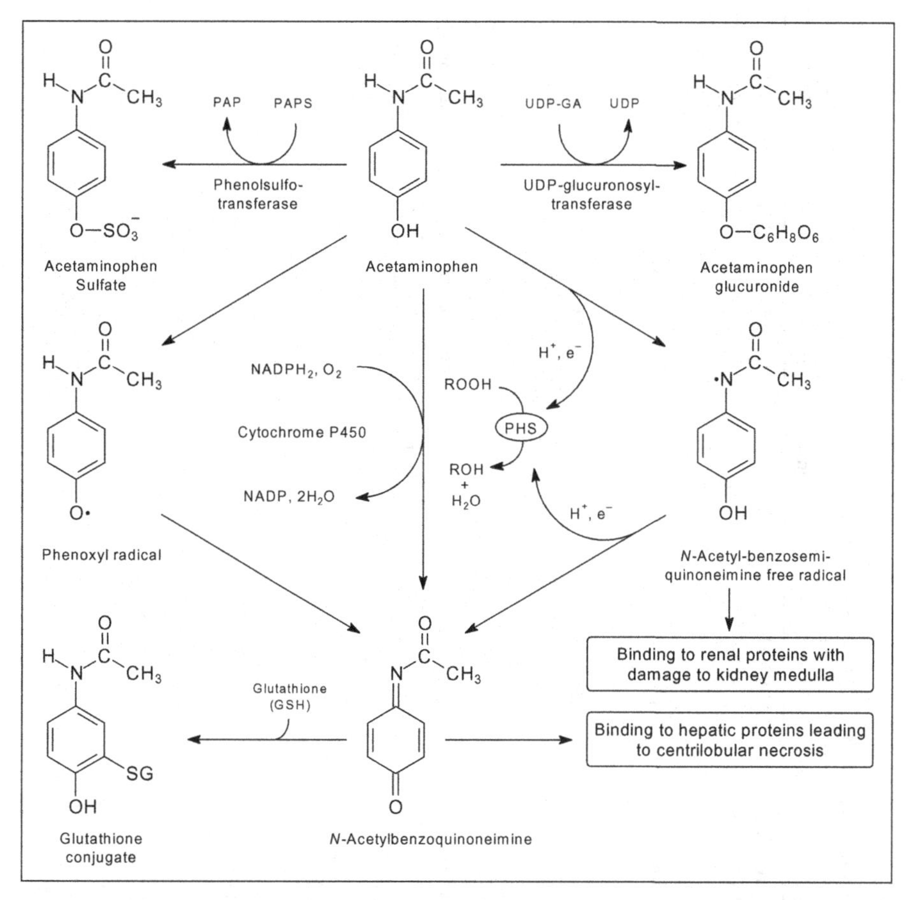

Figure 1.9 The metabolism of acetaminophen by Phase 1 and Phase 2 drug-metabolizing enzymes and the formation of hepatotoxic and nephrotoxic metabolites by cytochrome P450 and prostaglandin H synthase (PHS), respectively. Acetaminophen (*N*-acetyl-*para*-aminophenol or APAP) is conjugated directly with glucuronic acid and sulfonic acid by UDP-glucuronosyltransferase (UGT) and sulfotransferase (SULT), respectively. Cytochrome P450 oxidizes acetaminophen (by dehydrogenation) to the reactive metabolite *N*-acetyl-benzoquinoneimine, which causes liver toxicity if not detoxified by conjugation with glutathione. In the kidney medulla, acetaminophen can also be converted to *N*-acetyl-benzoquinoneimine (a two-electron oxidation product), but the process involves formation of *N*-acetyl-benzosemiquinoneimine radical (a one-electron oxidation product) PHS. Accordingly, cytochrome P450 and PHS play important roles in the hepatotoxic and nephrotoxic effects of acetaminophen, respectively.

Even when a drug undergoes extensive metabolism by CYP (see Fig. 1.10 for typical CYP reactions), it is possible that the rate-limiting step in its systemic clearance is transporter-mediated hepatic uptake, not its rate of metabolism by CYP, as in the case of bosentan (Tracleer™) (Treiber et al., 2004). Such drug candidates typically have a high liver-to-plasma ratio (which can be

TABLE 1.2 Common Chemical Groups and Enzymes Possibly Involved in Their Metabolism

	Chemical Group	Enzyme(s)	Reaction(s)		Chemical Group	Enzyme(s)	Reaction(s)
Alkane	$R{-}CH_2{-}R$	CYP	Hydroxylation, dehydrogenation	Aldehyde	$R{-}C(=O){-}H$	CYP, ALDH	Oxidative de-formylation, oxidation to carboxylic acid
Alkene	$R_2C{=}CR_2$	CYP, GST	Epoxidation, glutathione adduct formation	Amide	$R{-}C(=O){-}NH{-}R$	Amidase (esterase)	Hydrolysis
Alkyne	$R{-}C{\equiv}C{-}R$	CYP	Oxidation to carboxylic acid	Aniline	$C_6H_5{-}NH_2$	CYP, NAT, UGT, peroxidase, SULT	*N*-hydroxylation, *N*-acetylation, *N*-glucuronidation, *N*-oxidation, *N*-sulfonation
Aliphatic alcohol	$R{-}CH_2{-}OH$	CYP, ADH, catalase, UGT, SULT	Oxidation, glucuronidation, and sulfonation	Aromatic azaheterocycles		UGT, CYP, aldehyde oxidase	*N*-glucuronidation, hydroxylation, *N*-oxidation, ring cleavage, oxidation
Aliphatic amine	$R{-}NH_2$	CYP, FMO, MAO, UGT, SULT, MT, NAT, peroxidase	*N*-dealkylation, *N*-oxidation, deamination, *N*-glucuronidation, *N*-carbamoyl glucuronidation, *N*-sulfonation, *N*-methylation, *N*-acetylation	Carbamate	$R{-}NH{-}C(=O){-}O{-}R$	CYP, Esterase	Oxidative cleavage, hydrolysis

Amidine	HN=CR–NH_2	CYP	*N*-oxidation	Ester	R–CH_2–O–C(=O)–R	CYP, Esterase	Oxidative cleavage, hydrolysis
Arene	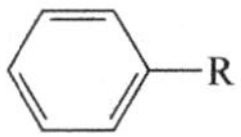 —R	CYP	Hydroxylation and epoxidation	Ether	R–CH_2–O–CH_2–R	CYP	*O*-dealkylation
Carboxylic acid	R–COOH	UGT, amino acid transferases	Glucuonidation, amino acylation	Ketone	R–CH_2–C(=O)–R	CYP, SDR, AKR	Baeyer–Villiger oxidation, reduction
Epoxide	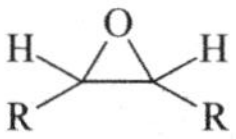 H, O, H, R, R	Epoxide hydrolase, GST	Hydrolysis, glutathione adduct formation	Phenol	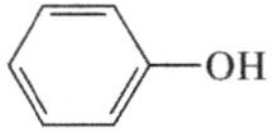 —OH	CYP, UGT, SULT, MT	Ipso-substitution, glucuronidation, sulfonation, methylation
Lactone	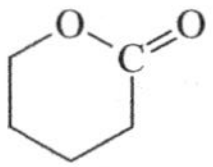 O, C=O	Lactonase (paraoxonase)	Hydrolysis (ring opening)	Thioether	R–CH_2–S–CH_2–R	CYP, FMO	*S*-dealkylation, *S*-oxidation

ADH, alcohol dehydrogenase; ALDH, aldehyde dehydrogenase; AKR, aldo-keto reductases; GST, glutathione *S*-transferase; MT, methyltransferase; SDR, short-chain dehydrogenases/reductases; SULT, sulfotransferase; UGT, UDP-glucuronosyltransferase.

Adapted from Williams et al. (2003).

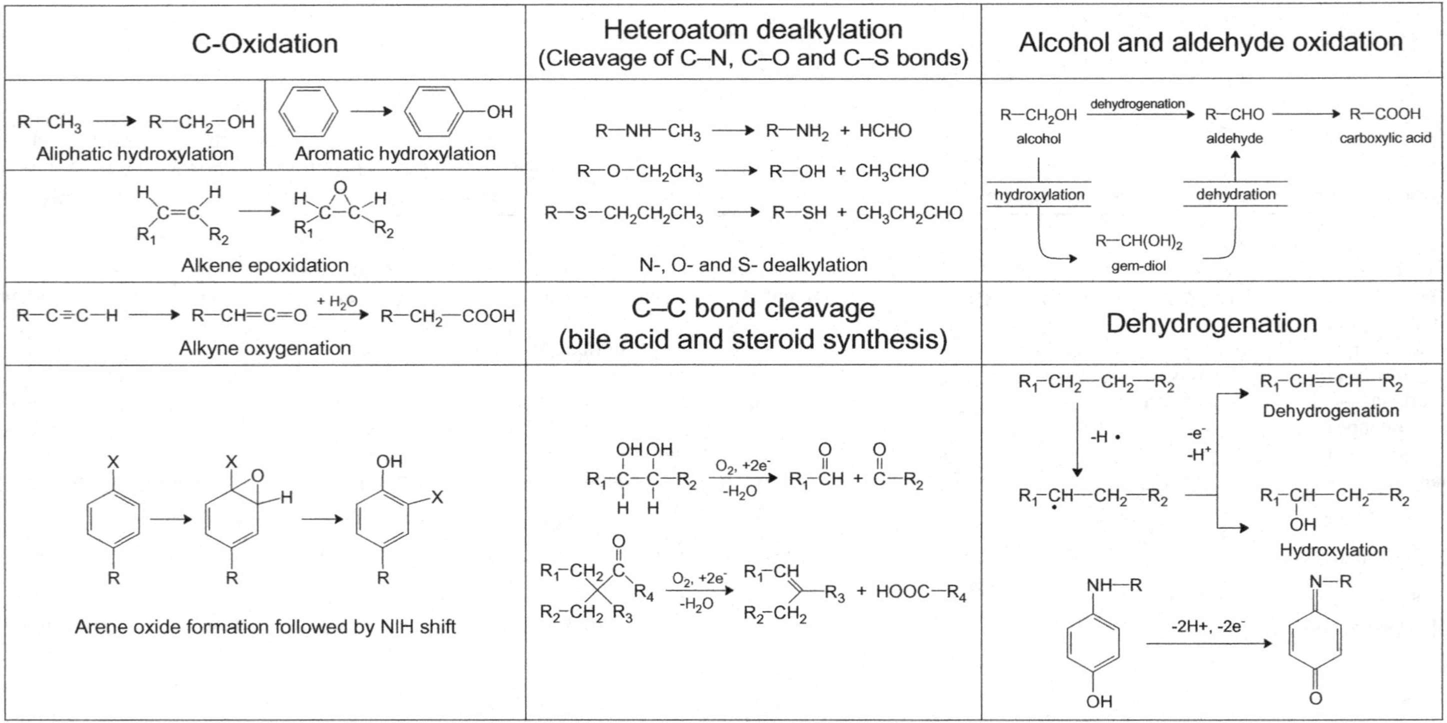

C-Oxidation

$R{-}CH_3 \longrightarrow R{-}CH_2{-}OH$

Aliphatic hydroxylation

Aromatic hydroxylation

Alkene epoxidation

$R{-}C{\equiv}C{-}H \longrightarrow R{-}CH{=}C{=}O \xrightarrow{+H_2O} R{-}CH_2{-}COOH$

Alkyne oxygenation

Arene oxide formation followed by NIH shift

Heteroatom dealkylation (Cleavage of C–N, C–O and C–S bonds)

$R{-}NH{-}CH_3 \longrightarrow R{-}NH_2 + HCHO$

$R{-}O{-}CH_2CH_3 \longrightarrow R{-}OH + CH_3CHO$

$R{-}S{-}CH_2CH_2CH_3 \longrightarrow R{-}SH + CH_3CH_2CHO$

N-, O- and S- dealkylation

C–C bond cleavage (bile acid and steroid synthesis)

$R_1{-}CH(OH){-}CH(OH){-}R_2 \xrightarrow[-H_2O]{O_2,\ +2e^-} R_1{-}CHO + R_2{-}CHO$

$\xrightarrow[-H_2O]{O_2,\ +2e^-}$ $+ HOOC{-}R_4$

Alcohol and aldehyde oxidation

$R{-}CH_2OH$ (alcohol) $\xrightarrow{\text{dehydrogenation}}$ $R{-}CHO$ (aldehyde) $\longrightarrow$ $R{-}COOH$ (carboxylic acid)

hydroxylation → $R{-}CH(OH)_2$ (gem-diol) → dehydration

Dehydrogenation

$R_1{-}CH_2{-}CH_2{-}R_2 \xrightarrow{-H\bullet} R_1{-}\dot{C}H{-}CH_2{-}R_2$

$-e^-$, $-H^+$ → $R_1{-}CH{=}CH{-}R_2$ Dehydrogenation

→ $R_1{-}CH(OH){-}CH_2{-}R_2$ Hydroxylation

$\xrightarrow{-2H+,\ -2e^-}$

Figure 1.10 Reactions catalyzed by cytochrome P450.

N-Oxidation		Oxidative deamination and dehalogenation	Dehydration
NH_2 → NHOH N-Hydroxylation	—N— → —N^+—O^- =N → =N^+—O^- N-Oxygenation	R_1–C(R_2)(H)–NH_2 → R_1–C(R_2)(OH)–NH_2 → R_1–C(R_2)=O + NH_3 R_1–C(R_2)(H)–X → R_1–C(R_2)(OH)–X → R_1–C(R_2)=O + HX	R_1–CH_2–CH(OOH)–R_2 —($-H_2O$)→ R_1–CH–CH–R_2 (epoxide, O) R–CH=N–OH —($-H_2O$)→ R–C≡N
S-Oxidation		**Reduction**	
R_1–S–R_2 → R_1–S^+(O^-)–R_2 → R_1–S(O^-)(O^+)–R_2 Sulfide Sulfoxide Sulfone		(epoxide, H, O, H) —($-2H+, -2e^-$; epoxide reduction)→ benzene + H_2O R_2–N^+(R_1)(R_3)–O^- —($-2H+, -2e^-$; N-oxide reduction)→ R_2–N(R_1)(R_3) + H_2O	R_2–C(R_1)(R_3)–OOH —($-2H+, -2e^-$; $-R_3H$; Alkyl peroxide reduction)→ R_1(R_2)C=O + H_2O —C—X —($-2H+, -2e^-$)→ —C• + HX

Figure 1.10 (*Continued*)

determined in nonclinical animals), and their plasma clearance may be impaired by cyclosporine (a broad and potent inhibitor of several hepatic uptake transporters) rather than 1-aminobenzotriazole (ABT) (an inactivator of most drug-metabolizing CYP enzymes) (Treiber et al., 2004; Strelevitz et al., 2006). There are occasions when the structure of a drug candidate leaves some doubt as to whether CYP will play the dominant role in its metabolism. In such cases, the selection of an appropriate *in vitro* test system can be guided by information on the routes of metabolism of the drug candidate in nonclinical animals (and in humans, of course, although information from clinical studies is not always available when reaction phenotyping studies are conducted in the course of drug development). Rats treated with the general CYP inhibitor ABT can provide valuable information on the overall importance of CYP (both hepatic and intestinal) to the disposition of a drug candidate, as demonstrated recently by Strelevitz et al. (2006) for midazolam. CYP is unlikely to play a major role in the disposition of a drug candidate whose pharmacokinetic profile in rats is unaffected by treatment with ABT. A similar approach can be taken with human hepatocytes: CYP is unlikely to play a major role in the disposition of a drug candidate whose metabolic stability in suspended human hepatocytes is unaffected by a 30- or 60-minute preincubation with 1 mM ABT. The mechanism of CYP inactivation by ABT and experimental approaches to exploit the broad inactivating effect of ABT on most drug-metabolizing CYP enzymes are discussed below (in this section).

Studies with various subcellular fractions are useful to ascertain which enzyme systems are involved in the metabolism of a drug candidate. In the absence of added cofactors, oxidative reactions such as oxidative deamination that are supported by mitochondria or by liver microsomes contaminated with mitochondrial membranes (as is the case with microsomes prepared from frozen liver samples) are likely catalyzed by MAO whereas oxidative reactions supported by cytosol are likely catalyzed by AO and/or xanthine oxidase (a possible role for these enzymes in the metabolism of a drug candidate can be deduced in large part from the structure of the drug candidate).

AO is a molybdozyme present in the cytosolic fraction of liver and other tissues of several mammalian species (Beedham, 2002; Beedham et al., 2003). It catalyzes both oxidation reactions (e.g., aldehydes to carboxylic acids, aza-heterocyclic aromatic compounds to lactams) and reduction reactions (e.g., nitrosoaromatics to hydroxylamines, isoxazoles to keto alcohols). In contrast to CYP, AO does not require a pyridine nucleotide cofactor; hence, it catalyzes reactions without the need to add a cofactor. In general, xenobiotics that are good substrates for AO are poor substrates for CYP, and vice versa (Rettie and Fisher, 1999). Naphthalene (with no nitrogen atoms) is oxidized by CYP, but not by AO, whereas the opposite is true of pteridine (1,3,5,8-tetraazanaphthalene), which contains four nitrogen atoms. The intermediate structure, quinazolone (1,3-diazanaphthalene) is a substrate for both enzymes. This complementarity in substrate specificity reflects the opposing preference of the two enzymes for oxidizing carbon atoms; CYP prefers to oxidize carbon atoms with high electron density, whereas AO (and xanthine oxidase) prefers to oxidize carbon atoms with low electron density. AO plays an important role in the metabolism of various drugs such as the hypnotic agent zaleplon (Kawashima et al., 1999; Lake et al., 2002, Renwick et al., 2002), the antiepileptic agent zonisamide (Sugihara et al., 1996), and the antipsychotic agent ziprasidone (Beedham et al., 2003). Thus, it is important to evaluate whether the metabolism of a drug is catalyzed only by CYP enzymes or if non-CYP enzymes such as AO are also involved. If the latter is the case, it is useful to assess the relative contribution of CYP enzymes and non-CYP enzymes. One method to assess the involvement of CYP enzymes in the metabolism of a drug candidate is to incubate the drug candidate with test systems in the presence of ABT.

As a general inactivator of CYP, ABT provides useful information on the role of CYP in the *in vitro* disposition of a drug candidate. It can also be used in nonclinical animals. Recently, ABT was used as an *in vivo* CYP inhibitor in rats to investigate the role of intestinal and hepatic metabolism in the disposition of midazolam. When rats were treated orally, ABT inhibited both intestinal and hepatic CYP enzymes involved in midazolam metabolism. However, only the hepatic CYP enzymes were inhibited when ABT was administered to rats intravenously (Strelevitz et al., 2006). The mechanism of CYP inactivation by ABT involves the destruction of the heme moiety by benzyne, as illustrated in Fig. 1.11. Although benzyne would be expected to bind to the heme moiety and inactivate all CYP enzymes, we have observed differential inactivation of CYP enzymes when NADPH-fortified human liver microsomes are incubated for 30 min with 2 mM ABT, as shown in Fig. 1.12. Although ABT caused extensive inactivation (>80%) of CYP3A4, 1A2, 2A6 and 2E1, it only caused a 50–60% loss of CYP2C9 activity, and it caused only a ~25% decrease in CYP4A11 activity (Ogilvie et al., 2008). Consequently, a reaction that is primarily catalyzed by CYP2C9 may be only partially inhibited by ABT.

Like CYP, FMO refers to a family of microsomal enzymes that require NADPH and oxygen (O_2) to catalyze the oxidative metabolism of drugs. Many of the reactions catalyzed by FMO can also be catalyzed by CYP enzymes (Rettie and Fisher, 1999; Cashman, 1999; Cashman and Zhang, 2006). FMO enzymes catalyze *N*- and *S*-oxidation reactions but not C-oxidation reactions. Therefore, CYP, but not FMO, would be suspected of converting a drug candidate to a metabolite whose formation involved aliphatic or aromatic hydroxylation, epoxidation, dehydrogenation, or heteroatom dealkylation (*N*-, *O*- or *S*-dealkylation), whereas both enzymes would be suspected of forming metabolites by *N*- or *S*-oxygenation. Therefore, if NADPH-fortified human liver microsomes convert a nitrogen-containing drug candidate to an *N*-oxide or *N*-hydroxy metabolite, or if they convert a sulfur-containing drug candidate to a sulfoxide or sulfone, it is advisable to determine the relative contribution of FMO and CYP enzymes.

CYP and FMO enzymes can be distinguished by their differential sensitivity to inactivation with detergent or heat. Treatment of human liver microsomes with the nonionic detergent Triton X-100 (final concentration 1% v/v) abolishes CYP activity with little or no loss of FMO activity. The converse is observed with heat inactivation. Incubating human liver microsomes at 50°C

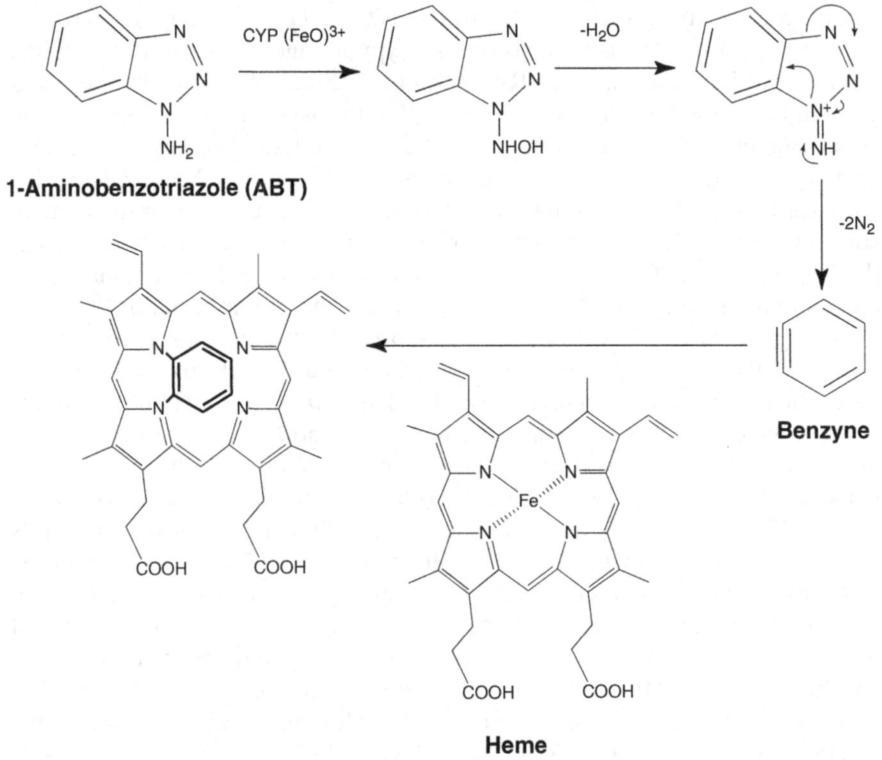

1-Aminobenzotriazole (ABT)

Benzyne

Heme

Figure 1.11 Mechanism of cytochrome P450 heme modification by the general cytochrome P450 inactivator 1-aminobenzotriazole (ABT).

for 1 or 2 min causes extensive (>80%) loss of FMO activity with little loss (<20%) of CYP activity. Detergent and heat inactivation provide a simple means to assess the relative contribution of CYP and FMO enzymes to oxidative reactions catalyzed by NADPH-fortified human liver microsomes. The general CYP inactivator ABT (see Figs. 1.11 and 1.12) can also be used to distinguish the role of CYP and FMO enzymes to reactions catalyzed by human liver microsomes or hepatocytes. Methimazole has been used to selectively inhibit FMO activity in liver microsomes, but it can also inhibit CYP enzymes. Recombinant human FMO1, FMO3 (the most active form in human liver microsomes (Cashman and Zhang, 2006) and FMO5 are commercially available and can be used to ascertain whether FMO enzymes are capable of catalyzing a given reaction. Finally, an important experimental consideration regarding FMO enzymes is that, in contrast to CYP enzymes, they cannot be inhibited by polyclonal antibodies, even when these antibodies recognize the FMO protein on a Western immunoblot.

If two kinetically distinct enzymes are involved in a given reaction, there is a high probability that only the high-affinity enzyme will contribute substan-

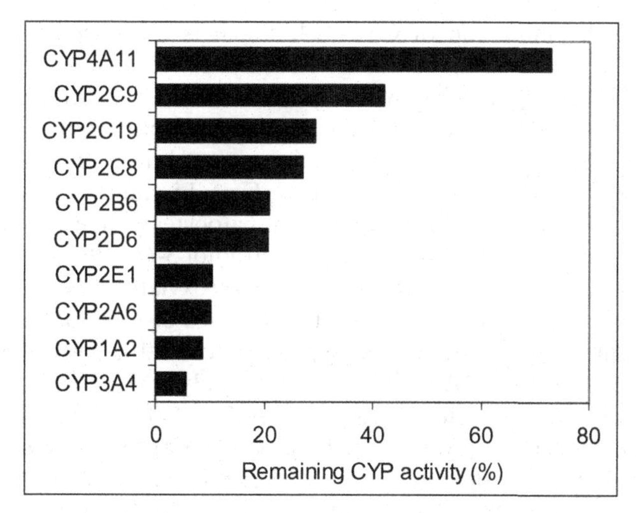

Figure 1.12 Effect of the CYP inactivator 1-aminobenzotriazle (1-ABT, 2 mM) on CYP enzyme activities in human liver microsomes. 1-Aminobenzotriazole (1-ABT) is a metabolism-dependent inhibitor of multiple cytochrome P450 enzymes based on its ability to bind covalently to the heme moiety. However, even at high concentrations, some human CYP enzymes (such as CYP2C9 and CYP4A11) are only partially inhibited by 1-ABT under conditions that result in extensive inhibition (>90%) of CYP3A4 and CYP1A2.

tially to drug clearance (unless the drug is administered at sufficiently high doses to achieve hepatic drug levels that allow even the low affinity enzyme to contribute significantly to metabolite formation). If two kinetically distinct enzymes are involved in a given reaction, the sample-to-sample variation in metabolite formation by a panel of human liver microsomes can be determined at several drug concentrations to identify the enzyme that is more relevant at a given substrate concentration. For example, the 5-hydroxylation of lansoprazole is mainly catalyzed by two CYP enzymes, CYP3A4 and CYP2C19 (Pearce et al., 1996). At high substrate concentrations (~100 μM) the 5-hydroxylation of lansoprazole by human liver microsomes is dominated by CYP3A4, a low-affinity, high-capacity enzyme. However, at pharmacologically relevant concentrations (~1 μM) the 5-hydroxylation of lansoprazole is catalyzed primarily by CYP2C19, as it is *in vivo*. In the study by Pearce et al. (Pearce et al., 1996), the correlation of CYP2C19 activity with lansoprazole 5-hydroxylation improved dramatically when the substrate concentration was lowered from 125 to 1 μM and, conversely, the correlation of the same activity with CYP3A4 progressively worsened.

Another example of misleading *in vitro* results caused by a low-affinity, high-capacity enzyme overshadowing the clinically relevant role of a high-affinity, low-capacity enzyme is the interaction between ethinyl estradiol-containing oral contraceptives and antibiotics such as rifampin. This interaction is often attributed to the induction of CYP3A4, which is the major CYP enzyme involved in the *oxidative* metabolism of ethinyl estradiol (e.g., Ortho-Evra® prescribing information, http://www.myortho360.com/myortho360/shared/pi/OrthoEvraPI.pdf#zoom=100 2009), but several lines of evidence suggest that induction of CYP3A4 is not the predominant mechanism by which rifampin increases the clearance of ethinyl estradiol. First, Li and colleagues reported that treatment of primary cultures of human hepatocytes with rifampin (33.3 μM) caused up to a 3.3-fold increase in ethinyl estradiol 3-*O*-sulfate formation (Li et al., 1999). Second, CYP3A4 catalyzes the 2-hydroxylation of ethinyl estradiol with a K_m value of approximately 3.4 μM (Shiraga et al., 2004), whereas the average plasma concentrations are approximately 1/10,000th of this value. Third, SULTs 1A1, 1A2, 1A3, 1E1, and 2A1 catalyze the 3-*O*-sulfonation of ethinyl estradiol with K_m values ranging from 6.7 to 4500 nM, near the pharmacologically relevant concentrations (Schrag et al., 2004). Finally, it is known that ethinyl estradiol is predominantly excreted in bile and urine as the 3-sulfate and, to a lesser extent, the 3-glucuronide (Li et al., 1999), which suggests that 3-sulfonation is the major pathway of ethinyl estradiol metabolism. Taken together, these data suggest that induction of SULTs can be clinically relevant at least for low-dose drugs that can be sulfonated with high affinity.

In summary, two of the most important criteria for the conduct of *in vitro* drug metabolism studies are (1) the choice of the test system (which should be selected to include those enzymes that will likely play a role in the metabolism of the drug candidate, which in turn is initially determined by the structure of the drug candidate) and (2) the concentration of the drug candidate, which should be pharmacologically relevant to prevent metabolism being dominated by a clinically irrelevant enzyme (one with low affinity but high capacity for metabolism of the drug candidate).

1.8 DRUG METABOLISM FROM AN ENZYME PERSPECTIVE

The biotransformation of drugs is catalyzed by various enzyme systems that can be divided into four categories based on the reaction they catalyze:

Hydrolysis (e.g., carboxylesterase)
Reduction (e.g., carbonyl reductase)
Oxidation (e.g., CYP)
Conjugation (e.g., UGT)

The mammalian enzymes involved in the hydrolysis, reduction, oxidation, and conjugation of xenobiotics are listed in Table 1.3, together with their principal subcellular location. The conjugation reactions include glucuronidation, sulfonation (often called sulfation), acetylation, methylation, conjugation with glutathione (mercapturic acid synthesis), and conjugation with amino acids (such as glycine, taurine, and glutamic acid).

TABLE 1.3 General Pathways of Drug Biotransformation and Their Major Subcellular Location

Reaction	Enzyme	Localization
Hydrolysis	Carboxylesterase	Microsomes, cytosol, lysosomes, blood
	Alkaline phosphatase	Plasma membrane
	Peptidase	Blood, lysosomes
	Epoxide hydrolase	Microsomes, cytosol
Reduction	Azo- and nitro-reduction	Microflora
	Carbonyl (aldo-keto) reduction	Cytosol, microsomes, blood
	Disulfide reduction	Cytosol
	Sulfoxide reduction	Cytosol
	Quinone reduction	Cytosol, microsomes
	Dihydropyrimidine dehydrogenase	Cytosol
	Reductive dehalogenation	Microsomes
	Dehydroxylation (cytochrome b5)	Microsomes
	Dehydroxylation (aldehyde oxidase)	Cytosol
Oxidation	Alcohol dehydrogenase	Cytosol
	Aldehyde dehydrogenase	Mitochondria, cytosol
	Aldehyde oxidase	Cytosol
	Xanthine oxidase	Cytosol
	Monoamine oxidase	Mitochondria
	Diamine oxidase	Cytosol
	Peroxidase	Microsomes, lysosomes, saliva
	Flavin-monooxygenases	Microsomes
	Cytochrome P450	Microsomes
Conjugation	Glucuronide (and glucose) conjugation	Microsomes
	Sulfate conjugation	Cytosol
	Glutathione conjugation	Cytosol, microsomes, mitochondria
	Amino acid conjugation	Mitochondria, microsomes
	Acetylation	Mitochondria, cytosol
	Methylation	Cytosol, microsomes, blood

In general, drug biotransformation is accomplished by a limited number of enzymes with broad substrate specificities. In humans, for example, two CYP enzymes, namely CYP2D6 and CYP3A4, metabolize over half the orally effective drugs in current use, but there is at least one drug predominantly metabolized by the other drug-metabolizing CYP enzymes expressed in human liver microsomes (namely CYP1A2, 2B6, 2C8, 2C9, 2C19, 2E1, and 2J2) (Gonzalez and Tukey, 2006; Parkinson and Ogilvie, 2008). Not all biotransformation reactions are catalyzed by the mammalian enzymes listed in Table 1.3. Some biotransformation reactions are catalyzed by enzymes in the gut microflora (largely anaerobic bacteria in the colon), whereas the biotransformation of other drugs is catalyzed by enzymes that participate in intermediary (endobiotic) metabolism.

Several drug-metabolizing enzymes are inducible, meaning their expression can be increased (upregulated) usually in response to exposure to high concentrations of drugs. Induction is mediated by ligand-activated receptors (so-called xenosensors) that are activated by xenobiotics (ligands) to DNA-binding proteins that upregulate the transcription of various genes encoding xenobiotic-biotransforming enzymes, especially CYP enzymes, which are usually induced to the greatest extent (in terms of fold increase) (Parkinson and Ogilvie, 2008). The major human xenosensors are the aryl hydrocarbon receptor (AhR), which induces CYP1A2, the CAR, and the PXR; the latter two xenosensors induces CYP2B6, CYP2C8, CYP2C9, CYP2C19, and CYP3A4. In rodents (but not humans or other primates), the peroxisome proliferator-activated receptor alpha (PPARα) mediates the induction of CYP4 enzymes.

Drug administration is largely through oral ingestion, and the small intestine and liver are highly developed to limit systemic exposure to orally ingested drugs, a process known as *first-pass elimination (or pre-systemic elimination)*. The enterocytes, at the tips of the small intestinal villi, express the efflux transporter P-glycoprotein (P-gp) (also known as MDR1 or ABCB1), which serves to limit drug absorption. The enterocytes express high levels of certain CYP and UGT enzymes, which biotransform a wide variety of drugs. The liver expresses a number of uptake transporters that actively remove drugs from the blood. They also express a number of efflux transporters that actively discharge drugs or their metabolites (especially conjugates) into the bile canaliculus for biliary excretion, or that actively discharge drug metabolites (especially conjugates) back into the blood for urinary excretion. The liver expresses the largest number and, with few exceptions, the highest concentrations of drug-metabolizing enzymes.

Although the liver contains higher concentrations of most xenobiotic-biotransforming enzymes, and because the number of hepatocytes in the liver exceeds the number of enterocytes in the small intestine, it might be assumed that, compared with the liver, the small intestine would make only a small contribution to first-pass metabolism, but this is not the case. Furanocoumarins in grapefruit juice inhibit intestinal but not hepatic CYP3A4 and yet grapefruit juice increases systemic exposure to felodipine and other drugs that undergo first-pass metabolism by CYP3A4 (Paine et al., 2006). The impact of the CYP3A4 inhibitor ketoconazole on the disposition of the CYP3A4 substrate midazolam depends on whether midazolam is given orally or intravenously. When midazolam is given intravenously, such that its clearance is dependent only on hepatic metabolism, ketoconazole causes a 3- to 5-fold increase in the AUC. However, when midazolam is given orally, ketoconazole causes a 10- to 15-fold increase in AUC, the difference reflecting the significant role of intestinal metabolism to the pre-systemic clearance of midazolam.

An overview of the function of the various drug-metabolizing enzymes listed in Table 1.3 can be found in Parkinson and Ogilvie (2008). Because CYP and UGT are the major Phase 1 and Phase 2 enzyme systems involved in drug metabolism, respectively, the following sections focus on the role of these two enzyme systems in drug metabolism and their role in the conversion of drugs to reactive metabolites.

1.8.1 CYP

Among all the drug-biotransforming enzymes, the CYP enzyme system ranks first in terms of catalytic versatility and the sheer number of drugs and other xenobiotics it detoxifies or activates to reactive intermediates. The highest levels of CYP enzymes involved in drug metabolism are found in liver endoplasmic reticulum (microsomes), but CYP enzymes are present in other organelles and in virtually all tissues. Microsomal and mitochondrial CYP enzymes play key roles in the biosynthesis or catabolism of steroid hormones, bile acids, fat-soluble vitamins such as vitamins A and D, fatty acids, and eicosanoids such as prostaglandins, thromboxane, prostacyclin, and leukotrienes, which underscores the catalytic versatility of CYP. In humans, the metabolism of the aforementioned endobiotics and the biotransformation of innumerable xenobiotics are catalyzed by 57 CYP enzymes, as shown in Table 1.4. In several

TABLE 1.4 Classification of the 57 Human Cytochrome P450 (CYP) Enzymes

Xenobiotics	Steroidogenic	Fatty Acids/Eicosanoids	Unknown
CYP1A1	CYP11A1	CYP4A11	CYP2A7
CYP1A2	CYP11B1	CYP4B1	CYP2S1
CYP1B1	CYP11B2	CYP4F2	CYP2U1
CYP2A6	CYP17A1	CYP4F8	CYP2W1
CYP2A13	CYP19A1	CYP4F12	CYP3A43
CYP2B6	CYP21A2	CYP5A1[e]	CYP4A22
CYP2C8[a]		CYP8A1[f]	CYP4F11
CYP2C9[a]	Bile Acid		CYP4F22
CYP2C18	CYP7A1	Vitamin D	CYP4V2
CYP2C19	CYP7B1	CYP24A1	CYP4X1
CYP2D6	CYP8B1	CYP26C1[g]	CYP4Z1
CYP2E1	CYP27A1[c]	CYP27B1	CYP20A1
CYP2F1	CYP39A1	CYP2R1	CYP27C1
CYP2J2[a]	CYP46A1		
CYP3A4[b]	CYP51A1[d]	Retinoic Acid	
CYP3A5		CYP26A1	
CYP3A7		CYP26B1	
CYP4F3[a]			

[a] Also involved in fatty acid and eicosanoid metabolism.
[b] Also involved in bile acid synthesis.
[c] Also involved in vitamin D metabolism.
[d] Also involved in cholesterol biosynthesis.
[e] Thromboxane A synthase (TBXAS1).
[f] Prostaglandin I2 (prostacyclin) synthase (PTGIS).
[g] Also involved in retinoic acid metabolism.

cases, there is no clear functional distinction in terms of endobiotic and xenobiotic metabolism because there are many examples of CYP enzymes playing an important role in the metabolism of both an endobiotic and a drug, or other xenobiotic.

The basic reaction catalyzed by CYP enzymes is monooxygenation (hydroxylation) in which one atom of oxygen is incorporated into a substrate, designated RH, and the other is reduced to water with reducing equivalents derived from NADPH, as follows:

$$\text{Substrate (RH)} + O_2 + \text{NADPH} + H^+ \rightarrow \text{Product (ROH)} + H_2O + \text{NADP}^+$$

Although CYP functions as a monooxygenase, the products are not limited to alcohols and phenols due to rearrangement reactions (Guengerich, 1991, 2001). A summary of the major reactions catalyzed by CYP appears in Fig. 1.10, from which is apparent that CYP can catalyze an impressive array of reactions. The major reactions catalyzed by CYP are listed below (most of which are illustrated in Fig. 1.10):

1. Hydroxylation of an aliphatic or aromatic carbon;
2. Epoxidation of a double bond;
3. Heteroatom (S-, N-, and I-) oxygenation and N-hydroxylation;
4. Heteroatom (O-, S-, and N-) dealkylation;
5. Oxidative group transfer;
6. Cleavage of esters;
7. Dehydrogenation.

Table 1.5 summarizes (1) *in vitro* reactions commonly used to measure the major drug-metabolizing CYP enzymes in human liver microsomes, (2) examples of positive controls for direct inhibition and metabolism-dependent inhibition of microsomal CYP enzymes, and (3) examples of positive controls for CYP induction in primary cultures of human hepatocytes. An expanded list of substrates, inhibitors, and inducers for each of the major drug-metabolizing CYP enzymes in human liver is given in Table 1.6. The drugs listed in Table 1.6 are largely based on examples cited by the FDA, which has provided, where possible, the following list of drugs to guide the conduct of *in vitro* and *in vivo* pharmacokinetic drug–drug interaction (DDI) studies:

1. preferred and acceptable CYP substrates and chemical inhibitors for reaction phenotyping *in vitro*;
2. sensitive probe substrates to monitor CYP activity *in vivo* (a sensitive substrate is one whose clearance is largely [>80%] determined by a single CYP enzyme such that loss of that CYP enzyme causes a fivefold or greater increase in exposure;
3. drugs that are strong or weak inhibitors of CYP enzymes *in vivo*;

TABLE 1.5 Commonly Used Marker Substrates to Support *In Vitro* Studies of Human CYP Enzymes and Commonly Used Positive Controls for CYP Inhibition and Induction Studies

Enzyme	Microsomal CYP Reactions			CYP Inhibitors		CYP Inducers		
	CYP Activity	Km	V_{max}	Direct	MDI	Inducible?	Xenosensor	Positive Control
CYP1A2	Phenacetin *O*-dealkylation	63	820	α-Naphthoflavone	Furafylline	Yes	AhR	Omeprazole
CYP2B6	Bupropion hydroxylation	77	570	Orphenadrine	Phencyclidine	Yes	CAR/PXR	Phenobarbital
CYP2B6	Efavirenz 8-hydroxylation	5.8	140	Orphenadrine	Phencyclidine	Yes	CAR/PXR	Phenobarbital
CYP2C8	Amodiaquine *N*-dealkylation	1.8	2500	Montelukast	Gem gluc.	Yes	PXR/CAR	Rifampin
CYP2C8	Paclitaxel 6α-hydroxylation	9.9[a]	650[a]	Montelukast	Gem gluc.	Yes	PXR/CAR	Rifampin
CYP2C9	Diclofenac 4′-hydroxylation	9.7	2800	Sulfaphenazole	Tienilic acid	Yes	PXR/CAR	Rifampin
CYP2C19	*S*-Mephenytoin 4′-hydroxylation	50	67	Modafinil	*S*-fluoxetine	Yes	PXR/CAR	Rifampin
CYP2D6	Dextromethorphan *O*-demethylation	7.7	220	Quinidine	Paroxetine	No	NA	NA
CYP2J2	Ebastine hydroxylation	3.0[b]	150[b]	Ketoconazole	Troleandomycin	No	NA	NA
CYP3A4/5	Testosterone 6β-hydroxylation	55[c]	3500	Ketoconazole	Troleandomycin	Yes	PXR/CAR	Rifampin
CYP3A4/5	Midazolam 1′-hydroxylation	2.5	1200	Ketoconazole	Troleandomycin	Yes	PXR/CAR	Rifampin

The kinetic constants, Km (μM) and Vmax (pmol/mg protein/min), were determined with XenoTech's XTreme 200 (pool of 200) except for paclitaxel 6α-hydroxylation and ebastine hydroxylation, which were determined with different pools of human liver microsomes.
MDI, metabolism-dependent inhibitor; Gem gluc., gemfibrozil glucuronide; NI, non-inducible enzyme.
[a] Km and Vmax were determined with a pool of 16.
[b] Km and Vmax were determined with a pool of 50.
[c] Testosterone 6β-hydroxylation by CYP3A4/5 shows positive (homotropic) cooperativity. Value represents S_{50}, with a Hill coefficient of 1.3.

TABLE 1.6 Examples of Clinically Relevant Substrates, Inhibitors, and Inducers of the Major Human Liver Microsomal Cytochrome P450 (CYP) Enzymes Involved in Xenobiotic Biotransformation

	CYP1A2	CYP2A6	CYP2B6	CYP2C8	CYP2C9	CYP2C19	CYP2E1
Substrates	Alosetron[g]	Coumarin[a]	Bupropion[a]	Amodiaquine[b]	Diclofenac[a]	Fluoxetine[b]	Aniline[b]
	Caffeine[b,c]	Nicotine[a]	Efavirenz[a,c]	Cerivastatin	Fluoxetine[b]	*S*-Mephenytoin[a,h]	Chlorzoxazone[a]
	Duloxetine[g]		Propofol[b]	Paclitaxel[a,h]	Flurbiprofen[b]	Lansoprazole[c]	Lauric acid[b]
	7-Ethoxyresorufin[b]		*S*-Mephenytoin[b]	Rosiglitazone[b,c]	Phenytoin[h]	Moclobemide	4-Nitrophenol[b]
	Phenacetin[a]		Cyclophosphamide	Repaglinide[c,g]	Tolbutamide[a,c]	Omeprazole[b,c,g]	
	Tacrine[b]		Ketamine		*S*-Warfarin[a,c,h]	Pantoprazole[c]	
	Tizanidine[h]		Meperidine				
	Theophylline[b,c,h]		Nevirapine				
Inhibitors	Acyclovir[f]	Methoxsalen[a]	Clopidogrel[b]	Gemfibrozil[b,c,d]	Amiodarone[c,e]	Fluvoxamine[c]	Clomethiazole[b]
	Cimetidine[f]	Pilocarpine[b]	3-Isopropenyl-3-	Montelukast[a]	Capecitabine	Moclobemide[c]	Diallyldisulfide[b]
	Ciprofloxacin[e]	Tranylcypromine[a]	methyl diamantane[b]	Pioglitazone[b]	Fluconazole[b,c,e]	Nootkatone[b]	Diethyldithiocarbamate[b]
	Famotidine[f]	Tryptamine[b]	2-Isopropenyl-2-	Quercetin[a]	Fluoxetine[b]	Omeprazole[c,d]	Disulfiram[c]
	Fluvoxamine[c,d]		methyladamantane[b]	Rosiglitazone[b]	Fluvoxamine[b]	Ticlopidine[b]	
	Furafylline[a]		Phencyclidine[b]	Rosuvastatin	Oxandrolone[e]		
	Mexilitene[e]		Sertraline[b]	Trimethoprim[b,f]	Sulfaphenazole[a]		
	α-Naphthoflavone[b]		Thio-TEPA[b]		Sulfinpyrazone[f]		
	Norfloxacin[f]		Ticlopidine[b]		Tienilic acid		
	Propafenone[e]		Phenylethylpiperidine				
	Verapamil[f]						
	Zileuton[e]						
Inducers	3-Methylcholanthrene[a]	Dexamethasone[a]	Phenobarbital[a]	Phenobarbital[b]	Phenobarbital	Phenobarbital	Ethanol[c]
	β-Naphthoflavone[a]	Pyrazole[b]	Phenytoin[b]	Rifampin[a,c]	Rifampin	Rifampin[a,c]	Isoniazid
	Omeprazole[a]		Rifampin				
	Lansoprazole[b]						
	TCDD						

	CYP2D6		CYP3A4				
Substrates	Atomoxetine[c]	(*R*)-Metoprolol	Alfentanil[h]	Clopidogrel	Fentanyl[h]	Midazolam[a,c,g]	Saquinavir[g]
	Amitriptyline	Methylphenidate	Alfuzosin	Cyclosporine[h]	Fluticasone[g]	Mifepristone	Sildenafil[c,g]
	Aripiprazole	Mexiletine	Alprazolam	Depsipeptide	Gallopamil	Mosapride	Sibutramine
	Brofaromine	Morphine	Amlodipine	Dexamethasone	Gefitinib	Nicardipine	Simvastatin[c,g]
	(±)-Bufuralol[a]	Nortriptyline	Amprenavir	Dextromethorphan[b]	Gepirone	Nifedipine[b]	Sirolimus[h]
	(*S*)-Chlorpheniramine	Ondansetron	Aprepitant	Diergotamine[h]	Granisetron	Nimoldipine	Sunitinib
	Chlorpromazine	Paroxetine	Artemether	α-Dihydroergocriptine	Gestodene	Nisoldipine	Tacrolimus[h]
	Clomipramine	Perhexilene	Astemiziole[h]	Disopyramide	Halofantrine	Nitrendipine	Tadalafil
	Codeine	Pimozide	Atazanavir	Docetaxel	Laquinimod	Norethindrone	Telithromycin
	Debrisoquine[b]	Propafenone	Atorvastatin	Domperidone	Imatinib	Oxatomide	Terfenadine[b,h]
	Desipramine[c,g]	(+)-Propranolol	Azithromycin	Dutasteride	Indinavir	Oxybutynin	Testosterone[a]
	Dextromethorphan[a,c]	Sparteine	Barnidipine	Ebastine	Isradipine	Perospirone	Tiagabine
	Dolasetron	Tamoxifen	Bexarotene	Eletriptan[c,g]	Itraconazole	Pimozide[h]	Tipranavir
	Duloxetine	Thioridazine[h]	Bortezomib	Eplerenone[g]	Karenitecin	Pranidipine	Tirilazad
	Fentanyl	Timolol	Brotizolam	Ergotamine[h]	Ketamine	Praziquantel	Tofisopam
	Haloperidol (reduced)	Tramadol	Budesonide[g]	Erlotinib	Levomethadyl	Quetiapine	Triazolam[b,c,g]
	Imipramine	(*R*)-Venlafaxine	Buspirone[c,g]	Erythromycin[b]	Lonafarnib	Quinidine[h]	Trimetrexate
	Loperamide		Capravirine	Eplerenone	Lopinavir	Quinine	Vardenafil[g]
			Carbamazepine	Ethosuximide	Loperamide	Reboxetine	Vinblastine
			Cibenzoline	Etoperidone	Lumefantrine	Rifabutin	Vincristine
			Cilastazol	Everolimus	Lovastatin[c,g]	Ritonavir	Vinorelbine
			Cisapride[h]	Ethinyl estradiol	Medroxyprogesterone	Rosuvastatin	Ziprasidone
			Clarithromycin	Etoricoxib	Methylprednisolone	Ruboxistaurin	Zonisamide
			Clindamycin	Felodipine[c,g]	Mexazolam	Salmetrol	

(*Contined*)

TABLE 1.6 (*CONTINED*)

	CYP2D6		CYP3A4				
Inhibitors	Amiodarone[f]	Fluoxetine[c,d]	Amiodarone	Cimetidine[f]	Fluvoxamine	Itraconazole[a,c,d]	Saquinavir[c,d]
	Buproprion	Methadone	Amprenavir[e]	Clarithromycin[c,d]	Fosamprenavir[e]	Mibefradil	St. John's wort
	Chlorpheniramine	Mibefradil	Aprepitant[e]	Diltiazem[e]	Gestodene	Nefazodone[c,d]	Telithromycin[c,d]
	Cimetidine	Paroxetine[c,d]	Atazanavir[c,d]	Erythromycin[e]	Grapefruit Juice	Nelfinavir[c,d]	Troleandomycin[b]
	Clomipramine	Quinidine[a,c]	Azamulin[b]	Felbamate	Ketoconazole[a,c,d]	Ritonavir[c,d]	Verapamil[b,e]
	Duloxetine[e]	Sertraline[f]	Bosentan	Fluconazole[e]	Indinavir[c,d]	Roxithromycin	
	Haloperidol	Terbinafine[e]					
Inducers	NA		Amprenavir	Efavirenz	Nifedipine	Rifampin[a,c]	Troglitazone[b]
			Avasimibe	Etoposide	Omeprazole	Rifapentine[b]	Troleandomycin
			Bosentan	Guggulsterone	Paclitaxel[b]	Ritonavir	Vitamin E
			Carbamazepine[c]	Hyperforin	PCBs	Simvastatin	Vitamin K2
			Clotrimazole	Lovastatin	Phenobarbital[b]	Spironolactone	Yin zhi wuang
			Cyproterone acetate	Mifepristone	Phenytoin[b]	Sulfinpyrazole	
			Dexamethasone[b]	Nelfinavir	Rifabutin	Topotecan	

Note: All FDA classifications are based on information available as of January 30, 2010 at the following URL: http://www.fda.gov/Drugs/DevelopmentApprovalProcess/DevelopmentResources/DrugInteractionsLabeling/ucm093664.htm.

[a] FDA-preferred *in vitro* substrate, inhibitor, or inducer.

[b] FDA-acceptable *in vitro* substrate, inhibitor, or inducer.

[c] FDA-provided examples of *in vivo* substrates, inhibitors, or inducers for oral administration (substrates in this category have plasma AUCs that are increased by at least 2-fold (5-fold for CYP3A4 substrates) when co-administered with inhibitors of the enzyme. Inhibitors in this category increase the AUC of substrates for that enzyme by at least 2-fold (5-fold for CYP3A4). Inducers in this category decrease the plasma AUC of substrates for that enzyme by at least 30%.

[d] Classified by the FDA as a "strong inhibitor" (i.e., caused a ≥5-fold increase in plasma AUC or ≥80% decrease in the clearance of CYP substrates in clinical evaluations).

[e] Classified by the FDA as a "moderate inhibitor" (i.e., caused a ≥2-fold but <5-fold increase in plasma AUC or 50–80% decrease in the clearance of *sensitive* CYP substrates when the inhibitor was given at the highest approved dose and the shortest dosing interval in clinical evaluations).

[f] Classified by the FDA as a "weak inhibitor" (i.e., caused a ≥1.25-fold but <2-fold increase in plasma AUC or 20–50% decrease in the clearance of *sensitive* CYP substrates when the inhibitor was given at the highest approved dose and the shortest dosing interval in clinical evaluations).

[g] Classified by the FDA as a "sensitive substrate" (i.e., drugs whose plasma AUC values have been shown to increase by ≥5-fold when co-administered with a known CYP inhibitor).

[h] Classified by the FDA as a "substrate with narrow therapeutic range" (i.e., drugs whose exposure–response indicates that increases in their exposure levels by concomitant use of CYP inhibitors may lead to serious safety concerns such as Torsades de Pointes).

4. preferred and acceptable inhibitors and inducers for use as positive controls for CYP inhibition and induction studies *in vitro*, and
5. drugs that are effective inducers of CYP enzymes *in vivo* (US FDA, 2006).

The global features of the major drug-metabolizing human CYP enzymes are as follows:

1. Two CYP enzymes, namely CYPD6 and CYP3A4, metabolize the majority of orally effective drugs, but they often metabolize different drugs and they do not metabolize all drugs; consequently there are many drugs whose clearance is largely determined by CYP enzymes other than CYP2D6 or CYP3A4, as shown in Table 1.6. Interindividual variation in CYP2D6, which is mainly confined to the liver, is largely determined by genetic factors whereas interindividual variation in CYP3A4, which is expressed in liver and small intestine, is largely determined by environmental factors (such as inhibitory and inducing drugs).
2. CYP2D6, CYP2C9, CYP2C19, and CYP3A5 are four drug-metabolizing CYP enzymes whose expression is significantly influenced by genetic polymorphisms that give rise to PMs and EMs, the incidence of which varies from one ethnic group to the next. Genetic polymorphisms do affect other CYP enzymes, but not to the same extent as these four CYP enzymes.
3. The induction of CYP3A4 is often associated with induction of CYP2B6, 2C8, 2C9, and 2C19, all of which are regulated by PXR/CAR. CYP1A2 and CYP2E1 are also inducible enzymes, but they are induced by different mechanisms and by different drugs. CYP1A2 is regulated by AhR. Although CYP2E1 is regulated by mechanisms other than those involving AhR, CAR, or PXR, its limited role in drug metabolism makes induction of CYP2E1 only a minor concern from a DDI perspective. Some drug candidates are CAR agonists that selectively induce CYP2B6. Not all CYP enzymes are inducible; CYP2D6, for example, is a noninducible enzyme.

The victim potential of a drug candidate is determined by identifying which CYP enzyme or enzymes are involved in its elimination, a process known as *reaction phenotyping* or *enzyme mapping*. Four *in vitro* approaches have been developed for reaction phenotyping. Each has its advantages and disadvantages, and a combination of approaches is usually required to identify which human CYP enzyme is responsible for metabolizing a xenobiotic (Ogilvie et al., 2008). The four approaches to reaction phenotyping are:

1. Correlation analysis, which involves measuring the rate of drug metabolism by several samples of human liver microsomes and correlating reaction rates with the variation in the level or activity of the individual

CYP enzymes in the same microsomal samples. This approach is successful because the levels of the CYP enzymes in human liver microsomes vary enormously from sample to sample (up to 100-fold) but, with judicious sample selection, they can vary independently from each other.

2. Chemical inhibition, which involves an evaluation of the effects of known CYP enzyme inhibitors, such as those listed in Table 1.5, on the metabolism of a drug by human liver microsomes. Chemical inhibitors of CYP must be used cautiously because most of them can inhibit more than one CYP enzyme and/or their effectiveness or specificity is influenced by the concentration of human liver microsomes. Some chemical inhibitors are metabolism-dependent inhibitors that require biotransformation to a metabolite that inactivates or noncompetitively inhibits CYP.
3. Antibody inhibition, which involves an evaluation of the effects of inhibitory antibodies against selected CYP enzymes on the biotransformation of a drug by human liver microsomes. Due to the ability of antibodies to inhibit selectively and noncompetitively, this method alone can potentially establish which human CYP enzyme is responsible for metabolizing a drug. Unfortunately, the utility of this method is limited by the availability of specific inhibitory antibodies.
4. Biotransformation by purified or recombinant human CYP enzymes, which can establish whether a particular CYP enzyme can or cannot biotransform a drug, but it does not address whether that CYP enzyme contributes substantially to reactions catalyzed by human liver microsomes. The information can potentially be obtained by adjusting rates of metabolism by recombinant CYP enzymes for (1) the specific content of each CYP enzyme in human liver microsomes, (2) the activity of each recombinant CYP enzyme relative to rates of metabolism (relative activity factor or RAF), or (3) a combination of both specific content and RAF, which is known as the intersystem extrapolation factor or ISEF (Proctor et al., 2004; Ogilvie et al., 2008).

As mentioned in the preceding section, reaction phenotyping *in vitro* should be carried out with pharmacologically relevant substrate concentrations so that the metabolism of the drug candidate *in vitro* is not dominated by a clinically irrelevant, high-capacity, low-affinity enzyme.

The perpetrator potential of a drug candidate is determined by identifying which CYP enzymes it can inhibit or induce. CYP inhibition studies are usually conducted with human liver microsomes or recombinant CYP enzymes and involve an evaluation of the drug candidate's ability to function as a direct-acting inhibitor and metabolism-dependent inhibitor of the major drug-metabolizing CYP enzymes in human liver microsomes (CYP1A2, 2B6, 2C8, 2C9, 2C19, 2D6, and 3A4) (Bjornsson et al., 2003a,b; US FDA, 2006; Huang et al., 2008; Grimm et al., 2009). Direct inhibition is evaluated by adding the inhibitor (drug candidate) and substrate (CYP marker substrate) simultaneously to human liver microsomes, whereas metabolism-dependent inhibition is evaluated by preincubating the drug candidate with NADPH-fortified human liver microsomes for 30 min prior to measuring residual CYP activity toward marker substrates. The 30-min preincubation allows the drug candidate to be converted to metabolites that inactivate CYP (or are more potent direct inhibitors than the parent compound). Experimental procedures are described in detail by Ogilvie et al. (2008). The potency of direct-acting inhibitors is expressed as the inhibitory constant K_i, whereas the efficiency of metabolism-dependent inhibition is expressed as k_{inact}/K_I (note that K_i for direct inhibition has a small '*i*' whereas K_I for metabolism-dependent inhibition has a large '*I*'). For direct-acting inhibitors, the FDA uses $[I]/K_i$ to assess the need for a clinical DDI study according to the following criteria:

$[I]/K_i > 1.0$: Clinically significant inhibition is probable.
$[I]/K_i = 0.1$ to 1.0: Clinically significant inhibition is possible.
$[I]/K_i < 0.1$: Clinically significant inhibition is unlikely.

where K_i is the experimentally determined inhibitory constant (determined *in vitro*) and $[I]$ is the mean C_{max} value at steady state for total drug (i.e., bound plus unbound). When the value of $[I]/K_i$ is >1.0, an *in vivo* DDI study is recommended. If the value is <0.1, the likelihood of a clinically relevant interaction is "remote," and an *in vivo* DDI study may not be necessary (Bjornsson et al., 2003a,b; US FDA, 2006; Huang et al., 2008; Zhang et al., 2009).

In the case of metabolism-dependent inhibitors, the FDA's draft guidance document does not describe the agency's approach to evaluating the need for clinical DDI studies other than stating that "*any time-dependent and concentration-dependent loss of initial product formation rate indicates mechanism-based inhibition*" and that this finding should be followed up with human *in vivo* studies (US FDA, 2006). A recent Pharma white paper on metabolism-dependent inhibitors summarizes several methods of predicting the likelihood of a drug causing *in vivo* inhibition based on *in vitro* studies, all of which are based on the following equation which relates changes in victim drug AUC (AUC_i-to-AUC_{ui} ratio) to the fraction of dose metabolized by the irreversibly inhibited enzyme:

$$\frac{AUC_i}{AUC_{ui}} = \frac{1}{\left(\frac{f_{m(CYP)}}{1+\left(\frac{k_{inact}\cdot[I]}{K_I\cdot k_{deg}}\right)}\right)+\left(1-f_{m(CYP)}\right)} \quad \text{(Eq. 1.2)}$$

where $[I]$ is the *in vivo* drug (inactivator) concentration, k_{inact} and K_I are kinetic parameters determined *in vitro* (i.e., the first-order rate constant of enzyme inactivation in the presence of saturating concentrations of inhibitor and the concentration of inhibitor supporting half-maximal rates of enzyme inactivation, respectively), $f_{m(\mathrm{CYP})}$ represents the fraction of total clearance of the victim drug to which the affected CYP enzyme contributes, and k_{deg} is the first-order rate constant of *in vivo* degradation of the affected enzyme (Grimm et al., 2009). Various "surrogate" concentrations can be used for [I]. The estimated $C_{\mathrm{max,u,inlet}}$ (estimated unbound steady-state C_{max} at the inlet to the liver) can be used in this equation in an attempt to approximate the actual unbound concentration in the liver, as described by (Kanamitsu et al., 2000). The estimated $C_{\mathrm{max,u,inlet}}$ is higher than the unbound systemic concentration, but less than the total systemic concentration. Equation 1.2 describes the impact of inactivating a liver CYP enzyme on systemic exposure to a victim drug. The Pharma white paper on metabolism-dependent inhibitors provides equations that take into account the simultaneous impact of CYP inactivation in both the liver and small intestine, which is particularly relevant for inactivators of CYP3A4 (Grimm et al., 2009).

Enzyme induction studies are usually conducted with primary cultures of human hepatocytes (Ogilvie et al., 2008). The FDA guidance document and a recent Pharma white paper (US FDA, 2006; Huang et al., 2008; Chu et al., 2009) describe the conduct of such studies, which are usually conducted in three different preparations of cultured human hepatocytes. A recent review article from the FDA recommends that *in vitro* induction studies include measurements of CYP1A2, CYP2B6, and CYP3A4 to assess the ability of the drug candidate to activate the major xenosensors AhR, CAR, and PXR, respectively. Activation of CAR and PXR generally results in induction of CYP2B6, 2C8, 2C9, 2C19, and 3A4, but there are drug candidates that selectively induce CYP2B6. The magnitude of CYP induction can vary from one preparation of human hepatocytes to the next, for which reason it is important to include positive controls in all *in vitro* induction studies (Madan et al., 2003). As shown in Table 1.5, omeprazole is the positive control for AhR-dependent CYP1A2 induction; phenobarbital is the positive control for CAR-dependent CYP2B6 induction, and rifampin is the positive control of PXR-dependent induction of CYP3A4 (and CYP2C8, 2C9, and 2C19). Because the magnitude of CYP induction by a given positive control can vary from one hepatocyte culture to the next, the recent Pharma white paper advocates the following acceptance criteria for the response to positive controls: (1) at least a twofold induction of CYP enzyme activity, or (2) at least a sixfold induction of CYP mRNA levels (Chu et al., 2009).

An assessment of the clinical relevance of the results of *in vitro* enzyme induction studies in human hepatocytes is based on the degree of CYP induction by the drug candidate (at pharmacologically relevant concentrations) relative to the appropriate positive controls (i.e., it is based on percent relative effectiveness). The FDA's draft guidance document indicates that a clinical induction study is required if, at pharmacologically relevant concentrations, a drug candidate induces a CYP enzyme to 40% or more of the appropriate positive control (US FDA, 2006). A more conservative approach is advocated in the recent Pharma white paper (Chu et al., 2009), which recommends the following three-tier system for assessing the clinical relevance of a drug candidate that causes enzyme induction in human hepatocytes at pharmacologically relevant concentrations (based on total [bound plus unbound] plasma concentration):

If induction is ≥40% of positive control, then clinical induction is probable;
If induction is 15–40% of positive control, then clinical induction is possible;
If induction is ≤15% of positive control, then clinical induction is unlikely.

In addition to their use in the aforementioned studies (reaction phenotyping, CYP inhibition, and CYP induction), human liver microsomes and hepatocytes are also used to support *in vitro* studies of metabolite profiling and intrinsic clearance. Metabolite profiling studies with microsomes and hepatocytes from humans and nonclinical species are recommended in the FDA's guidance on Metabolites in Safety Testing as a valuable *in vitro* approach to identify the potential for a drug candidate to be converted to a human-specific or human-abundant metabolite (US FDA, 2008). Such *in vitro* studies are particularly valuable in guiding the selection of the non-rodent species (dog or monkey) to support nonclinical safety studies. Such studies (in conjunction with an evaluation of the drug candidate's structure) can also guide the selection of the appropriate test system (microsomes or hepatocytes) for *in vitro* studies of species differences in metabolism and reaction phenotyping. The use of microsomes and hepatocytes to assess drug clearance is outlined below.

When the clearance of a drug is largely determined by its rate of metabolism by CYP, its rate of metabolism by human liver microsomes or hepatocytes *in vitro* generally provides a reasonable estimate of its intrinsic hepatic clearance *in vivo*. Whole liver intrinsic clearance ($CL_{int\text{-}whole\ liver}$) can be estimated from k_{el}, the rate constant of elimination (min^{-1}), which is determined by measuring the rate of disappearance of the drug candidate (at a pharmacologically relevant concentration [one that is well below K_m]) from incubations with NADPH-fortified human liver microsomes or human hepatocytes based on single exponential decay ($y = A_0e^{-kel}$). *In vitro* intrinsic clearance can also be determined from V_{max}/K_m (the quotient of the Michaelis–Menten kinetic constants, which are determined by measuring rates of metabolism over a wide range of substrate concentrations), but this approach requires considerably more work than an analysis of the time course of substrate disappearance at a low, pharmacologically relevant concentration of drug candidate.

For microsomal incubations, the rate constant of elimination is used to calculate the *in vitro* intrinsic clearance in microsomes, $CL_{int\text{-}mcs}$ (in units of mL/min/mg microsomal protein), according to the following equation (Youdim et al., 2008):

$$CL_{int\text{-}mcs}\ (\text{mL/min/mg microsomal protein}) = k_{el}\ (\text{min}^{-1}) \times \frac{\text{incubation volume (mL)}}{\text{mg protein per incubation}} \quad \text{(Eq. 1.3)}$$

The rate of *in vitro* intrinsic clearance in microsomes ($CL_{int\text{-}mcs}$) is used to estimate the *in vivo* rate of hepatic clearance, $CL_{int\text{-}whole\ liver}$ (in units of mL/min) based on the following equation.

$$CL_{int\text{-}whole\ liver}\,(mL/min) = CL_{int\text{-}mcs}\,(mL/min/mg\ protein) \times MPPGL\,(mg/g) \times liver\ weight\,(g)$$

A value of 47 mg microsomal protein/g liver is used for human MPPGL (milligram of microsomal protein per gram of liver) (Hakooz et al., 2006).

For hepatocyte incubations, the rate constant of elimination is used to calculate the *in vitro* intrinsic clearance $CL_{int\text{-}hepatocytes}$ (in units of mL/min/10^6 cells per incubation) according to the following equation (Barter et al., 2007):

$$CL_{int\text{-}hepatocytes}\,(mL/min/10^6\ cells) = k_{el}\,(min^{-1}) \times \frac{incubation\ volume\,(mL)}{number\ of\ million\ cells\ per\ incubation} \quad \text{(Eq. 1.4)}$$

The rate of *in vitro* intrinsic clearance in hepatocytes ($CL_{int\text{-}hepatocytes}$) is used to estimate the *in vivo* rate of hepatic clearance, $CL_{int\text{-}whole\ liver}$ (in units of mL/min) based on the following equation:

$$CL_{int\text{-}whole\ liver}\,(mL/min) = CL_{int\text{-}hepatocytes}\,(mL/min/10^6\ cells) \times HPPGL\,(10^6\ cells/g) \times liver\ weight\,(g) \quad \text{(Eq. 1.5)}$$

A value of 99×10^6 cells/g is used for human HPPGL (hepatocytes per gram of liver) (Griffiths et al., 2005; Barter et al., 2007). The average weight of human liver is 1650 g (Price et al., 2003; Johnson et al., 2005).

Estimations of *in vitro* intrinsic clearance are usually based on the following assumptions:

1. the pharmacokinetic behavior of the drug candidate conforms to the well-stirred model;
2. the free concentration of drug candidate in the microsomal/hepatocyte incubation is represented by the total concentration of drug candidate (i.e., its nominal concentration). *In vitro* clearance can be based on the free concentration of drug candidate if this is measured experimentally or estimated from physiochemical properties as described by Hallifax and Houston (2006) and Gertz et al. (2008). The unbound fraction of drug candidate in a microsomal incubation (fu_{inc} or fu_{mic}) can be predicted reasonably well based from logP or $logD_{7.4}$, according to the following equation:

$$fu_{inc} = \frac{1}{1 + C \cdot 10^{0.072 \cdot \log P/D^2 + 0.067 \cdot \log P/D - 1.126}} \quad \text{(Eq. 1.6)}$$

where *C* is the concentration of microsomal protein (mg/mL), logP/D is the logP of the molecule if it is a base (basic pKa > 7.4), and $\log D_{7.4}$ of the molecule if it is neutral or an acid (acidic pKa < 7.4).

In vitro rates of intrinsic clearance can overestimate and underestimate *in vivo* rates of systemic clearance for several reasons. First, for drugs that are rapidly metabolized *in vitro*, the estimated *in vivo* rate of clearance may exceed hepatic blood flow. Therefore, care should be taken not to estimate rates of hepatic intrinsic clearance ($CL_{\text{int-whole liver}}$) that exceed hepatic blood flow, which is 1780 mL/min in humans (Boxenbaum, 1980). Second, two drugs with the same *in vitro* intrinsic clearance may have different *in vivo* clearance rates because of differences in Vd. Acidic drugs and drugs containing a sulfonamide moiety tend to bind tightly to plasma proteins but not to microsomes (Brunton et al., 2006). Accordingly, unless microsomes/hepatocytes are incubated *in vitro* in the presence of plasma, the free concentration of acids/sulfonamides *in vitro* will be much greater than that *in vivo*. Accordingly, *in vitro* estimates of clearance of acidic and sulfonamide-containing drugs tend to overestimate their rate of clearance *in vivo*. Conversely, rates of microsomal glucuronidation *in vitro* generally underestimate *in vivo* rates of glucuronidation (Soars et al., 2002; Miners et al., 2006; Ogilvie et al., 2008), which is discussed in the next section on glucuronidation.

1.8.2 Glucuronidation

Like CYP-catalyzed oxidations, UGT-catalyzed reactions involving UDP-glucuronic acid (UDPGA) (and occasionally UDP-glucose, UDP-xylose, and UDP-galactose) are a major pathway of drug metabolism in humans. The site of glucuronidation is generally an electron-rich nucleophilic oxygen, nitrogen, or sulfur heteroatom. Therefore, substrates for glucuronidation contain such functional groups as aliphatic alcohols and phenols (which form *O*-glucuronide ethers), carboxylic acids (which form *O*-glucuronide esters, also known as acyl glucuronides), primary and secondary aromatic and aliphatic amines (which form *N*-glucuronides), and free sulfhydryl groups (which form *S*-glucuronides). Certain xenobiotics, such as phenylbutazone, sulfinpyrazone, suxibuzone, ethchlorvynol, Δ^6-tetrahydrocannabinol, and feprazone contain carbon atoms that are sufficiently nucleophilic to form *C*-glucuronides. The *C*-glucuronidation of the enolic form of phenylbutazone is catalyzed specifically by UGT1A9 (Nishiyama et al., 2006).

A common pathway of drug metabolism involves hydroxylation by CYP (Phase 1) followed by glucuronidation by UGT (Phase 2). C-hydroxylation by CYP results in formation of a phenol (in the case of aromatic hydroxylation) or an alcohol (in the case of aliphatic hydroxylation). In general, phenols are conjugated with glucuronic acid (and sulfonic acid) much more rapidly than are alcohols. This principle is illustrated in Fig. 1.13. 17β-Estradiol and morphine both contain a non-sterically hindered phenol and alcohol (i.e., they

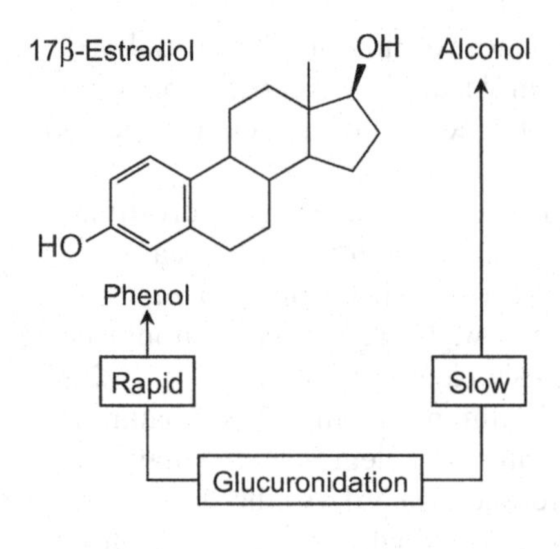

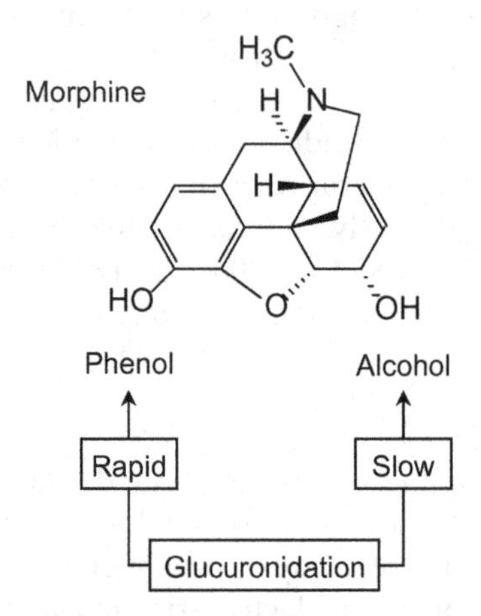

Figure 1.13 An illustration indicating that rates of glucuronidation proceed much faster with phenols (aromatic OH groups) than with alcohols (aliphatic OH groups). 17β-Estradiol (top) and morphine (bottom) both contain a non-sterically hindered phenol and alcohol (i.e. they both contain an exposed aromatic and aliphatic OH group). Both are conjugated with glucuronic acid, but the rate of conjugation of the phenol greatly exceeds (by at least fivefold) the rate of glucuronidation of the alcohol. In human liver microsomes, glucuronidation of the phenolic OH group in 17β-estradiol is catalyzed by UGT1A1, whereas glucuronidation of the phenolic OH group in morphine is catalyzed by UGT2B7.

both contain an exposed aromatic and aliphatic OH group). Both are conjugated with glucuronic acid, but the rate of conjugation of the phenol greatly exceeds (by >fivefold) the rate of glucuronidation of the alcohol. In human liver microsomes, glucuronidation of the phenolic OH group in 17β-estradiol is catalyzed by UGT1A1 whereas glucuronidation of the phenolic OH group in morphine is catalyzed by UGT2B7.

The major human UGT enzymes are summarized in Table 1.7. Probe substrates have been identified for some but not all of the human UGTs, including UGT1A1 (17β-estradiol 3-glucuronidation and bilirubin), UGT1A3 (hexafluoro-1α,25-trihydroxyvitamin D3), UGT1A4 (trifluoperazine), UGT1A6 (serotonin and 1-naphthol), UGT1A9 (propofol), UGT2B7

TABLE 1.7 Major Human UDP-Glucuronosyltransferase Enzymes (UGTs)

UGT	Present in Liver?	Tissue	Example Substrates
1A1	Yes	Liver, small intestine, colon	**Bilirubin**, **17β-estradiol (3-glucuronidation)**, raloxifene, 17α-ethinyl estradiol, carvedilol, levothyroxine, acetaminophen
1A3	Yes	Liver, small intestine, colon	**Hexafluoro-1α,25-trihydroxyvitamin D3**, tertiary amines, antihistamines, 17β-estradiol, ketotifen, naproxen, ketoprofen, ibuprofen, fenoprofen, valproic acid, ezetimibe, norbuprenorpine
1A4	Yes	Liver, small intestine, colon	**Trifluoperazine**, tertiary amines, antihistamines, lamotrigine, amitriptyline, cyclobenzaprine, olanzapine
1A5	Yes	Liver	Unknown
1A6	Yes	Liver, small intestine, colon, stomach	**1-Naphthol, serotonin**, 4-nitrophenol, 4-methylumbelliferone, ibuprofen, acetaminophen, SN-38, diclofenac
1A7	No	Esophagus, stomach, lung	Octylgallate, arylamines, 4-hydroxybiphenyl, 4-hydroxyestrone, mycophenolic acid, SN-38
1A8	No	Colon, small intestine, kidney?	Entacapone, troglitazone, anthraquinone, 8-hydroxyquinoline, furosemide, raloxifene, niflumic acid, ciprofibric acid, clofibric acid, valproic acid, mycophenolic acid, diflunisal, furosemide
1A9	Yes	Liver, colon, kidney	**Propofol**, thyroid hormones, entacapone, salicylic acid, scopoletin, fenofibrate, acetaminophen, ketoprofen, ibuprofen, fenoprofen, naproxen, furosemide, diflunisal, diclofenac, bumetanide
1A10	No	Stomach, small intestine, colon	1-Naphthol, mycophenolic acid, estrogen, raloxifene, troglitazone, furosemide
2A1	No	Olfactory	Valproic acid, ibuprofen
2A2	Unknown	Unknown	Unknown
2A3	Unknown	Unknown	Unknown
2B4	Yes	Liver, small intestine	Hyodeoxycholate, estriol, codeine, androsterone, carvedilol

(*Contined*)

TABLE 1.7 (*Contined*)

UGT	Present in Liver?	Tissue	Example Substrates
2B7	Yes	Kidney, small intestine, colon, liver	**Zidovudine (AZT), morphine (6-glucuronidation[a])**, ibuprofen, ketoprofen, diclofenac, opioids, oxazepam, carvedilol, clofibric acid, naloxone, valproic acid, tiprofenic, zomepirac, benoxaprofen
2B10	Yes	Liver, ileum, prostate	Nicotine, imipramine, amitriptyline
2B11	Yes	Mammary, prostate, others	4-Nitrophenol, naphthol, estriol, 2-aminophenol, 4-hydroxybiphenyl
2B15	Yes	Liver, small intestine, prostate	***S*-Oxazepam**, androgens, flavonoids, 4-hydroxytamoxifen, estriol, entacapone, SN-38, tolcapone, diclofenac
2B17	Yes	Liver, prostate	Androgens, eugenol, scopoletin, galangin, ibuprofen
2B28	Yes	Liver, mammary	17β-Estradiol, testosterone

Bold text represents selective substrates (or reactions).
[a] UGT2B7 also catalyzes the 3-glucuronidation of morphine, but this reaction is also catalyzed by other UGTs.
Adapted from Miners et al. (2006), Kiang et al. (2005), Williams et al. (2004), and Fisher et al. (2000b).

(morphine 6-glucuronidation and zidovudine [AZT]), and UGT2B15 (*S*-oxazepam) (Miners et al., 2006). Selective inhibitors have only been identified for UGT1A4 (hecogenin) and 2B7 (fluconazole) (Miners et al., 2006).

In contrast to the situation with CYP enzymes, there are fewer clinically relevant inhibitory DDIs caused by inhibition of UGT enzymes, and AUC increases are rarely greater than twofold (Williams et al., 2004), whereas dramatic AUC increases have been reported for CYP enzymes, such as the 190-fold increase in AUC reported for the CYP1A2 substrate ramelteon (Rozerem™) upon coadministration of fluvoxamine (Rozerem prescribing information: http://general.takedapharm.com/content/file/PI.pdf?applicationCode=2bcc07ca-d9c0-4704-9a28-963127115641&fileTypeCode=ROZEREMPI 2008). For instance, plasma levels of indomethacin are increased approximately twofold upon coadministration of diflunisal, and *in vitro* studies indicate that this interaction is due in part to inhibition of indomethacin glucuronidation in the intestine (Gidal et al., 2003; Mano et al., 2006). Valproic acid coadministration increases the AUC of lorazepam and lamotrigine by 20 and 160%, respectively (Williams et al., 2004). Clinically relevant DDIs due to induction of UGT enzymes have also been observed, but again the magnitude of the interaction is less than that observed for CYP induction. Rifampin coadministration increases mycophenolic acid clearance by 30%, and increases the AUC of its acyl glucuronide (formed by UGT2B7), and its 7-*O*-glucuronide (formed by various UGT1 enzymes) by more than 100 and 20%, respectively (Naesens et al., 2006). In spite of the limited number of reports of UGT-mediated DDIs, UGT enzymes are increasingly involved in the metabolism of new drug candidates. The apparent increased involvement of UGT enzymes is due in part to the selection process for new drug candidates, which is often biased against chemicals that interact with CYP enzymes. The selection of chemicals with little potential of inhibiting CYP enzymes, chemicals that are metabolically stable in NADPH-fortified human liver microsomes, and chemicals with high aqueous solubility are chemicals that are more likely to be metabolized by UGT than CYP. The FDA acknowledges that "there are few documented cases of clinically significant drug-drug interactions related to non-CYP enzymes … ," but goes on to say that "the identification of drug metabolizing enzymes of this kind (i.e., glucuronosyltransferases, sulfotransferases, and *N*-acetyl transferases) is encouraged" (US FDA, 2006).

Some UGT enzymes (e.g., UGT1A7, 1A8, 1A10, and 2A1) are not expressed in human liver (see Table 1.7). Of the latter enzymes, UGT1A7, 1A8, and 1A10 are expressed in the gastrointestinal tract where they appear to be important for pre-hepatic elimination of various orally administered drugs. Of the hepatically expressed UGT enzymes, UGT1A1, 1A3, 1A4, 1A6, 1A9, 2B7, and 2B15 are considered to be the UGT enzymes most important for hepatic drug metabolism (Miners et al., 2006). The current UGT nomenclature system may be found at: http://www.flinders.edu.au/medicine/sites/clinical-pharmacology/ugt-homepage.cfm.

Typical reactions catalyzed by UGT enzymes require the cofactor UDPGA. However, the C-terminus of all UGT enzymes contains a membrane-spanning domain that anchors the enzyme in the endoplasmic reticulum, and the enzyme faces the lumen of the endoplasmic reticulum, where it is ideally placed to conjugate lipophilic xenobiotics and their metabolites generated by oxidation, reduction, or hydrolysis. When human liver is homogenized, the endoplasmic reticulum appears to vesiculate with roughly one-half of the microsomes "inside-out" and the other half "outside-in." Those that vesiculate with the lumen oriented toward the middle cannot catalyze glucuronidation reactions because the water-soluble cofactor UDPGA cannot cross the membrane in order to access the UGT enzymes. Access of these "latent" UGT enzymes to UDPGA can be achieved by solubilizing the membranes with detergent (such as the zwitterionic detergent l-naphthol, 4-hydroxybiphenyl, [3-[3-chlolamidopropyl]dimethylammonio]- 1-propanesulfonic acid [CHAPS]) or by permeabilizing the membrane with the pore-forming peptide alamethicin (Fisher et al., 2000a). Treatment of microsomes with detergents virtually eliminates CYP activity, hence, detergents cannot be used in studies designed to investigate the possible coupling of oxidation reactions catalyzed by CYP with conjugation reactions catalyzed by UGT. This is not a limitation

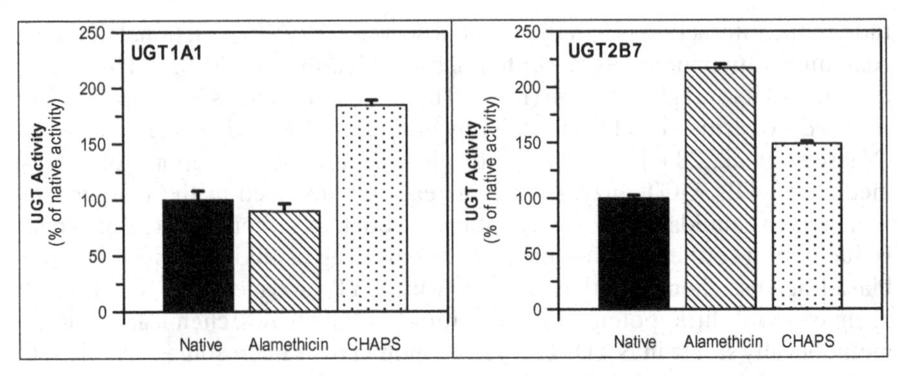

Figure 1.14 The activating effects of CHAPS and alamethicin on rates of glucuronidation (UGT1A1 and UGT2B7 activity) in human liver microsomes. The activity UGT1A1 (17β-estradiol 3-glucuronidation) and UGT2B7 (morphine 6-glucuronidation) was measured in 16 samples of human liver microsomes in the presence and absence of the zwitterionic detergent CHAPS and the pore-forming peptide alamethicin. Values are mean ± standard deviation (n = 16).

of alamethicin, which activates UGT activity without inhibiting CYP (Fisher et al., 2000a). The effects of CHAPS and alamethicin on UGT1A1 and UGT2B7 activity in 16 samples of human liver microsomes is shown in Fig. 1.14. CHAPS, but not alamethicin caused a twofold increase in UGT1A1 activity (17β-estradiol 3-glucuronidation) whereas alamethicin, but not CHAPS, caused slightly more than a twofold increase in UGT2B7 activity (morphine 6-glucuronidation). Fig 1.14 illustrates that the magnitude of the increase in UGT activity with membrane disrupters is modest (less than threefold), and the best activator for one UGT enzyme may not be the best activator of another UGT enzyme.

The *in vitro* activity of UGT enzymes is highly dependent not only on the substance used to activate the microsomal membranes, but also on incubation conditions. The kinetic properties of UGT enzymes have been demonstrated to vary with the concentration of cofactor, membrane composition, type of buffer, ionic strength and pH (Miners et al., 2006). *In vitro* intrinsic clearance (CL_{int}) values (measured as V_{max}/K_m) for zidovudine (AZT) glucuronidation in human liver microsomes were shown to vary sixfold depending on incubation conditions, but even under conditions that produced the greatest CL_{int}, the *in vivo* clearance rate was underpredicted by three- to fourfold (Miners et al., 2006). This *in vitro* underprediction of the *in vivo* rate of clearance of drugs that are glucuronidated is typical when kinetic data based on experiments with human liver microsomal are used to assess CL_{int}, and is likely due to a number of factors including the presence or absence of albumin, nonspecific binding, atypical *in vitro* kinetics, active uptake of the aglycone into hepatocytes, and significant extrahepatic expression of various UGT enzymes (Miners et al., 2006). The prediction of the *in vivo* clearance of drugs that are glucuronidated by hepatocytes appears to be more accurate than predictions made with microsomes, but underprediction is still the likely outcome. The case of zidovudine appears to be an exception to this rule, as *in vivo* clearance values are predicted well with hepatocytes (Miners et al., 2006). The use of either microsomes or recombinant UGT2B7 also appears to underpredict the *in vivo* magnitude of the *inhibitory* interaction between fluconazole and zidovudine by 5- to 10-fold (Miners et al., 2006). However, when 2% bovine serum albumin (BSA) is added to either system, there is a decrease in the K_i value of 85% which results in a much improved prediction of the *in vivo* interaction (Miners et al., 2006). The effect of BSA is not due to nonspecific binding, but the exact mechanism remains to be elucidated.

The kinetic properties of UGT enzymes are possibly influenced by the formation of homo- and heterodimers among UGT enzymes and by the formation of heterodimers with other microsomal enzymes such as various CYP enzymes or epoxide hydrolase (Miners et al., 2006). For instance, it has been demonstrated that the ratio of morphine 3-glucuronide to morphine 6-glucuronide formed by UGT2B7 is altered by the presence of CYP3A4 (Miners et al., 2006). It is not known whether dimerization between UGT2B7 and CYP3A4, if it occurs, also alters the substrate specificity of other individual UGT enzymes, which would have implications for studies designed to determine the substrate specificity of recombinant UGT enzymes, which are invariably expressed individually. Additionally, there is currently no universally accepted method to quantify UGT enzymes in a recombinant preparation, which precludes the accurate determination of RAFs, as for recombinant CYP enzymes. Finally, posttranslational modifications to UGT enzymes that occur *in vivo* in humans (e.g., phosphorylation and *N*-glycosylation) may not occur in the cell expression system chosen to produce the recombinant UGT enzymes (i.e., bacterial systems), which can impact activity in a substrate-dependent manner (Miners et al., 2006). All of these findings suggest that the use of recombinant human UGT enzymes may not provide accurate indications of the extent to which a given UGT can glucuronidate a given drug candidate.

Some primary amines, or the demethylated metabolites of secondary and tertiary amines, such as carvedilol, sertraline, varenicline, mofegiline, garenoxacin, tocainide, and sibutramine, among others, have been reported to be converted to *N*-carbamoyl glucuronides (Gipple et al., 1982; Tremaine et al., 1989; Beconi et al., 2003; Hayakawa et al., 2003; Obach et al., 2005; Link et al., 2006; Obach et al., 2006).

$$\text{R–NH}_2 + \text{HCO}_3^- + \text{UDP–C}_6\text{H}_9\text{O}_6 \rightarrow \text{R–NH–CO–O–C}_6\text{H}_9\text{O}_6 + \text{UDP} + \text{OH}^-$$

Marked species difference have been found in the formation of *N*-carbamoyl glucuronides, and humans have been found to produce these conjugates from even fewer drugs, including varenicline, sertraline, and mofegiline. To form this type of conjugate *in vitro*, the incubation must be performed under a CO_2

atmosphere, and include a carbonate–bicarbonate buffer. Although not directly demonstrated, it has been hypothesized that a transient carbamic acid intermediate is formed by the interaction of the amine with the dissolved CO_2, followed by its glucuronidation (Obach et al., 2005). Given that the *in vitro* formation of *N*-carbamoyl glucuronides occurs only under special incubation conditions that are not typically employed *in vitro*, it is possible that many other primary and secondary amines or their oxidative metabolites can be converted to such conjugates but have not been detected.

Glucuronide conjugates formed in the liver are exported by transporters on the canalicular or sinusoidal membrane for excretion in bile or urine, respectively. Transport of glucuronides across the sinusoidal membrane into the blood is mediated by MRP1 (ABCC1), MRP3 (ABCC3), and MRP4 (ABCC4). The transport of glucuronides across the canalicular membrane into bile is mediated by MRP2 (ABCC2), and BCRP (ABCG2). Glucuronides can also be taken up by hepatocytes through the action of OATP1B1 (OATP2), and OATP1B3 (OATP8) on the sinusoidal membrane (Giacomini and Sugiyama, 2006). Whether glucuronides are excreted in bile or urine depends, in part, on the size of the aglycone (the parent compound or unconjugated metabolite). In rat, glucuronides are preferentially excreted in urine if the molecular weight of the aglycone is less than 250 g/mol, whereas glucuronides of larger molecules (aglycones with molecular weight >350 g/mol) are preferentially excreted in bile. Molecular weight cutoffs for the preferred route of excretion vary among mammalian species. The carboxylic acid moiety of glucuronic acid, which is ionized at physiological pH, promotes excretion because (1) it increases the aqueous solubility of the xenobiotic and (2) it is recognized by the biliary and renal organic anion transport systems, which enable glucuronides to be secreted into urine and bile.

1.9 REACTIVE METABOLITES

The terms "Phase 1" and "Phase 2" drug-metabolizing enzymes were originally coined by R.T. Williams (1959) to distinguish those enzymes that could convert xenobiotics to reactive (toxic) metabolites from those that could not. This classification indicates that R.T. Williams fully appreciated the potential for CYP to convert drugs and other xenobiotics to toxic metabolites, but he apparently underappreciated the ability of conjugating enzymes to do likewise. Today, *all* drug-metabolizing enzymes, including the glutathione *S*-transferases, are recognized as playing a role in the metabolic activation of one or more xenobiotics to reactive metabolites (Mahajan and Evans, 2008; Parkinson and Ogilvie, 2008). CYP is recognized as the enzyme system responsible for converting many diverse drugs to reactive metabolites; as evidenced by the fact that CYP (in the form of NADPH-fortified liver microsomes) is included as the activating system in the Ames bacterial mutagenicity assay, and that covalent binding of drug candidates to microsomal protein or glutathione has been used to assess the potential of drug candidates to be converted to reactive metabolites. However, some drugs, including the NSAID class of carboxylic acid-containing drugs, are converted to reactive metabolites by UGTs. The following two sections discuss the metabolic activation of drugs by CYP-dependent oxidation and by glucuronidation of acid-containing drugs to acyl glucuronides with an emphasis on covalent binding to microsomal protein and hepatotoxicity.

1.9.1 CYP-Dependent Activation

All of the major drug-metabolizing CYP enzymes in human liver microsomes (and certain CYP enzymes expressed in extrahepatic tissues) have been implicated in the metabolic activation of drugs or other xenobiotics to reactive metabolites associated with various adverse events (such as liver toxicity). A representative sampling is listed in Table 1.8. In some cases, more than one CYP enzyme is capable of activating a drug to a reactive metabolite. Acetaminophen, for example, is converted to its reactive metabolite *N*-acetylbenzoquinoneimine by three hepatic CYP enzymes, namely CYP1A2, CYP2E1, and CYP3A4. Furthermore, as shown in Fig. 1.9, acetaminophen can be activated to nephrotoxic metabolites by prostaglandin H synthase (PHS). Many of the CYP-catalyzed reactions shown in Fig. 1.10 have been implicated in the formation of reactive metabolites (ultimate toxicants) or their precursors (proximate toxicant). CYP-catalyzed activation reactions include (but are not limited to):

1. The *para*-hydroxylation of aromatic amines to form *para*-aminophenols (like acetaminophen) that can be further activated by dehydrogenation to reactive quinoneimines.
2. The *N*-hydroxylation of aromatic amines to hydroxylamines that undergo conjugation by *O*-acetylation or *O*-sulfonation. The conjugate (X) represents a good leaving group that promotes formation of a reactive nitrenium ion ($R\text{–}NH^+$) as follows:

$$R\text{–}NH_2 \rightarrow R\text{–}NH\text{–}OH \rightarrow R\text{–}NH\text{–}OX \rightarrow R\text{–}NH^+ + XO^-$$

3. The epoxidation of alkenes to reactive epoxides.
4. The conversion of alkynes to ketocarbenes (R=C=O), possibly via an initial epoxidation reaction.
5. The conversion of aromatic hydrocarbons to oxiranes (epoxides).
6. Oxidative and reductive dehalogenation of drugs like halothane (to form reactive acyl halides and carbon-centered radicals, respectively)
7. *S*-oxygenation of thioamides resulting in the formation of iminosulfinic acids.

TABLE 1.8 Examples of Xenobiotics Activated by Human Cytochrome P450 (CYP) Enzymes

CYP1A2	CYP2E1
Acetaminophen	Acetaminophen
2-Acetylaminofluorene	Acrylonitrile
4-Aminobiphenyl	Benzene
2-Aminofluorene	Carbon tetrachloride
2-Naphthylamine	Chloroform
NNK	Dichloromethane
Amino acid pyrolysis products	1,2-Dichloropropane
(DiMeQx, MelQ, MelQx, Glu P-1,	Ethylene dibromide
Glu P-2, IQ, PhlP, Trp P-1, Trp P-2)	Ethylene dichloride
Tacrine	Ethyl carbamate
CYP2A6 and 2A13	Halothane
NNK and bulky nitrosamines	*N*-Nitrosodimethylamine
N-Nitrosodiethylamine	Styrene
Aflatoxin B1	Trichloroethylene
CYP2B6	Vinyl chloride
6-Aminochrysene	CYP3A4
Cyclophosphamide	Acetaminophen
Ifosfamide	Aflatoxin B_1 and G_1
CYP2C8, 9, 18, 19	6-Aminochrysene
Tienilic acid	Benzo[*a*]pyrene 7,8-dihydrodiol
Phenytoin	Cyclophosphamide
Valproic acid	Ifosfamide
CYP2D6	1-Nitropyrene
NNK	Sterigmatocystin
CYP2F1	Senecionine
3-Methylindole	*Tris*(2,3-dibromopropyl) phosphate
Acetaminophen	CYP4B1
Valproic acid	Ipomeanol
CYP1A1 and 1B1	3-Methylindole
Benzo[*a*]pyrene and other polycyclic aromatic hydrocarbons	2-Aminofluorene

NNK, 4-(methylnitrosamino)-1-(3-pyridyl)-1-butanone, a tobacco-specific nitrosamine.
Adapted from Guengerich and Shimada (1991).

Safety concerns over the metabolic activation of drugs by CYP are founded on several observations. First, experiments in nonclinical species treated with CYP inhibitors or inducers and experiments in CYP knockout mice have established a key role for CYP in mediating the toxicity of several drugs and xenobiotics. Second, metabolic activation plays a key role in the carcinogenesis of many genotoxic chemicals such as polycyclic aromatic hydrocarbons (which are converted to DNA-reactive diol-epoxides). It is for this reason that NADPH-fortified liver microsomes are included as the activating system on the Ames bacterial mutagenicity assay. Third, the metabolic activation of certain drugs, like mibefradil (Posicor), leads to inactivation of CYP. Such irreversible, metabolism-dependent inhibition can cause DDIs due to extensive and prolonged inhibition of CYP activity. The toxicity of certain drugs, such as halothane and tienilic acid (a metabolism-dependent inhibitor of CYP2C9) involves an immune reaction with production of antibodies against proteins covalently modified with the drug metabolite (hapten formation).

Concern over CYP-dependent metabolic activation led to the practice of assessing the potential of drug candidates to be converted to reactive (hepatotoxic) metabolites based on their covalent binding to microsomal protein (reviewed in Evans et al., 2004). Scientists at Merck developed the procedure of basing (in part) the decision to discontinue the development of a drug candidate on whether covalent binding to microsomal protein exceeds 50 pmol equivalents/mg protein (hereafter abbreviated to 50 pmol adducts/mg protein). This threshold stemmed from studies of the covalent binding of hepatotoxic drugs to rat or mouse liver protein *in vivo*, and was then adapted to an *in vitro* test system. The hepatotoxicants evaluated were acetaminophen, phenacetin, amodiaquine, bromobenzene, ethinyl estradiol, furosemide, 4-iopmeanol, isoniazid, and iproniazid. From these studies, Merck established that hepatic necrosis *in vivo* is associated with a threshold of covalent binding of 1 nmol/mg protein. Interestingly, only four of the nine compounds tested *in vivo* actually resulted in covalent binding that exceeded this threshold (namely, acetaminophen, bromobenzene, furosemide, and 4-ipomeanol). In the case of acetaminophen, this level of covalent binding (1.3-1.6 nmol adducts/mg protein) was observed in phenobarbital-induced mice (which are more susceptible to the hepatotoxic effects of acetaminophen than noninduced mice).

The *in vivo* threshold of 1 nmol adducts/mg protein was reduced by a safety factor of 20 to establish the *in vivo* threshold of 50 pmol adducts/mg liver protein, which also corresponds to a level of protein binding that, when measured by liquid scintillation counting, is approximately 10 times background (for ^{14}C-labeled compounds). This value (50 pmol adducts/mg liver protein) is also used as the *in vitro* threshold. It should be noted, however, that binding *in vivo* is based on binding to total liver protein (covalent binding to total liver homogenate), whereas binding *in vitro* is based on binding only to microsomal protein.

Merck's decision tree is illustrated in Fig. 1.15. It is based on both *in vivo* and *in vitro* studies: The former study involves measuring covalent binding to total liver and plasma protein at 2, 6, and 24 h after orally administering the radioactive drug candidate to rats at 20 mg/kg, whereas the latter involves measuring covalent binding to microsomal protein following a 1-hour incubation of the radioactive drug candidate with NADPH-fortified rat and human liver microsomes (1 mg microsomal protein/mL). The results are not intended to be used on a strict go/no-go basis but are considered in the context of numerous qualifying considerations, as summarized in Fig. 1.15.

The key question surrounding the use of covalent binding as a means of assessing the potential of a drug candidate to cause liver toxicity in the clinic

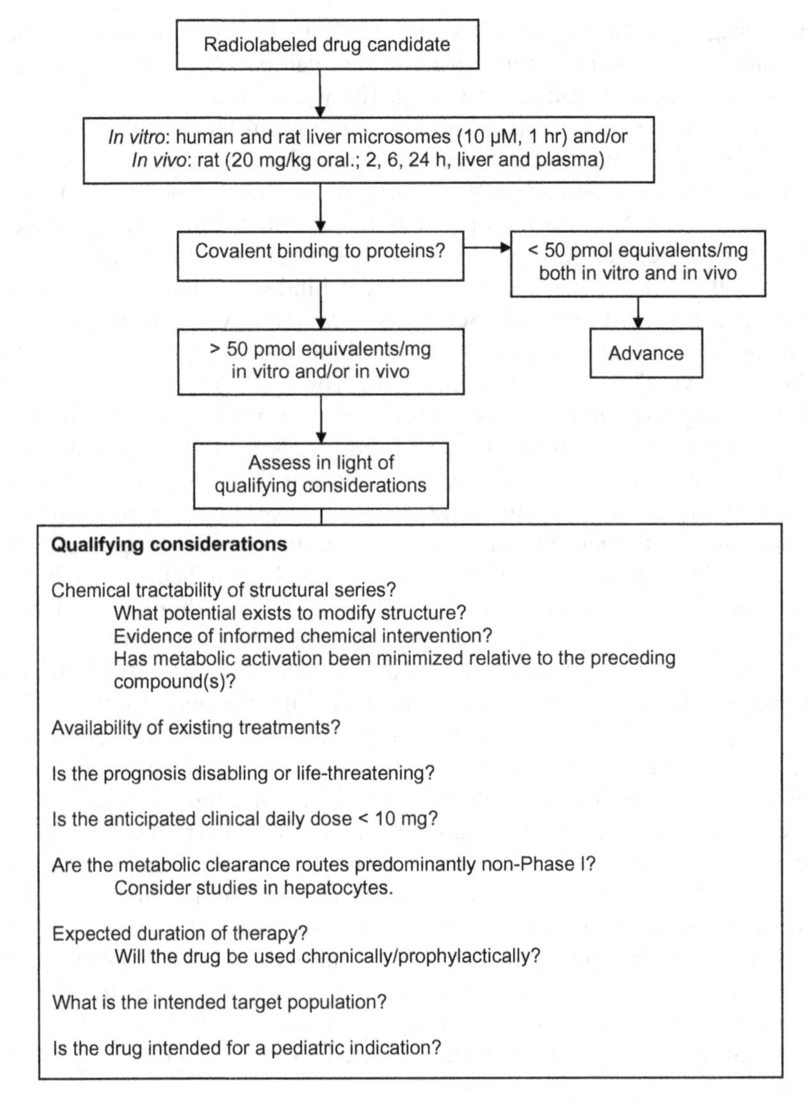

Figure 1.15 Merck's decision tree for assessing the suitability of lead compounds for development based, in part, on metabolic activation to metabolites that bind covalently to protein. From Evans et al. (2004).

is: how well does covalent binding *in vivo* (in rats) or *in vitro* (in liver microsomes) correlate with liver toxicity in the clinic? The answer to this question is: the correlation is weak, and there are many false positives and false negatives. Covalent binding would be expected to identify only a subset of hepatotoxicants, namely those that cause liver necrosis by overwhelming the hepatocyte's defense system and by binding to cellular protein (either with or without the need for conversion to reactive metabolites by CYP or some other drug-metabolizing enzyme). There is the potential for both false negative and false positive results. In terms of false negatives, drugs that cause cholestatic liver injury by inhibiting the transport of bile acids will not necessarily be identified as hepatotoxic drugs based on covalent binding to protein, which likely accounts for the negative results obtained by Evans et al. (2004) for cholestatic steroids like ethinyl estradiol.

Likewise, drugs that cause cellular damage indirectly, by causing oxidative stress for example, will not necessarily be identified as hepatotoxic drugs based on covalent binding to protein. Even drugs that are converted to reactive metabolites and cause liver necrosis may not be detected on the basis of binding to liver microsomal protein if the metabolic activation involves non-microsomal enzymes, as in the case of felbamate (whose activation involves metabolism by cytosolic alcohol dehydrogenase). For this reason, Leone et al. (2007) advocated the use of *both* covalent binding and measures of oxidative stress (Nrf2 activation) to identify drug candidates with the potential to cause liver toxicity, as discussed later in this section.

Merck's covalent binding assay, like many others (Leone et al., 2007; Masubuchi et al., 2007), are based on studies that focus on drugs and other chemicals that are known to cause hepatotoxicity (especially liver necrosis). This bias is somewhat understandable, but it leaves open the possibility that many non-hepatotoxic drugs may bind covalently to protein to the same degree as hepatotoxic drugs. Indeed, the following examples illustrate that there is evidence to suggest that this is indeed the case.

Tirmenstein and Nelson (1989) compared the hepatotoxicity and degree of covalent binding of acetaminophen (*N*-acetyl-4-aminophenol) and its regio-isomer 3′-hydroxyacetanilide (*N*-acetyl-3-aminophenol) in phenobarbital-induced mice. They treated mice with 250 mg/kg acetaminophen or 600 mg/kg of 3′-hydroxyacetanilide. Under these dosage regimens, acetaminophen and 3′-hydroxyacetanilide caused the same degree of covalent binding to total liver protein but only treatment with acetaminophen caused liver toxicity.

Obach et al. (2008) compared the degree of covalent binding of nine hepatotoxic drugs and nine non-hepatotoxic drugs to NADPH-fortified human liver microsomes. The hepatotoxic drugs included acetaminophen, benoxaprofen, carbamazepine, diclofenac, felbamate, indomethacin, nefazadone, sudoxicam, and tienilic acid. The non-hepatotoxic drugs included buspirone, diphenyhydramine, ibuprofen, meloxicam, paroxetine, propranolol, raloxifene, simvastatin, and theophylline. Seven of the nine non-hepatotoxic drugs bound covalently to human liver microsomes; the two that did not were ibuprofen and theophylline. Likewise, seven of the hepatotoxic drugs bound covalently to human liver microsomes; the two that did not were benoxaprofen and felbamate. The two groups of drugs could not be categorized as hepatotoxic or non-hepatotoxic on the basis of covalent binding to human liver microsomes. Based on the kinetics of protein binding (i.e., based on the *in vitro* intrinsic clearance attributable to covalent binding [V_{max}/K_m]), only two hepatotoxic drugs bound more extensively to protein than acetaminophen compared with five non-hepatotoxic drugs (one of which, paroxetine, exhibited the greatest degree of covalent binding).

In a follow-up to their study of covalent binding of nine hepatotoxic and nine non-hepatotoxic drugs to liver microsomal protein (Obach et al., 2008), Obach's group published the results of related studies in human liver S9 fraction and hepatocytes (Bauman et al., 2009). The overall results were the same: covalent binding alone could not discriminate hepatotoxic drugs from non-hepatotoxic drugs, and although the discrimination improved by examining covalent binding to liver S9 fraction in the presence of glutathione or taking into account the total daily body burden of covalent binding, these methods all produced multiple false positive and false negative results.

Takakusa et al. (2008a) also compared the covalent binding of 16 hepatotoxic and 4 non-hepatotoxic drugs to human liver microsomes. The *in vitro* conditions were those recommended by Evans et al. (2004), as summarized in Fig. 1.15. For 10 of the 16 hepatotoxic drugs, the degree of covalent binding to human liver microsomes was less than 50 pmol adducts/mg protein. This was also true with rat liver microsomes; that is, 10 out of 16 drugs produced less than 50 pmol adducts/mg protein. The results were reasonably consistent across species inasmuch as 8 of the 16 hepatotoxic drugs produced less than 50 pmol adducts/mg protein in *both* rat and human liver microsomes. Three of the non-hepatotoxic drugs (amlodipine, caffeine, and warfarin) bound to human liver microsomes in a predominantly NADPH-dependent manner (suggesting formation of reactive metabolites by CYP). The degree of covalent binding of these non-hepatotoxic drugs (9–16 pmol adducts/mg protein) was greater than that of five of the hepatotoxic drugs (namely, phenytoin, procainamide, sulfamethoxazole, valproic acid, and zomepirac). Two of the hepatotoxic drugs that are known to be converted to reactive metabolites by CYP (namely, amodiaquine and procainamide) produced *less* covalent binding in the presence of NADPH than in its absence, whereas the covalent binding of all four non-hepatotoxic drugs increased twofold or more in the presence of NADPH. Overall, the results of the study by Takakusa et al. (2008a) tend to support the conclusions reached by Obach et al. (2008) and Bauman et al. (2009). All three studies illustrate the lack of a simple relationship between covalent binding and liver toxicity. It can be argued that the application of covalent binding criteria would have prevented the development of hepatotoxic drugs that were subsequently withdrawn from the market (true positives such as nefazodone and tienilic acid), but application of these same criteria would have prevented the development of several non-hepatotoxic drugs, including paroxetine, raloxifene, propranolol, buspirone, and diphenhydramine (false positives).

When a drug is found to cause clinically relevant hepatotoxicity, studies are usually conducted to assess its conversion by CYP to one or more reactive metabolites based on covalent binding of drug metabolites to microsomes, glutathione, or other trapping agents. Reactive metabolites are invariably identified, but the role of these reactive metabolites in drug-induced liver injury is seldom established. The aforementioned studies with hepatotoxic and non-hepatotoxic drugs provided evidence that *most* drugs bind covalently to microsomal protein (although the extent is often less than 50 pmol adducts/mg protein). Accordingly, when a drug is converted to a reactive metabolite, it is not possible to conclude with any degree of certainty that the drug will necessarily cause liver toxicity. Conversely, the formation of a reactive metabolite does not preclude the possibility that the parent drug is responsible for any clinically observed hepatotoxicity. The relative toxicity of the parent drug and its metabolites can be assessed in nonclinical species following treatment of rodents with CYP inhibitors or inducers. A drug could be administered to rats (or mice) that were pretreated with various CYP inducers (to increase metabolite formation) or a broad-spectrum CYP inhibitor (to decrease metabolite formation) to ascertain whether liver toxicity was potentiated by CYP induction and abrogated by CYP inhibition or vice versa. The rodent enzyme inducers might include β-naphthoflavone (CYP1A inducer), phenobarbital (CYP2B inducer), and pregnenolone-16α-carbonitrile (CYP3A inducer), but this group could be extended to include isoniazid (CYP2E inducer) and clofibric acid (CYP4A inducer). The broad-spectrum CYP inhibitor could be clotrimazole (or a related azole-antimycotic drug) or ABT. A related approach involves treating rats with lipopolysaccharide (LPS) which causes a pro-inflammatory condition (and suppression of CYP) that predisposes rats to the hepatotoxic effects of certain drugs (briefly reviewed in Guengerich and MacDonald, 2007).

The information that can be obtained from studies in CYP-induced and CYP-inhibited rats could be clinically relevant to the point of influencing the design of clinical studies. If such experiments identify the parent drug as being more hepatotoxic than its metabolites, then the clinical population at risk (and the population that requires special attention, monitoring, and protection) includes any genetically determined PMs and/or those patients on co-medications that inhibit the metabolism of the drug. As shown in Fig. 1.3, this scenario is represented by perhexilline, which causes hepatotoxicity predominantly in CYP2D6 PMs (poor metabolizers of perhexilline) which led to the withdrawal of perhexilline in Europe and its regulatory disapproval in the United States (Parkinson and Ogilvie, 2008).

If such experiments identify the metabolite(s) as being more hepatotoxic than the parent drug, then the clinical population at risk (and the population that requires special attention, monitoring and protection) includes genetically determined EMs and/or those patients on co-medications that induce the metabolism of the drug. This scenario is represented by acetaminophen, a drug that is converted to a reactive quinoneimine by several CYP enzymes including CYP2E1 and CYP3A4, and whose hepatotoxicity is enhanced by CYP2E1 inducers (alcohol and isoniazid) and CYP3A4 inducers (rifampin) (Parkinson and Ogilvie, 2008). It should be noted that, regardless of whether hepatotoxicity is due to the parent drug or its metabolites, numerous other factors may introduce interindividual differences in susceptibility to drug-induced liver injury, such as nutritional status (which can impact drug conjugation), underlying disease processes (such as inflammation), and polymorphisms in the expression of glutathione *S*-transferases and other protective enzymes.

Studies of liver toxicity in rodents should arguably involve an evaluation of whether drug treatment causes an upregulation of genes controlled by Nrf2 (Leone et al., 2007). Nrf2 (nuclear factor E2 p45-related factor-2) functions as a key xenosensor in the antioxidant response (ARE) signaling pathway. It is activated by drugs that are converted to electrophilic metabolites and/or cause oxidative stress, both of which decrease glutathione levels. Nrf2 is normally sequestered in the cytoplasm and targeted for ubiquitination and proteolytic degradation by its binding to KEAP-1 (Kelch-like ECH-related protein), a sulfhydryl (thiol)-rich protein. Alkylation or oxidation of the sulfhydryl groups in KEAP-1 leads to the release of Nrf2, which dimerizes with small Maf proteins in the nucleus to form a DNA-binding protein (active transcription factor). The Nrf2 transcription factor binds to the region of DNA known as the antioxidant response element (also known as the electrophilic response element) and thereby induces enzymes that detoxify electrophiles and drug metabolites that generate reactive oxygen species (ROS), including glutathione *S*-transferase (GSTA1), microsomal epoxide hydrolase, aldo-keto reductase (AKR7A, also known as aflatoxin aldehyde reductase), NAD(P)H-quinone oxidoreductase (NQO1; also known as DT-diaphorase), and glutamate-cysteine ligase (GCL), which catalyzes the rate-limiting step in glutathione synthesis. Induction of GCL increases the rate of glutathione synthesis under conditions of oxidative stress and a decrease in glutathione concentration (Parkinson and Ogilvie, 2008).

Another *in vivo* experimental approach to investigate the hepatotoxic potential of a drug involves manipulating glutathione levels in rats to ascertain if glutathione depletion potentiates hepatotoxicity and if glutathione augmentation has a protective effect. Hepatic glutathione levels can be lowered by treating rats with diethyl maleate (which depletes glutathione) or buthionine-*S*-sulfoximine (which inhibits glutathione synthesis). Glutathione levels and the levels of some forms of glutathione *S*-transferase can be induced by ethoxyquin, butylated hydroxyanisole (BHA), and oltipraz (all of which activate Nrf2).

In summary, recent studies established that hepatotoxic drugs cannot be distinguished from non-hepatotoxic drugs based on their activation by CYP to metabolites that bind covalently to microsomal protein. Studies in nonclinical animals that combine an evaluation of both covalent binding and Nrf2 activation can potentially improve the discrimination between hepatotoxic and non-hepatotoxic drugs, but it is not common for the hepatotoxicity of a drug candidate to be evaluated in rodents following manipulation of CYP activity (with inhibitors and inducers) or glutathione levels. The situation is more promising for carboxylic acid-containing drugs that are converted to potentially toxic acyl glucuronides. In the case of acyl glucuronides, covalent binding to protein does appear to be a good predictor of hepatotoxicity, as outlined in the next section.

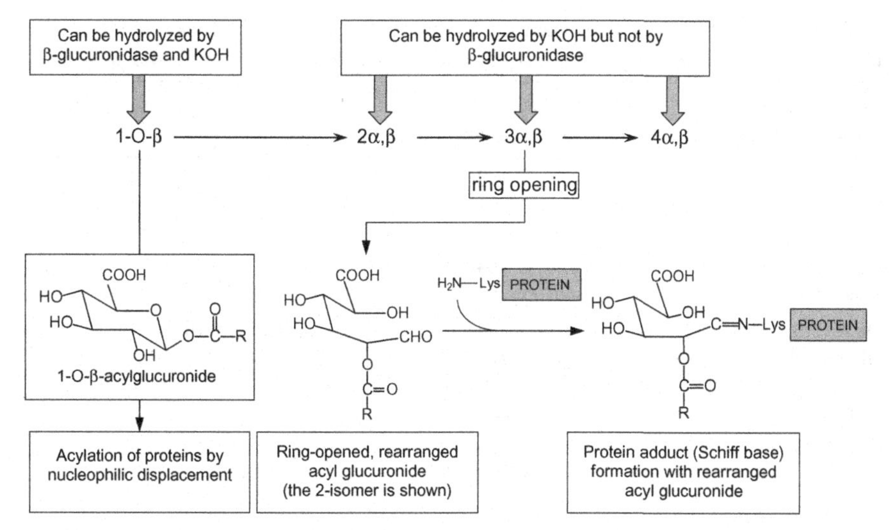

Figure 1.16 Activation of carboxylic-acid containing drugs to reactive metabolites by formation and rearrangement (isomerization) of an acyl glucuronide. A carboxylic acid-containing drug (R-COOH) can potentially be glucuronidated by UDP-glucuronosyltransferase to a 1-*O*-β-acyl glucuronide. The acyl moiety can potentially transfer from the 1-position to the 2-, 3-, and 4-positions. This rearrangement (isomerization) allows the glucuronic acid ring system to open up to form an aldehyde that can potentially form a Schiff base with lysine residues, which is the main mechanism (although not the only mechanism) by which acyl glucuronides bind covalently to protein. Consequently, there is a strong relationship between the rate of isomerization of 1-*O*-β-acyl glucuronides (as measured by their half-life in aqueous buffer at pH 7.4) and hepatotoxicity, as shown in Table 1.9.

1.9.2 Acyl Glucuronidation

Numerous carboxylic acid-containing drugs are glucuronidated to form an acyl glucuronide, which are recognized as potentially reactive metabolites capable of binding covalently to hepatic and plasma proteins and causing hepatotoxicity (Benet et al., 1993; Ebner et al., 1999; Boelsterli, 2002). As shown in Fig. 1.16, glucuronidation of a carboxylic acid-containing drug by UDP-glucuronosyltransferase initially results in the formation of the 1β-isomer. Fig. 1.16 also shows that the initial 1β-isomer can rearrange (by internal migration of the glucuronide moiety) to the 2-, 3- and 4-isomers (and it can also undergo nonenzymatic hydrolysis back to the aglycone). The properties of the 1β-isomer differ significantly from those of the 2-, 3-, and 4-isomers. Although all isomeric acyl glucuronides can be hydrolyzed under strong alkaline conditions, only the 1β-isomer can be hydrolyzed with β-glucuronidase under slightly acidic conditions. Furthermore, whereas the 1β-isomer is relatively inert, the 2-, 3-, and 4-isomers can all undergo ring opening to relatively reactive aldehydes; as such, they can form a Schiff base with amines (such as the

TABLE 1.9 Relationship between the Hepatotoxicity of Acyl Glucuronides and Their Chemical Reactivity as Determined by Half-Life in Aqueous Buffer at pH 7.4

Acidic Drug (*Withdrawn)	Half-Life of Acyl Glucuronide at pH 7.4 (min)	Hepatotoxicity
Tolmetin*	0.26	Toxic acyl-glucuronide
Isoxepac*	0.29	
Probenecid	0.40	
Zenarestat*	0.42	
Zomepirac*	0.45	
Diclofenac	0.51	Intermediate
Diflunisal	0.67	
(*R*)-Naproxen	0.92	
Salicylic acid	1.3	
Indomethacin	1.4	
(*S*)-Naproxen	1.8	
Ibuprofen	3.3	
Bilirubin	4.4	
Furosemide	5.3	Nontoxic acyl-glucuronide
Flufenamic acid	7.0	
Clofibric acid	7.3	
Mefenamic acid	17	
Telmisartan	26	
Gemfibrozil	44	
Valproic acid	79	

Adapted from Ebner et al. (1999) and Boelsterli et al. (2002).

ε-NH2 group of lysine), which is the principal mechanism by which acyl glucuronides bind covalently to protein, which is thought to be a key step in their ability to cause toxicity (such as hepatotoxicity). Because rearrangement of the 1β-isomer to the 2-, 3-, and 4-isomers precedes the covalent binding of acyl glucuronides to protein and because covalent binding to protein appears to be a key step in the hepatotoxicity of acyl glucuronides, there is a good relationship between the hepatotoxicity of acyl glucuronides and their rate of isomerization. As shown in Table 1.9, most of the drugs that can be converted to acyl glucuronides with a half-life in aqueous buffer (pH 7.4) of 30 min or less have been withdrawn from the market or carry black box warnings for hepatotoxicity (Ebner et al., 1999; Boelsterli, 2002). An exception to this general rule is probenecid, a benzoic acid-containing drug. Although the acyl glucuronide of probenecid undergoes rapid rearrangement/nonenzymatic hydrolysis, such that its half-life in aqueous buffer at pH 7.4 is 16–24 min (Hansen-Moller and Schmit, 1991; Akira et al., 2002), this uricosuric and renal tubular blocking agent (a penicillin-sparing agent) does not carry a warning for liver toxicity, and is generally considered to be a non-hepatotoxic drug. However, probenecid does cause other adverse events, such as hypersensitivity reactions, which may be a consequence of its sulfonamide structure or possibly the covalent binding of its acyl glucuronide to proteins. Although probenecid may be an exception, the data in Table 1.9 argue in favor of (bio) synthesizing the 1β-*O*-acyl-glucuronide of a carboxylic acid-containing drug and measuring its half-life in 100 mM potassium phosphate buffer (pH 7.4) to ascertain whether its rate of rearrangement/hydrolysis is comparable to that of known hepatotoxic acyl-glucuronides (which have half-lives of 30 min or less) or comparable to that of nontoxic acyl-glucuronides (which have considerably longer half-lives).

When a carboxylic acid-containing drug is known or suspected of being converted to an acyl glucuronide, it is prudent to acidify plasma and other biological samples (to pH 3 to 5) to prevent rearrangement of the 1β-*O*-acyl-glucuronide to its isomeric forms, namely 2-, 3-, and 4-acyl glucuronides. The isomeric acyl glucuronides can often be separated chromatographically and analyzed individually. Furthermore, 1β-*O*-acyl-glucuronide can be distinguished from its 2-, 3-, and 4-isomers based on their sensitivity to hydrolysis by β-glucuronidase, which hydrolyzes 1β-*O*-acyl-glucuronide but not its isomers (Fig. 1.16). This difference can be used to assess the relative amounts of the 1β-*O*-acyl-glucuronide and its isomeric forms. The mass fragmentation pattern of 1β-*O*-acyl-glucuronide differs from that of the isomeric acyl-glucuronides. In the mass spectrometer, the former (1β-*O*-acyl-glucuronide) loses 176 amu due to loss of glucuronic acid (with formation of the aglycone). However, this does not happen (to a significant extent) with the isomeric glucuronides; they lose water (−18 amu). It may be possible, therefore, to use differences in fragmentation patterns to determine how much acyl-glucuronide is present as 1β-*O*-acyl-glucuronide and how much is present as a combination of 2-, 3-, and 4-glucuronide (based on the relativel size of the −176 and −18 amu signals).

Reactive acyl glucuronides include several unsubstituted acetic acid derivatives, like tolmetin (withdrawn due to hepatotoxicity) (Bolze et al., 2002). Mono- and di-substituted acetic acid derivatives (like ibuprofen and gemfibrozil) are considerably more stable and less reactive. The isomerization of acyl glucuronides of unsubstituted acetic acid derivatives is presumably free of steric hindrance, which is consistent with their short half-lives and rapid rate of covalent binding to protein. The difference in the hepatotoxicity of ibufenac and ibuprofen (shown below) underscores the influence of acetic

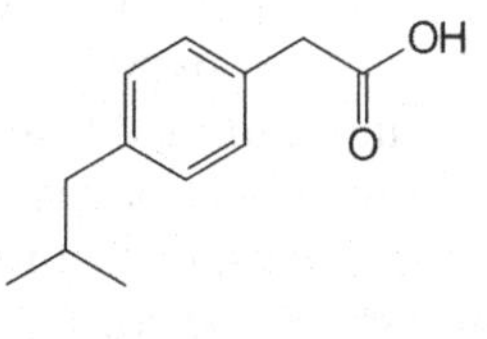

OH
O
Ibufenac

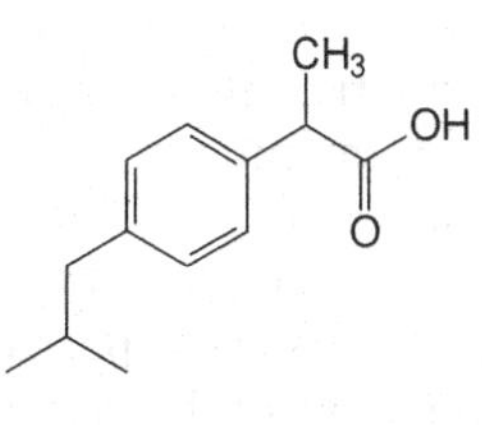

CH_3
OH
O
Ibuprofen

acid substitution on acyl glucuronide reactivity. These two NSAIDs are structurally related and differ by a single methyl group, as shown above. Ibufenac, an unsubstituted acetic acid derivative, is hepatotoxic, whereas ibuprofen is relatively nontoxic. Addition of a single methyl group on the α-position of the acetic acid moiety appears sufficient to lessen the toxicity of the acyl glucuronide of ibuprofen to the point where this NSAID is an over-the-counter drug.

1.9.3 Drug Transporters

Uptake and efflux transporters influence the absorption, distribution, and elimination of numerous drugs, as summarized in Figs. 1.1 and 1.8. Transporters do not biotransform drugs, but they can influence the extent of drug metabolism, and they are important determinants of DDIs. Benet (2009) provided evidence that, by recycling a drug in the intestine, the efflux transporter P-gp (MDR1 or ABCB1) can increase the extent to which that drug is metabolized by intestinal drug-metabolizing enzymes like CYP3A4. It is beyond the scope of this chapter to describe the function and regulation of drug transporters and their role in drug disposition and drug metabolism. However, Table 1.10 lists the major uptake and efflux transporters together with lists of inhibitors and substrates.

Drugs that are substrates or inhibitors of transporters can be involved in DDIs as victims or perpetrators, respectively. The 2006 FDA guidance document recommends performing *in vitro* assays to determine if drug candidates are substrates or inhibitors of P-glycoprotein (P-gp/MDR1/ABCB1) and describes criteria for evaluating the need for subsequent P-gp interactions *in vivo* (US FDA, 2006; Zhang et al., 2009). To determine whether a drug candidate is a P-gp substrate, its bidirectional permeability is determined in a monolayer of polarized, P-gp-expressing cells such as Caco-2 or MDCKII-MDR1 cells, which are grown on a permeable membrane separating the apical- and basolateral-facing chambers. "Bidirectional" means the permeability of the drug candidate is assessed in both directions: from the basolateral to the apical chamber (B → A) and *vice versa* (A → B). P-gp is expressed on the apical membrane of the polarized cells; hence, P-gp will facilitate the passage of a drug added to the basolateral compartment (B → A) but retard its passage from the apical compartment (A → B). The efflux ratio is determined by dividing the permeability of the drug candidate in the B → A direction by its permeability in the A → B direction (BA/AB). If the efflux ratio is ≥2, the experiment is repeated in the presence of a P-gp inhibitor. If the efflux ratio in the presence of a P-gp inhibitor is reduced by 50% or close to unity, then it is reasonable to conclude that the drug candidate is a P-gp substrate and a clinical interaction study may be warranted to evaluate its victim potential. Factors such as permeability, solubility, and nonspecific binding may affect the outcome of the *in vitro* experiments and should be considered when determining if *in vivo* experiments are needed.

TABLE 1.10 A Summary of Substrates and Inhibitors of Transporters Involved in the Uptake and Efflux of Drugs

Uptake Transporters						
Name	Gene	Alias	Relevant Polymorphisms	Location (Order of Expression)	Substrates	Inhibitors
OATP1B1	SLC1B1	OATP-C, OATP2	[*1A, *1B, *5, *15, *16, *17][3]	Liver	Organic anions[2], **atorvastatin**[3], atrasentan[3], benzylpenicillin[3], bosentan[3], caspofungin[3], cerivastatin[3], enalapril[3], fluvastatin[3], irinotecan (SN-38 metabolite)[3], methotrexate[3], olmesartan[3], **pitavastatin**[3] **pravastatin**[3], **repaglinide**[3], rifampin[3], **rosuvastatin**[3] , **simvastatin acid**[3], temocapril[3], troglitazone sulfate[3], valsartan[3], thyroxine[8], nategrinide[6], fexofenadine[9], bromosulphophthalein[14], Estrone-3-sulfate[14], estradiol-17ß-glucuronide[14], bilirubin glucuronide[14], bilirubin[14], bile acids[14]	**Cyclosporine**[3], gemfibrozil[3], gemfibrozil glucuronide[3], **rifampin**[3], clarithromycin[3], erythromycin[3], roxithromycin[3], telithromycin[3], indinavir[3], ritonavir[3], saquinavir[3], **Lopinavir**[14]
OATP1B3	SLC01B3	OATP8		Liver	Organic anions[2], telmisartan[6], fexofenadine[6], digoxin[8], methotrexate[8], rifampin[8], **pitavastatin**[3] **rosuvastatin**[3], bromosulphophthalein[14], cholecystokinin 8[14], telmisartan glucuronide[14], calsartan[14], olmesartan[14], estradiol-17ß-glucuronide[14], bile acids[14]	Cyclosporine[3], gemfibrozil[3], gemfibrozil glucuronide[3], rifampin[3], clarithromycin[3], erythromycin[3], roxithromycin[3], telithromycin[3], ritonavir[14], lopinavir[14]
OATP2B1	SLC02B1	OATP-B, SLC21A9		Liver, intestine, kidney, brain	Organic anions[2], **pravastatin**[8], fexofenadine[9], **atorvastatin**[3] , **rosuvastatin**[3], estrone-3-sulfate[14], bromosulphophthalein[14], taurocholate[14], glyburide[14]	Cyclosporine[3], gemfibrozil[3], gemfibrozil glucuronide[3], rifampicin[14]

(*Contined*)

TABLE 1.10 (*Contined*)

Uptake Transporters						
Name	Gene	Alias	Relevant Polymorphisms	Location (Order of Expression)	Substrates	Inhibitors
OATP1A2	SLC02A1			Liver, brain, kidney, intestine	Fexofenadine[3], levofloxacin[3], pitavastatin[3], rocuronium[3], rosuvastatin[3], saquinavir[3], thyroxin[3], estrone-3-sulfate[14], dehyroepiandrosterone sulphate[14], bile salts[14], methotrexate[14], bromosulphophthalein[14], ouabain[14], digoxin[14], **statins**[14]	Grapefruit juice[3], orange juice[3], apple juice[3], **naringin**[3], hesperidin[3], **rifampicin**[3], rifamycin SV[3], ritonavir[14], lopinavir[14], saquinavir[14]
OAT1	SLC22A6			Kidney, choroid plexus, brain	Polyspecific organic anions[1], acyclovir[8], adefovir[8], methotrexate[8], zidovudine[8], cidofovir[9], cefadroxil[9], cefamandole[9], cefazolin[9], cimetidine[9], para-aminohippurate[14], cidofovir[14], **lamivudine**[14], **zalcitabine**[14], **tenofovir**[14], **ciprofloxacin**[14], **methotrexate**[14]	Probenecid[8], cefadroxil[8], cefamandole[8], cefazolin[8], novobiocin[14]
OAT2	SLC22A7			Kidney, Liver	Polyspecific organic anions[1], salicylic acid[6], indomethacin[6], zidovudine[8]	
OAT3	SLC22A8			Kidney, choroid plexus, brain	Polyspecific organic anions[1], cimetidine[8], methotrexate[8], zidovudine[8], cefadroxil[9], cefamandole[9], cefazolin[9], pravastatin[9], estrone-3-sulfate[14], non-steroidal anti-inflammatory drugs[14], cefaclor[14], ceftizoxime[14], **furosemide**[14], **bumetanide**[14]	Probenecid[8], cefadroxil[8], cefamandole[8], cefazolin[8], novobiocin[14]

OCT1	SLC22A1	Liver	Organic cations[2], **metformin**[6], cimetidine[6], acyclovir[8], amantadine[8], desipramine[8], ganciclovir[8], adefovir[9], cidofovir[9],1-methyl-4-phenylpyridinium (MPP+)[10], tetraethyl ammonium[14], oxaliplatin[14]	Disopyramide[8], midazolam[8], phenformin[8], phenoxy-benzamine[8], quinidine[8], quinine[8], ritonavir[8], verapamil[8]
OCT2	SLC22A2	Kidney, brain	Amantadine[8], cimetidine[8], memantine[8], Tetraethylammonium[14], procainamide[14], ranitidine[14], amiloride[14], oxaliplatin[14], **varenicline**[14]	Desipramine[8], phenoxy-benzamine[8], quinidine[8], **cimetidine**[14], pilsicainide[14], **cetrizine**[14], testosterone[14]
OCT3	SLC22A3	Prostate, heart, muscle, liver, placenta, kidney, heart	Cimetidine[8]	Desipramine[8], prazosin[8], phenoxy-benzamine[8] quinidine[8]
NTCP	SLC10A1	Liver, pancreas	Bile acids[2], **rosuvastatin**[8]	Cyclosporine[3], gemfibrozil[3], gemfibrozil glucuronide[3]
PEPT1	SLC15A1	Intestine, kidney	Di-, and tripeptides[1], β-lactams[1], ACE inhibitors[1], renin inhibitors[1], valacyclovir[1], ampicillin[8], amoxicillin[8], captopril[8], enalapril[9], cefadroxil[9], glycylsarcosine[14], cephalexin[14], bestatin[14], aminolevulinic acid[14]	Glycyl-proline[14]
PEPT2	SLC15A2	Kidney	Di-, and tripeptides[1], ampicillin[8], amoxicillin[8], captopril[8], valacyclovir[8], cefadroxil[9], glycylsarcosine[14], cephalexin[14], bestatin[14], enalapril[14], aminolevulinic acid[14]	Zofenopril[14], fosinopril[14]

(*Contined*)

TABLE 1.10 (*Contined*)

Efflux Transporters						
Name	Gene	Alias	Relevant Polymorphisms	Location (Order of Expression)	Substrates	Inhibitors
MDR1	ABCB1	P-gp	[C3435T]	Intestine, liver, kidney, brain, placenta, adrenal, testes	Lipid cations[2], docetaxel[4], etoposide[4], imatinib[4], paclitaxel[4], teniposide[4], vinblastine[4], vincristine[4], bunitrolol[4], carvedilol[4], celiprolol[4], talinolol[4], diltiazem[4], mibefradil[4], verapamil[4], digitoxin[4], **digoxin**[4], quinidine[4], amprenavir[4], indinavir[4], nelfinavir[4], saquinavir[4], ritonavir[4], dexamethasone[4], methylprednisolone[4], cyclosporin A[4], sirolimus[4], tacrolimus[4], ondansetron[4], erythromycin[4], levofloxacin[4], **atorvastatin**[4], lovastatin[4], fexofenadine[4], terfenadine[4], amitryptiline[4], colchicine[4], Debrisoquine[4], itraconazole[4], losartan[4], morphine[4], phenytoin[4], rifampin[4], **loperamide**[8], topotecan[8], **pravastatin**[3] **pitavastatin**[3] **simvastatin**[3], berberine[14], irinotecan[14], doxorubicin[14], vinblastine[14], paclitaxel[14]	Verapamil[8], ketoconazole[8], itraconazole[8], elacridar (GF120918)[8], LY335979[8], valspodar (PSC833)[8], **quinidine**[14]
MDR3	ABCB4			Liver	Phospholipids[2], digoxin[8], paclitaxel[8], vinblastine[8], phosphatidylcholine[14]	Verapamil[14], cyclosporine[14]
BSEP	ABCB11			Liver	Bile acids[2], vinblastine[8], taurocholic acid[14], pravastatin[14]	Cyclosporine[14], rifampicin[14], glibenclamide[14]
BCRP	ABCG2	MXR, ABCP BMDP	[Q141K, F208S, S248P, E334stop, F489L, S441N, Q126stop][5]	Intestine, liver, kidney, skeletal muscle, brain breast, placenta, hematopoietic stem cells, tumors	Sulfoconjugates[2], SN-38[5], methotrexate[5], mitoxantrone[5], **topotecan**[5] **imatinib**[5], **gefitinib**[5], hematoporphyrin[5], pheophorbide A[5], genistein[5], quercetin[5], estrone-3-sulfate[5], Hoechst 33342, 2-amino-1-methyl-6-phenylimidazo[4,5,β]pyridine[5], fluoroquinolones[6], **pitavastatin**[6], **rosuvastatin**[6], **sulfasalazine**[6], daunorubicin[8], doxorubicin[8], **atorvastatin**[3] **pravastatin**[3], irinotecan[14], porphyrins[14]	Novobiocin[5], tacrolimus[5], fumitremorgin C[5], Ko143[5], elacridar (GF120918)[8], gefitinib[8], estrone-17ß-estradiol[14]

MRP1	ABCC1		Intestine, liver (basolateral), kidney, brain, tumor cells	Adefovir[8], indinavir[8]	
MRP2	ABCC2	CMOAT	Intestine, liver (apical), kidney, brain	Conjugates[2], GSH[2], **pravastatin**[8], cerivastatin[8], valsartan[8], olmesartan[8], methotrexate[8], SN-38[8], cefodizime[8], glucuronides[8], indinavir[8], cisplatin[8], glutathione[14], etoposide[14], mitoxantrone[14], olmesartan[14]	Cyclosporine[8], delaviridine[14], efavirenz[14], emtricitabine[14]
MRP3	ABCC3	CMOAT2	Intestine, liver (basolateral), kidney, placenta, adrenal	Glucuronides[2], etoposide[8], methotrexate[8], tenoposide[8], estradiol-17ß-glucuronide[14], fexofenadine[14]	Delaviridine[14], efavirenz[14], emtricitabine[14]
MRP4	ABCC4		Liver (basolateral)	Bile acids + GSH[2], prostanoids[2], urate[2], adefovir[14], tenofovir[14], cyclic AMP[14], dehydoepiandrosterone sulphate[14], methotrexate[14], topotecan[14], furosemide[14], cyclic GMP[14]	Celecoxib[14], diclofenac[14]
MRP6	ABCC6		Liver (basolateral), kidney	Organic anions[2], cisplatin[8], daunorubicin[8]	

(*Contined*)

TABLE 1.10 (*Contined*)

Other Transporters						
Name	Gene	Alias	Relevant Polymorphisms	Location (Order of Expression)	Substrates	Inhibitors
MATE1	SLC47A1			Liver (apical), kidney	Organic cations[2], tetraethyl ammonium[12], metformin[14], N-methylpyridinium (MPP+)[14]	Quinidine[14], cimetidine[14], procainamide[14]
MATE2-K	SLC47A2			Kidney proximal tubule	Metformin[14], N-methylpyridinium (MPP+)[14], tetraethylammonium[14]	Cimetidine[14], quinidine[14], pramipexole[14]
OSTα& OSTβ	OSTA & OSTB			Liver (basolateral membrane)	Bile acids[2]	
hCNT1	SLC28A1				Pyrimidines[11], uridine[13]	
hCNT2	SLC28A2				Purines[11]	
hCNT3	SLC28A3				Pyrimidines[11], purines[11], uridine[13]	
hENT1	SLC29A1			Hepatic mitochondria	Pyrimidines[11], purines[11]	
hENT2	SLC29A2				Pyrimidines[11], purines[11]	

[1]FDA CP Presentation by Kathleen Hillgren, Ph.D: Overview of Transporter Families and Importance: Intestine and Kidney.
[2]FDA CP Presentation by Dietrich Keppler, Professor, MD: Overview of Transporter Families and Importance: Focus on Liver and Brain.
[3]FDA CP Presentation by Mikko Niemi, MD, Ph.D: OATP Basic and Clinical Studies: Drug Interactions; Probe Substrates.
[4]FDA CP Presentation by Martin F. Fromm, Professor, MD: Basic Research and Current Knowledge of MDR1.
[5]FDA CP Presentation by Toshihisa Ishikawa, Ph.D: Overview: Focus on Key Transporters. Basic Research and Current Knowledge of BCRP (ABCG2).
[6]FDA CP Presentation by Yuichi Sugiyama, Ph.D: Non-selective Inhibitors of Hepatic Influx and Efflux Transporters: Implications to Pharmacokinetics and Hepatic Drug Disposition.
[7]http://pharmacogenetics.ucsf.edu/
[8]*US FDA*, 2006.
[9]Shitara et al., 2006.
[10]Shu et al., 2007.
[11]Presentation by Jash Unadkat PhD: Mitochondrial and Plasma Membrane Nucleoside Transporters: Role in Toxicity of Nucleoside Drugs. 4th World Conference on Drug Absorption, Transport and Delivery. Kanazawa, Ishikawa, Japan. June 20–22, 2007.
[12]Otsuka et al., 2005.
[13]Zhang, et al., 2003.
[14]The International Transporter Consortium. Membrane transporters in drug development. *Nature Reviews* 9:215–236 (2010).
Boldfaced drugs can be used for *in vivo* DDI studies.

Source: FDA Critical Path Transporter Workshop. Bethesda, MD, USA. October 2–3, 2008 (FDA CP).

Drug candidates are evaluated as P-gp inhibitors *in vitro* by evaluating their effect on the bidirectional permeability of a marker P-gp substrate (e.g., digoxin) across a monolayer of P-gp expressing cells (such as Caco-2 or MDCKII-MDR1 cells). Inhibition of P-gp activity toward a marker substrate like digoxin is reflected in a decrease of the efflux ratio by 50% or close to unity. An IC_{50} or K_i value is determined as a measure of inhibitory potency. A clinical drug interaction study is recommended if $[I]/IC_{50} > 0.1$ or $[I]_2/IC_{50} > 10$, where $[I]$ is the maximum concentration of total drug in plasma (C_{max} for bound plus unbound drug) at the highest clinical dose and where $[I]_2$ is the theoretical maximum gastrointestinal concentration of drug after oral dosing, which is estimated by dividing the highest clinical dose by 250 mL. (Note: Recent guidance suggests that $[I]$ may be the free concentration.)

Based on the paper published by the International Transporter Consortium (ITC) (Giacomini et al., 2010) the FDA is expected to extend the current requirements for evaluating drug candidates as P-gp substrates and inhibitors to BCRP as well as adding requirements for testing if compounds are substrates or inhibitors of the hepatic uptake transporters OATP1B1 (OATP2,OATP-C) and OATP1B3 (OATP8) and renal uptake transporters OCT2, OAT1, and OAT3. Furthermore, the European Medicines Agency (EMA) in a guidance document published in April 2010 (European Medicines Agency, Guideline on the Investigation of Drug Interactions, 2010) suggested that compounds should also be assessed as inhibitors of the hepatic uptake transporter OCT1 and the bile salt efflux pump (BSEP) bringing the total number of transporters considered relevant by regulatory agencies for having the potential to cause DDIs to nine.

REFERENCES

Akira K, Uchijima T, Hashimoto T. 2002. Rapid internal acyl migration and protein binding of synthetic probenecid glucuronides. *Chem Res Toxicol* 15:765–772.

Backman J, Kyrklund C, Neuvonen M, Neuvonen P. 2002. Gemfibrozil greatly increases plasma concentrations of cerivastatin. *Clin Pharmacol Ther* 72:685–691.

Barter ZE, Bayliss MK, Beaune PH, Boobis AR, Carlile DJ, Edwards RJ, Houston JB, Lake BG, Lipscomb JC, Pelkonen OR, Tucker GT, Rostami-Hodjegan A. 2007. Scaling factors for the extrapolation of in vivo metabolic drug clearance from in vitro data: Reaching a consensus on values of human microsomal protein and hepatocellularity per gram of liver. *Curr Drug Metab* 8:33–45.

Bauman JN, Kelly JM, Tripathy S, Zhao SX, Lam WW, Kalgutkar AS, Obach RS. 2009. Can in vitro metabolism-dependent covalent binding data distinguish hepatotoxic from nonhepatotoxic drugs? An analysis using human hepatocytes and liver S-9 fraction. *Chem Res Toxicol* 22(2):332–340.

Beconi MG, Mao A, Liu DQ, Kochansky C, Pereira T, Raab C, Pearson P, Lee Chiu S-H. 2003. Metabolism and pharmacokinetics of a dipeptidyl peptidase IV inhibitor in rats, dogs, and monkeys with selective carbamoyl glucuronidation of the primary amine in dogs. *Drug Metab Dispos* 31:1269–1277.

Beedham C. 2002. Molybdenum hydroxylases. In *Enzyme Systems That Metabolise Drugs and Other Xenobiotics*, ed. Ioannides C, 147–187. New York: Wiley.

Beedham C, Miceli JJ, Obach RS. 2003. Ziprasidone metabolism, aldehyde oxidase, and clinical implications. *J Clin Psychopharmacol* 23:229–232.

Benet LZ. 2009. The drug transporter-metabolism alliance: Uncovering and defining the interplay. *Mol Pharm* 6:1631–1643.

Benet LZ, Spahn-Langguth H, Iwakawa S, Volland C, Mizuma T, Mayer S, Mutschler E, Lin ET. 1993. Predictability of the covalent binding of acidic drugs in man. *Life Sci* 53:PL141–PL146.

Bjornsson TD, Callaghan JT, Einolf HJ, Fischer V, Gan L, Grimm S, Kao J, King SP, Miwa G, Ni L, Kumar G, McLeod J, Obach RS, Roberts S, Roe A, Shah A, Snikeris F, Sullivan JT, Tweedie D, Vega JM, Walsh J, Wrighton SA. 2003a. The conduct of in vitro and in vivo drug-drug interaction studies: A Pharmaceutical Research and Manufacturers of America (PhRMA) perspective. *Drug Metab Dispos* 31:815–832.

Bjornsson TD, Callaghan JT, Einolf HJ, Fischer V, Gan L, Grimm S, Kao J, King SP, Miwa G, Ni L, Kumar G, McLeod J, Obach SR, Roberts S, Roe A, Shah A, Snikeris F, Sullivan JT, Tweedie D, Vega JM, Walsh J, Wirghton SA. 2003b. The conduct of in vitro and in vivo drug-drug interaction studies: A PhRMA perspective. *J Clin Pharmacol* 43:443–469.

Boelsterli UA. 2002. Xenobiotic acyl glucuronides and acyl CoA thioesters as protein-reactive metabolites with the potential to cause idiosyncratic drug reactions. *Curr Drug Metab* 3:439–445.

Bolze S, Bromet N, Gay-Feutry C, Massiere F, Boulieu R, Hulot T. 2002. Development of an in vitro screening model for the biosynthesis of acyl glucuronide metabolites and the assessment of their reactivity toward human serum albumin. *Drug Metab Dispos* 30:404–413.

Boxenbaum H. 1980. Interspecies variation in liver weight, hepatic blood flow, and antipyrine intrinsic clearance: Extrapolation of data to benzodiazapines and phenytoin. *J Pharmacokinet Biopharm* 8:165–176.

Brunton LL, Lazo JS, Parker KL. 2006. *Goodman & Gilman's The Pharmacological Basic of Therapeutics*, 11th ed. New York: McGraw-Hill.

Cashman J. 1999. In vitro metabolism: FMO and related oxygenations. In *Handbook of Drug Metabolism*, ed. Woolf EJ, 477–505. New York: Marcel Dekker.

Cashman JR, Zhang J. 2006. Human flavin-containing monooxygenases. *Annu Rev Pharmacol Toxicol* 46:65–100.

Chu V, Einolf HJ, Evers R, Kumar G, Moore D, Ripp S, Silva J, Sinha V, Sinz M, Skerjanec A. 2009. In vitro and in vivo induction of cytochrome p450: A survey of the current practices and recommendations: A Pharmaceutical Research and Manufacturers of America perspective. *Drug Metab Dispos* 37:1339–1354.

Collins C, Levy R, Ragueneau-Majlessi I, Hachad H. 2006. Prediction of maximum exposure in poor metabolizers following inhibition of nonpolymorphic pathways. *Curr Drug Metab* 7:295–299.

Curran PG, DeGroot LJ. 1991. The effect of hepatic enzyme-inducing drugs on thyroid hormones and the thyroid gland. *Endocr Rev* 12:135–150.

Ebner T, Heinzel G, Prox A, Beschke K, Wachsmuth H. 1999. Disposition and chemical stability of telmisartan 1-*O*-acylglucuronide. *Drug Metab Dispos* 27:1143–1149.

Evans DC, Watt AP, Nicoll-Griffith DA, Baillie TA. 2004. Drug-protein adducts: An industry perspective on minimizing the potential for drug bioactivation in drug discovery and development. *Chem Res Toxicol* 17:3–16.

Fisher M, Campanale K, Ackermann B, Vandenbranden M, Wrighton S. 2000a. In vitro glucuronidation using human liver microsomes and the pore-forming peptide alamethicin. *Drug Metab Dispos* 28:560–566.

Fisher MB, VandenBranden M, Findlay K, Burchell B, Thummel KE, Hall SD, Wrighton SA. 2000b. Tissue distribution and interindividual variation in human UDP-glucuronosyltransferase activity: Relationship between UGT1A1 promoter genotype and variability in a liver bank. *Pharmacogenetics* 10:727–739.

Fromm MF, Busse D, Kroemer HK, Eichelbaum M. 1996. Differential induction of prehepatic and hepatic metabolism of verapamil by rifampin. *Hepatology* 24: 796–801.

Gasche Y, Daali Y, Fathi M, Chiappe A, Cottini S, Dayer P, Desmeules J. 2004. Codeine intoxication associated with ultrarapid CYP2D6 metabolism. *N Engl J Med* 351:2827–2831.

Giacomini KM, Sugiyama Y. 2006. Membrane transporters and drug response. In *Goodman & Gilman's The Pharmacological Basis of Therapeutics*, ed. Goodman LS, Gilman A, Brunton LL, Lazo JS, Parker KL, 41–70. New York: McGraw-Hill.

Giacomini KM, Huang SM, Tweedie DJ, Benet LZ, Brouwer KL, Chu X, Dahlin A, Evers R, Fischer V, Hillgren KM, Hoffmaster KA, Ishikawa T, Keppler D, Kim RB, Lee CA, Niemi M, Polli JW, Sugiyama Y, Swaan PW, Ware JA, Wright SH, Yee SW, Zamek-Gliszczynski MJ, Zhang L. 2010. Membrane transporters in drug development. *Nat Rev Drug Discov* 9:215–236.

Gertz M, Kilford PJ, Houston JB, Galetin A. 2008. Drug lipophilicity and microsomal protein concentration as determinants in the prediction of the fraction unbound in microsomal incubations. *Drug Metab Dispos* 36:535–542.

Gidal BE, Sheth R, Parnell J, Maloney K, Sale M. 2003. Evaluation of VPA dose and concentration effects on lamotrigine pharmacokinetics: Implications for conversion to lamotrigine monotherapy. *Epilepsy Res* 57:85–93.

Gipple KJ, Chan KT, Elvin AT, Lalka D, Axelson JE. 1982. Species differences in the urinary excretion of the novel primary amine conjugate: Tocainide carbamoyl O-beta-D-glucuronide. *J Pharm Sci* 71:1011–1014.

Goetz MP, Rae JM, Suman VJ, Safgren SL, Ames MM, Visscher DW, Reynolds C, Couch FJ, Lingle WL, Flockhart DA, Desta Z, Perez EA, Ingle JN. 2005. Pharmacogenetics of tamoxifen biotransformation is associated with clinical outcomes of efficacy and hot flashes. *J Clin Oncol* 23:9312–9318.

Gonzalez FJ, Tukey RH. 2006. Drug metabolism. In *Goodman & Gilman's The Pharmacological Basis of Therapeutics*, ed. Goodman LS, Gilman A, Brunton LL, Lazo JS, Parker KL, 71–91. New York: McGraw-Hill.

Gorski J, Vannaprasaht S, Hamman M, Ambrosius W, Bruce M, Haehner-Daniels B, Hall S. 2003. The effect of age, sex, and rifampin administration on intestinal and hepatic cytochrome P450 3A activity. *Clin Pharmacol Ther* 74:275–287.

Gorski JC, Huang SM, Pinto A, Hamman MA, Hilligoss JK, Zaheer NA, Desai M, Miller M, Hall SD. 2004. The effect of echinacea (*Echinacea purpurea* root) on cytochrome P450 activity in vivo. *Clin Pharmacol Ther* 75:89–100.

Griffiths HH, Jones OO, Smith R. 2005. Determination of microsomal and hepatocyte scaling factors for in vitro in vivo extrapolation. *Drug Metab Rev* 37(Suppl. 1): 71.

Grimm SW, Einolf HJ, Hall SD, He K, Lim H-K, Ling K-HJ, Lu C, Nomeir AA, Seibert E, Skordos KW, Tonn GR, Van Horn R, Wang RW, Wong YN, Yang TJ, Obach RS. 2009. The conduct of in vitro studies to address time-dependent inhibition of drug-metabolizing enzymes: A perspective of the Pharmaceutical Research and Manufacturers of America. *Drug Metab Dispos* 37:1355–1370.

Guengerich FP. 1991. Reactions and significance of cytochrome P-450 enzymes. *J Biol Chem* 266:10019–10022.

Guengerich FP. 2001. Uncommon P450-catalyzed reactions. *Curr Drug Metab* 2:93–115

Guengerich FP, MacDonald JS. 2007. Applying mechanisms of chemical toxicity to predict drug safety. *Chem Res Toxicol* 20:344–369.

Hakooz N, Ito K, Rawden H, Gill H, Lemmers L, Boobis AR, Edwards JE, Carlile DJ, Lake BG, Houston BJ. 2006. Determination of a human hepatic microsomal scaling factor for predicting in vivo drug clearance. *Pharm Res* 23:533–539.

Hansen-Moller J, Schmit U. 1991. Rapid high-performance liquid chromatography assay for the simultaneous determination of probenecid and its glucuronide in urine. Irreversible binding of probenecid to serum albumin. *J Pharm Biomed Anal* 9: 65–73.

Hallifax D, Houston JB. 2006. Binding of drugs to hepatic microsomes: Comment and assessment of current prediction methodology with recommendation for improvement. *Drug Metab Dispos* 34:724–726.

Hallifax D, Houston JB. 2007. Saturable uptake of lipophilic amine drugs into isolated hepatocytes: Mechanisms and consequences for quantitative clearance prediction. *Drug Metab Dispos* 35:1325–1332.

Hayakawa H, Fukushima Y, Kato H, Fukumoto H, Kadota T, Yamamoto H, Kuroiwa H, Nishigaki J, Tsuji A. 2003. Metabolism and disposition of novel des-fluoro quinolone garenoxacin in experimental animals and an interspecies scaling of pharmacokinetic parameters. *Drug Metab Dispos* 31:1409–1418.

Holtbecker N, Fromm MF, Kroemer HK, Ohnhaus EE, Heidemann H. 1996. The nifedipine-rifampin interaction: Evidence for induction of gut wall metabolism. *Drug Metab Dispos* 24:1121–1123.

Howgate EM, Rowland YK, Proctor NJ, Tucker GT, Rostami-Hodjegan A. 2006. Prediction of in vivo drug clearance from in vitro data: I. Impact of inter-individual variability. *Xenobiotica* 36:473–497. http://www.ema.europa.eu/ema/index.jsp?curl=pages/medicines/human/referrals/Cerivastatin/human_referral_000108.jsp&mid=WC0b01ac0580024e9a&murl=menus/regulations/regulations.jsp&jsenabled=true# (accessed July 22, 2010).

Huang SM, Strong JM, Zhang L, Reynolds KS, Nallani S, Temple R, Abraham S, Habet SA, Baweja RK, Burckart GJ, Chung S, Colangelo P, Frucht D, Green MD, Hepp P, Karnaukhova E, Ko HS, Lee JI, Marroum PJ, Norden JM, Qiu W, Rahman

A, Sobel S, Stifano T, Thummel K, Wei XX, Yasuda S, Zheng JH, Zhao H, Lesko LJ. 2008. New era in drug interactions evaluation: US Food and Drug Administration update on CYP enzymes, transporters, and the guidance process. *J Clin Pharmacol* 48:662–670.

Johnson TN, Tucker G, Tanner MS, Rostami-Hodjegan A. 2005. Changes in liver volume from birth to adulthood: A meta analysis. *Liver Transplantation* 12: 1481–1493.

Kanamitsu SI, Ito K, Sugiyama Y. 2000. Quantitative prediction of in vivo drug-drug interactions from in vitro data based on physiological pharmacokinetics: Use of maximum unbound concentration of inhibitor at the inlet to the liver. *Pharm Res* 17:336–343.

Kawashima K, Hosoi K, Naruke T, Shiba T, Kitamura M, Watabe T. 1999. Aldehyde oxidase-dependent marked species difference in hepatic metabolism of the sedative-hypnotic, zaleplon, between monkeys and rats. *Drug Metab Dispos* 27:422–428.

Kiang TKL, Ensom MHH, Chang TKH. 2005. UDP-glucuronosyltransferases and clinical drug-drug interactions. *Pharmacol Ther* 106:97–132.

Koren G, Cairns J, Chitayat D, Gaedigk A, Leeder SJ. 2006. Pharmacogenetics of morphine poisoning in a breastfed neonate of a codeine-prescribed mother. *Lancet* 368:704.

Kumar S, Samual K, Subramanian R, Braun MP, Stearns RA, Chiu SL, Evans DC, Baillie TA. 2002. Extrapolation of diclofenac clearance from in vitro microsomal metabolism data: Role of acyl glucuronidation and sequential oxidative metabolism of the acyl glucuronide. *J Pharmacol Exp Ther* 303:969–978.

Lake BG, Ball SE, Kao J, Renwick AB, Price RJ, Scatina JA. 2002. Metabolism of zaleplon by human liver: Evidence for involvement of aldehyde oxidase. *Xenobiotica* 32:835–847.

Leone AM, Kao LM, McMillian MK, Nie AY, Parker JB, Kelley MF, Usuki E, Parkinson A, Lord PG, Johnson MD. 2007. Evaluation of felbamate and other antiepileptic drug toxicity potential based on hepatic protein covalent binding and gene expression. *Chem Res Toxicol* 20:600–608.

Li AP, Hartman NR, Lu C, Collins JM, Strong JM. 1999. Effects of cytochrome P450 inducers on 17alpha-ethinyloestradiol (EE2) conjugation by primary human hepatocytes. *Br J Clin Pharmacol* 48:733–742.

Link M, Hakala KS, Wsol V, Kostiainen R, Ketola RA. 2006. Metabolite profile of sibutramine in human urine: A liquid chromatography-electrospray ionization mass spectrometric study. *J Mass Spectrom* 41:1171–1178.

MacIntyre AC, Cutler DJ. 1988. The potential role of lysosomes in tissue distribution of weak bases. *Biopharm Drug Dispos* 9:513–526.

Madan A, Graham RA, Carroll KM, Mudra DR, Burton LA, Krueger LA, Downey AD, Czerwinski M, Forster J, Ribadeneira MD, Gan L-S, LeCluyse EL, Zech K, Robertson P Jr., Koch P, Antonian L, Wagner G, Yu L, Parkinson A. 2003. Effects of prototypical microsomal enzyme inducers on cytochrome P450 expression in cultured human hepatocytes. *Drug Metab Dispos* 31:421–431.

Mahajan MK, Evans CA. 2008. Dual negative precursor ion scan approach for rapid detection of glutathione conjugates using liquid chromatography/tandem mass spectrometry. *Rapid Commun Mass Spectrom* 22:1032–1040.

Majerus PWT, Douglas M. 2006. Blood conjugation and anticoagulant, thrombolytic, and antiplatelet drugs. In *Goodman & Gilman's The Pharmacological Basis of Therapeutics*, 11th ed., ed. Goodman LS, Gilman A, Brunton LL, Lazo JS, Parker KL, 1200–1256. New York: McGraw-Hill.

Mano Y, Usui T, Kamimura H. 2006. In *vitro* drug interaction between diflunisal and indomethacin via glucuronidation in humans. *Biopharm Drug Dispos* 27:267–273.

Masubuchi N, Makino C, Murayama N. 2007. Prediction of in vivo potential for metabolic activation of drugs into chemically reactive intermediate: Correlation of in vitro and in vivo generation of reactive intermediates and in vitro glutathione conjugation formation in rats and humans. *Chem Res Toxicol* 20:455–464.

Mikus G, Schowel V, Drzewinska M, Rengelshausen J, Ding R, Riedel KD, Burhenne J, Weiss J, Thomsen T, Haefeli WE. 2006. Potent cytochrome P450 2C19 genotype-related interaction between voriconazole and the cytochrome P450 3A4 inhibitor ritonavir. *Clin Pharmacol Ther* 80:126–135.

Miners JO, Knights KM, Houston JB, Mackenzie PI. 2006. In vitro-in vivo correlation for drugs and other compounds eliminated by glucuronidation in humans: Pitfalls and promises. *Biochem Pharmacol* 71:1531–1539.

Naesens M, Kuypers DR, Streit F, Armstrong VW, Oellerich M, Verbeke K, Vanrenterghem Y. 2006. Rifampin induces alterations in mycophenolic acid glucuronidation and elimination: Implications for drug exposure in renal allograft recipients. *Clin Pharmacol Ther* 80:509–521.

Nishiyama T, Kobori T, Arai K, Ogura K, Ohnuma T, Ishii K, Hayashi K, Hiratsuka A. 2006. Identification of human UDP-glucuronosyltransferase isoform(s) responsible for the C-glucuronidation of phenylbutazone. *Arch Biochem Biophys* 454: 72–79.

Obach RS, Cox LM, Tremaine LM. 2005. Sertraline is metabolized by multiple cytochrome P450 enzymes, monoamine oxidases, and glucuronyl transferases in human: An in vitro study. *Drug Metab Dispos* 33:262–270.

Obach RS, Reed-Hagen AE, Krueger SS, Obach BJ, O'Connell TN, Zandi KS, Miller S, Coe JW. 2006. Metabolism and disposition of varenicline, a selective alpha4beta2 acetylcholine receptor partial agonist, in vivo and in vitro. *Drug Metab Dispos* 34:121–130.

Obach RS, Kalgutkar AS, Soglia JR, Zhao SX. 2008. Can in vitro metabolism-dependent covalent binding data in liver microsomes distinguish hepatotoxic from nonhepatotoxic drugs? An analysis of 18 drugs with consideration of intrinsic clearance and daily dose. *Chem Res Toxicol* 21:1814–1822.

Ogilvie BW, Zhang D, Li W, Rodrigues AD, Gipson AE, Holsapple J, Toren P, Parkinson A. 2006. Glucuronidation converts gemfibrozil to a potent, metabolism-dependent inhibitor of CYP2C8: Implications for drug-drug interactions. *Drug Metab Dispos* 34:191–197.

Ogilvie BW, Usuki E, Yerino P, Parkinson A. 2008. In vitro approaches for studying the inhibition of drug-metabolizing enzymes responsible for the metabolism of drugs (reaction phenotyping) with emphasis on cytochrome P450. In *Drug-Drug Interactions*, 2nd ed., ed. Rodrigues AD, 231–358. New York: Informa Healthcare USA.

Otsuka M, Matsumoto T, Morimoto R, Arioka S, Omote H, Moriyama Y. 2005. A human transporter protein that mediates the final excretion step for toxic organic cations. *PNAS* 102:17923–17928.

Ozdemir O, Boran M, Gokce V, Uzun Y, Kocak B, Korkmaz S. 2000. A case with severe rhabdomyolysis and renal failure associated with cerivastatin-gemfibrozil combination therapy—A case report. *Angiology* 51:695–697.

Paine MF, Widmer WW, Hart HL, Pusek SN, Beavers KL, Criss AB, Brown SS, Thomas BF, Watkins PB. 2006. A furanocoumarin-free grapefruit juice establishes furanocoumarins as the mediators of the grapefruit juice-felodipine interaction. *Am J Clin Nutr* 83:1097–1105.

Parkinson A, Ogilvie BW. 2008. Biotransformation of xenobiotics. In *Casarett & Doull's Toxicology: The Basic Science of Poisons*, 7th ed., ed. Klaassen CD, 161–304. New York: McGraw-Hill.

Pearce RE, Rodrigues AD, Goldstein JA, Parkinson A. 1996. Identification of the human P450 enzymes involved in lasoprazole metabolism. *J Pharmacol Exp Ther* 277:805–816.

Pope LE, Khalil MH, Berg JE, Stiles M, Yakatan GJ, Sellers EM. 2004. Pharmacokinetics of dextromethorphan after single or multiple dosing in combination with quinidine in extensive and poor metabolizers. *J Clin Pharmacol* 44:1132–1142.

Price PS, Conolly RB, Chaisson CF, Gross EA, Young JS, Mathis ET, Tedder DR. 2003. Modeling interindividual variation in physiological factors used in PBPK models of humans. *Crit Rev Toxicol* 33:469–503.

Proctor NJ, Tucker GT, Rostami-Hodjegan A. 2004. Predicting drug clearance from recombinantly expressed CYPs: Intersystem extrapolation factors. *Xenobiotica* 34:151–178.

Renwick AB, Ball SE, Tredger JM, Price RJ, Walters DG, Kao J, Scatina JA, Lake BG. 2002. Inhibition of zaleplon metabolism by cimetidine in the human liver: In vitro studies with subcellular fractions and precision-cut liver slices. *Xenobiotica* 32:849–862.

Rettie AE, Fisher MB. 1999. Transformation enzymes: Oxidative; non-P450. In *Handbook of Drug Metabolism*, ed. Woolf T, 131–151. New York: Marcel Dekker, Inc.

Schrag ML, Cui D, Rushmore TH, Shou M, Ma B, Rodrigues AD. 2004. Sulfotransferase 1E1 is a low km isoform mediating the 3-O-sulfation of ethinyl estradiol. *Drug Metab Dispos* 32:1299–1303.

Shiraga T, Niwa T, Ohno Y, Kagayama A. 2004. Interindividual variability in 2-hydroxylation, 3-sulfation, and 3-glucuronidation of ethynylestradiol in human liver. *Biol Pharm Bull* 27:1900–1906.

Shitara Y, Hirano M, Sato H, Sugiyama Y. 2004. Gemfibrozil and its glucuronide inhibit the organic anion transporting polypeptide 2 (OATP2/OATP1B1:SLC21A6)-mediated hepatic uptake and CYP2C8-mediated metabolism of cerivastatin: Analysis of the mechanism of the clinically relevant drug-drug interaction between cerivastatin and gemfibrozil. *J Pharmacol Exp Ther* 311:228–236.

Shitara Y, Horie T, Sugiyama Y. 2006. Transporters as a determinant of drug clearance and tissue distribution. *Eur J Pharm Sci* 27:425–446.

Shu Y, Sheardown SA, Brown C, Owen RP, Zhang S, Castro RA, Ianculescu AG, Yue L, Lo JC, Burchard EG, Brett CM, Giacomini KM. 2007. Effect of genetic variation in the organic cation transporter 1 (OCT1) on metformin action. *J Clin Invest* 117:1422–1431.

Soars MG, Burchell B, Riley RJ. 2002. In vitro analysis of human drug glucuronidation and prediction of in vivo metabolic clearance. *J Pharmacol Exp Ther* 301: 382–390.

Strelevitz TJ, Foti RS, Fisher MB. 2006. In vivo use of the P450 inactivator 1-aminobenzotriazole in the rat: Varied dosing route to elucidate gut and liver contributions to first-pass and systemic clearance. *J Pharm Sci* 95:1334–1341.

Sugihara K, Kitamura S, Tatsumi K. 1996. Involvement of mammalian liver cytosols and aldehyde oxidase in reductive metabolism of zonisamide. *Drug Metab Dispos* 24:199–202.

Takakusa H, Masumoto H, Yukinaga H, Ma kino C, Nakayama S, Okazaki O, Sudo K. 2008a. Covalent binding and tissue distribution/retention assessment of drugs associated with idiosyncratic drug toxicity. *Drug Metab Dispos* 36:1770–1779.

Tirmenstein MA, Nelson SD. 1989. Subcellular binding and effects on calcium homeostasis produced by acetaminophen and a nonhepatotoxic regioisomer, 3′-hydroxyacetanilide, in mouse liver. *J Biol Chem* 264:9814–9819.

Treiber A, Schneiter R, Delahaye S, Clozel M. 2004. Inhibition of organic anion transporting polypeptide-mediated hepatic uptake is the major determinant in the pharmacokinetic interaction between bosentan and cyclosporin A in the rat. *J Pharmacol Exp Ther* 308:1121–1129.

Tremaine LM, Stroh JG, Ronfeld RA. 1989. Characterization of a carbamic acid ester glucuronide of the secondary amine sertraline. *Drug Metab Dispos* 17:58–63.

Tsunoda SM, Velez RL, von Moltke LL, Greenblatt DJ. 1999. Differentiation of intestinal and hepatic cytochrome P450 3A activity with use of midazolam as in vivo probe: Effect of ketoconazole. *Clin Pharmacol Ther* 66:461–471.

Tucker GT, Houston JB, Huang SM. 2001. Optimizing drug development: Strategies to assess drug metabolism/transporter interaction potential—Toward a consensus. *Pharm Res* 18:1071–1080.

US Food and Drug Administration (US FDA). 2006. Draft Guidance for Industry: Drug Interaction Studies—Study Design, Data Analysis and Implications for Dosing and Labeling. http://www.fda.gov/downloads/Drugs/GuidanceCompliance RegulatoryInformation/Guidances/UCM072101.pdf.

US Food and Drug Administration (US FDA). 2008. Guidance for Industry Safety Testing of Drug Metabolites. http://www.fda.gov/downloads/Drugs/Guidance ComplianceRegulatoryInformation/Guidances/ucm079266.pdf.

Williams RT. 1959. *Detoxication Mechanisms: The Metabolism and Detoxication of Drugs, Toxic Substances, and Other Organic Compounds*, 2nd ed. New York: Wiley.

Williams JA, Hurst SI, Bauman J, Jones BC, Hyland R, Gibbs JP, Obach RS, Ball SE. 2003. Reaction phenotyping in drug discovery: Moving forward with confidence? *Curr Drug Metab* 4:527–534.

Williams JA, Hyland R, Jones BC, Smith DA, Hurst S, Goosen TC, Peterkin V, Koup JR, Ball SE. 2004. Drug-drug interactions for UDP-glucuronosyltransferase substrates: A pharmacokinetic explanation for typically observed low exposure (AUCI/AUC) ratios. *Drug Metab Dispos* 32:1201–1208.

Wu CY, Benet LZ. 2005. Predicting drug disposition via application of BCS: Transport/absorption/elimination interplay and development of a biopharmaceutics drug disposition classification system. *Pharm Res* 22(1):11–23.

Youdim KA, Zayed A, Dickins M, Phipps A, Griffiths M, Darekar A, Hyland R, Fahmi O, Hurst S, Plowchalk DR, Cook J, Guo F, Obach RS. 2008. Application of CYP3A4 in vitro data to predict clinical durg-durg interactions; predictions of compounds as objects of interaction. *Br J Clin Pharmacol* 65:680–692.

Zhang J, Visser F, Vickers MF, Lang T, Robins MJ, Nielsen LPC, Nowak I, Baldwin SA, Young JD, Cass CE. 2003. Uridine binding motifs of human concentrative nucleoside transporters 1 and 3 produced in *Saccharomyces cerevisiae*. *Mol Pharmacol* 64:1512–1520.

Zhang L, Zhang YD, Zhao P, Huang SM. 2009. Predicting drug-drug interactions: An FDA perspective. *AAPS J* 11:300–306.

Zineh I, Beitelshees AL, Gaedigk A, Walker JR, Pauly DF, Eberst K, Leeder JS, Phillips MS, Gelfand CA, Johnson JA. 2004. Pharmacokinetics and CYP2D6 genotypes do not predict metoprolol adverse events or efficacy in hypertension. *Clin Pharmacol Ther* 76:536–544.

CHAPTER 2

ANALYTICAL TOOLS AND APPROACHES FOR METABOLITE IDENTIFICATION IN DRUG METABOLISM

YONGMEI LI
Boehringer Ingelheim Pharmaceuticals, Inc., Drug Metabolism and Pharmacokinetics, Ridgefield, CT

2.1 INTRODUCTION

The identification of drug metabolites is an integral part of contemporary drug discovery and development processes in order to assess the metabolic fate of a drug candidate in animals and humans (Baillie et al., 2002; Watt et al., 2003). Knowledge of the metabolic fate of a drug is very important, as metabolites could be pharmacologically active (Gad, 2003; Fura et al., 2004), toxic (Guengerich, 2001; Evans et al., 2004; Kalgutkar et al., 2005), or involved in drug–drug interactions via inhibition or induction of drug-metabolizing enzymes (Obach, 2003).

Samples for metabolite identification can be generated using *in vitro* systems, such as recombinant drug-metabolizing enzymes (e.g., cytochrome P450 [CYP]), subcellular fractions (e.g., liver microsomes, S9 fractions, or cytosol), whole cells (e.g., hepatocytes), tissues, or organs (Pelkonen et al., 2005). Alternatively, samples such as plasma, blood, bile, urine, or feces can come from *in vivo* studies after dosing a test compound.

During the early discovery stage, simple *in vitro* systems are used to determine the metabolic fate of new chemical entities (NCEs). The focus of metabolite identification is to identify "metabolic soft spots" of NCEs in order to provide feedback to medicinal chemists for further lead optimization, with the aim of blocking metabolism to optimize the pharmacokinetic and

Biotransformation and Metabolite Elucidation of Xenobiotics, Edited by Ala F. Nassar

safety profiles of newly synthesized drug candidates. There is also a need to compare metabolite exposure in *in vitro* incubations and *in vivo* samples from different species. These studies can help the selection of appropriate toxicology species for safety evaluation. At the development stage, a more complete metabolite profile is needed. In addition to more extensive studies with the NCEs, the availability of radiolabeled compounds significantly reduces the complexity of identifying metabolites. Radiolabeled compounds can be incubated with *in vitro* systems or dosed to the appropriate toxicology species and humans (ADME studies) to provide much more complete information on metabolic pathways, excretory routes, and their relative quantitative importance.

The identification of metabolites in complex biological matrices is a challenging task due to the presence of endogenous components, which can mask the detection of metabolites. In addition, metabolite identification can be further complicated by the fact that the metabolites may be present at relatively low concentrations. Recent progress in analytical techniques has made it easier to conduct metabolite identification in early drug discovery and development stages. Liquid chromatography coupled with tandem mass spectrometry (LC-MS/MS) has become the most powerful tool for analysis of drug metabolites because of its sensitivity, selectivity, fast speed, compatibility with sample matrices, and flexibility (Lee and Kerns, 1999; Clarke et al., 2001; Hop et al., 2002; Kostiainen et al., 2003). A tremendous amount of structural information can be obtained from LC-MS/MS analyses. However, mass spectrometric analysis often cannot define absolute positions of metabolic substitution, especially for metabolites formed through aromatic or aliphatic oxidation. In those cases, nuclear magnetic resonance (NMR), LC-NMR, other complementary chemical techniques, or synthesis of authentic standards are necessary for definitive structural assignments. This chapter will focus on the modern techniques for metabolite identification and quantitation. The strategies for metabolite identification are also covered in this chapter.

2.2 SAMPLE PREPARATION

Although it is often considered as the rate-limiting step, sample preparation is still essential to remove or reduce the amount of proteins, salts, and endogenous matrix compounds contained in the biologic matrices and/or increase the concentrations of the trace analytes reaching the detector. Appropriate sample preparation before analyses not only improves the specificity, sensitivity, and reliability of the analyses, but also reduces capillary and column clogging and therefore, improves overall analytical efficiency.

The most common sample preparation methods include protein precipitation, solid-phase extraction, and liquid–liquid extraction. Direct injection may be used occasionally. Solid samples, such as feces and tissues, need additional pretreatments, including homogenization, filtration, centrifugation, solvent evaporation, pH adjustment, and other procedures as necessary to significantly reduce endogenous components.

2.2.1 Direct Injection

Direct injection is the simplest way of preparing a sample but is typically only suitable for samples such as urine and bile. Complex biological matrices, such as serum and plasma, that contain large amounts of proteins and cellular components, if introduced to the instruments directly, can rapidly reduce instrument performance and clog the columns. However, recent development in techniques in special and selective extraction supports, including restricted access media (RAM) (Souverain et al., 2004) and turbulent flow chromatography (TFC) (Mullett, 2007; Xu et al., 2007), have enabled direct and repetitive injection of complex biologic matrices for analysis. The RAM approach separates small molecules from macromolecules in biologic samples through a physical barrier (pore size of the packing particles) or a chemical diffusion barrier (a bonded polymer/protein network on the outside of the particle). Only small molecules are able to reach the absorption sites within the RAM particles and be retained, while macromolecules are limited to the outer surface of the particles and can be easily removed. TFC is another frequently used method for online sample preparation. This method separates small molecules from macromolecules based on their difference in diffusion coefficient in a turbulent flow condition generated by high flow rate in short, narrow columns packed with large-sized particles (50–150 μm). Following separation and removal of macromolecules, small molecules are then eluted to the LC-MS/MS for metabolite identification. It should be noted that these online extraction methods have the risk of losing metabolites due to variable recoveries during the extraction process, which limits their effectiveness in metabolite identification.

2.2.2 Protein Precipitation

Protein precipitation is a common method to denature and remove proteins from biological samples. The basic protein precipitation procedure involves two steps: addition of a precipitant to a biological sample to denature the proteins, which effectively breaks the bonds between protein and drug or metabolite(s), and centrifugation to remove the denatured proteins (Polson et al., 2003). The metabolite(s) of interest remain in the supernatant layer for analysis or further sample preparation. Commonly used precipitants include organic solvent, acid, salt, and metal ions (Polson et al., 2003), among which organic solvents, such as acetonitrile and methanol, are the most widely used. The ratio of organic solvents to the sample is usually between 2:1 and 3:1; for example, an acetonitrile to plasma ratio of 3:1 is necessary for complete protein precipitation. One should be aware that the use of acids as precipitant may not be ideal in some cases as they can catalyze the hydrolysis of certain conjugates such as glucuronides and sulfates. After protein precipitation, the sample can be further processed by solvent evaporation and reconstituted for analysis.

Protein precipitation is simple and fast and can retain all analytes, thus being widely used in sample preparation. The method has also been successfully automated using either centrifugation or filtration in a 96-well format (Biddlecombe and Pleasance, 1999; Watt et al., 2000). The major limitation of the protein precipitation approach is that the residual salts and other endogenous materials in the supernatant may cause ion suppression or enhancement, which could lead to variation in response between samples. When ion suppression occurs, analytes would have lower detection sensitivities and may be missed in metabolite identification.

2.2.3 Liquid–Liquid Extraction (LLE)

LLE separates the analytes of interest from other unwanted components in a biological sample based on their differential solubility and partitioning between the aqueous and organic phases. The usual procedure is to mix an aqueous sample solution with an immiscible organic solvent where the analytes are extracted from one phase to the other. Repeated LLE can maximally extract the analytes or remove interfering substances, which results in a cleaner sample. The analyte extract can be further treated (concentrated and reconstituted in an appropriate solution) to make it suitable for subsequent analysis. LLE is most effective for samples with non-polar or moderately polar analytes of interest. Beside liquid–liquid partitioning and solubility properties, many other factors also influence the extraction outcome, including solution pH, pKa, and ionic strength.

LLE generally produces cleaner sample than protein precipitation and solid phase extraction. However, LLE is labor-intensive and time-consuming. In addition, LLE requires large volumes of solvents and is difficult for full automation due to multiple disjointed steps involved, such as mixing, centrifugation, and evaporation (O'Connor, 2002). Another issue with the use of LLE is the recovery of polar metabolites, especially when there are no authentic standards of the metabolites to ensure sufficient extraction recovery. Therefore, in general, the use of LLE has waned.

2.2.4 Solid Phase Extraction (SPE)

SPE is a routine technique to extract and concentrate analytes in complex biological matrices. It is also considered as a more efficient and environmentally friendlier approach than LLE (Fedeniuk and Shand, 1998). Compared to protein precipitation, SPE produces a cleaner sample but is not considered as cost-effective because of higher labor and material costs associated with this approach.

The basic SPE follows a generic protocol of four steps (Fedeniuk and Shand, 1998) and can be further optimized as needed to improve sample cleanup: (1) sorbent conditioning to prepare the sorbent bed for reproducible retention; (2) application of sample to the sorbent to selectively retain analytes of interest; (3) rinsing the sorbent bed to remove interferences and other unwanted components while retaining the analytes on the sorbent; and (4) elution of the retained analytes using a strong solvent to disrupt the analyte–sorbent binding. The four general mechanisms utilized for SPE include nonpolar, polar, ion-exchange, and covalent interactions. SPE has been automated in a 96-well plate format with a robotic sample processor (O'Connor, 2002; Xu et al., 2007), which greatly improves its accuracy, precision, and productivity, and makes it well-suited for LC-MS applications.

2.3 INSTRUMENTATION

Gas chromatography (GC), radioimmunoassay, high-performance liquid chromatography (HPLC) with ultraviolet (UV), fluorescence, or electrochemical detection were the traditional techniques for analysis of biological samples applied in the past. While GC and HPLC are currently still the primary techniques to separate the analytes of interest from interfering and background substances, the traditional detection capabilities cannot keep up with the demand for high sensitivity and specificity of metabolite identification. The current analytical method of choice for trace-level metabolite identification and quantification is to couple GC or HPLC with mass spectrometers because of their exceptional sensitivity and selectivity and their additional capacities in molecular weight determination and structural characterization of compounds in complex mixtures.

2.3.1 Gas Chromatography–Mass Spectrometry (GC-MS)

GC was coupled with MS in the 1960s because of the requirements of analytes to be in a gas phase for MS analysis. GC-MS is suitable for separating and quantifying trace analytes that are organically extractable, nonpolar, volatile, and highly thermally stable (Pasikanti et al., 2008). The capability of GC-MS in analyzing such compounds in complex matrices has been well documented (Rubiolo et al., 2008; Hoh et al., 2009; Vizcaino et al., 2009). However, the use of GC-MS for analysis of polar and nonvolatile compounds is very difficult because the analytes have to be extracted to volatile organic solvent either directly or through chemical derivatization prior to analysis, which is a very time-consuming and tedious process and sometimes may not be feasible, especially when dealing with large molecules (Christie, 1998; Rufer et al., 2006; Pasikanti et al., 2008). This problem has significantly limited the use of GC-MS and has led to the choice of LC-MS over GC-MS as the preferred analytical technique.

2.3.2 LC-MS

Compared to GC, HPLC operates at ambient temperatures and therefore does not require analytes of interest to be volatile or thermally stable. The development of atmospheric pressure ionization (API, discussed later) technologies was the breakthrough that made it possible for the direct combination of liquid chromatography and mass spectrometry. LC-MS has now become the method of choice for metabolite identification because of its fast speed and high sensitivity. General components of an LC-MS system are highlighted in Fig. 2.1.

In addition, LC-MS does not necessarily require complex sample preparation. Crude extracts of *in vitro* incubations or *in vivo* samples, involving complex matrices, can be directly applied to LC-MS for metabolite identification and structure determination.

Sample Introduction In addition to appropriate sample preparation methods, as discussed earlier, chromatographic performance is also essential for achieving sufficient specificity and sensitivity of LC-MS analysis. Together, sample preparation and chromatographic performance can help reduce noise level and thus facilitate metabolite detection and identification.

HPLC HPLC has been widely accepted and used for separation of analytes in pharmaceutical research worldwide. It is the most widely applied technology in metabolite identification because of its universality and compatibility with API-MS. HPLC used in metabolite analysis generally involves slow LC gradients with columns of >2 mm in internal diameter and 10–20 cm in length. Particle size of the packing material is generally greater than 2 μm.

Fast chromatographic gradient methodology has been used to reduce analysis time and increase throughput for metabolite identification. For example,

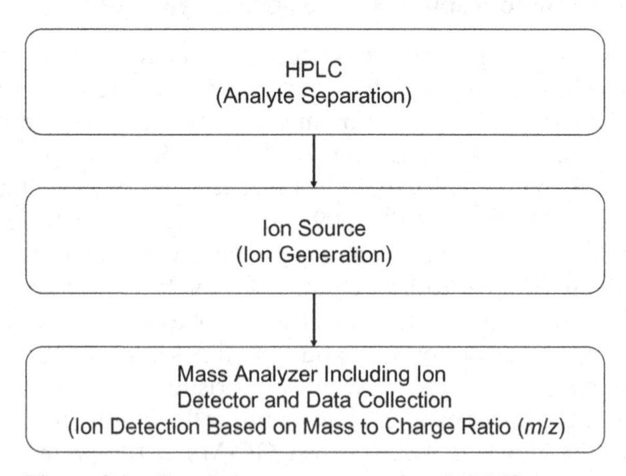

Figure 2.1 General components of an LC-MS system.

Hop et al. explored the use of short HPLC columns (2 cm) containing 3 μm particles for fast gradient elution of under 2 min in analysis of several compounds in liver microsomal incubations, without sacrificing chromatographic resolution (Hop et al., 2002). In addition, a significant improvement in chromatographic performance has been achieved by the recent emergence of ultra performance liquid chromatography (UPLC) (Castro-Perez, 2007).

UPLC The first commercially available UPLC (the ACQUITY UPLC system) was from Waters (Milford, MA) which can work at pressures up to 15,000 psi (Xu et al., 2007), followed by UPLC systems from other manufactures, including Accela™ High-Speed Chromatographic System (up to 15,000 psi) from Thermo Scientific (West Palm Beach, FL), 1200 series Rapid Resolution LC System (up to 8702 psi), and the newest version 1290 Infinity LC System (up to 17,404 psi) from Agilent (Santa Clara, CA). Compared to HPLC, UPLC technology pushes chromatographic peak capacity (number of peaks resolved per unit time of chromatography) and separation speeds to new limits, making it especially useful for complex matrices such as plasma, tissues, and feces. The improved chromatographic performance of UPLC is achieved by smaller particle size (sub 2 μm) packing materials. Smaller particles can effectively shorten the diffusion path of an analyte and thus significantly improve the efficiency of separation. In addition, it allows for a wider range in the van Deemter curve for increased flow rates without losing chromatographic resolution. The overall result of UPLC is higher efficiency at high flow rates (Castro-Perez, 2007). Due to the smaller particle used, UPLC requires technology that can handle the significant increase in back pressure of LC systems.

Because of the better resolution of UPLC, structurally similar metabolites that co-elute in traditional HPLC can now be separated under UPLC. In addition, ion suppression is reduced because of better separation of metabolites from endogenous components. Furthermore, throughput can be improved significantly due to reduced chromatographic run-times. Castro-Perez et al. (2005) demonstrated significantly improved sensitivity, chromatographic resolution, and speed of analysis by using UPLC-MS to analyze a number of *in vitro* samples. For example, a greater number of metabolites of prochlorperazine were identified in rat liver microsomal incubatons using UPLC-MS compared with those identified using regular HPLC-MS. HPLC-MS and UPLC-MS chromatograms of the prochlorperazine sample are shown in Fig. 2.2. Eight doubly hydroxylated metabolites of prochlorperazine (*m/z* 406) were detected with the UPLC analysis; while in the HPLC separation, just three of the doubly hydroxylated metabolites were detected.

UPLC usually produces chromatographic peaks of width from 1 to 3 s, which requires the mass spectrometer to acquire data quickly enough to keep up with the chromatographic output. Therefore, UPLC is ideally coupled with mass spectrometers capable of high-speed scanning for metabolite identification, such as time-of-flight (TOF) instruments (described in the section Time-of-Flight and Quadrupole Time-of-Flight) (Castro-Perez, 2007), in order

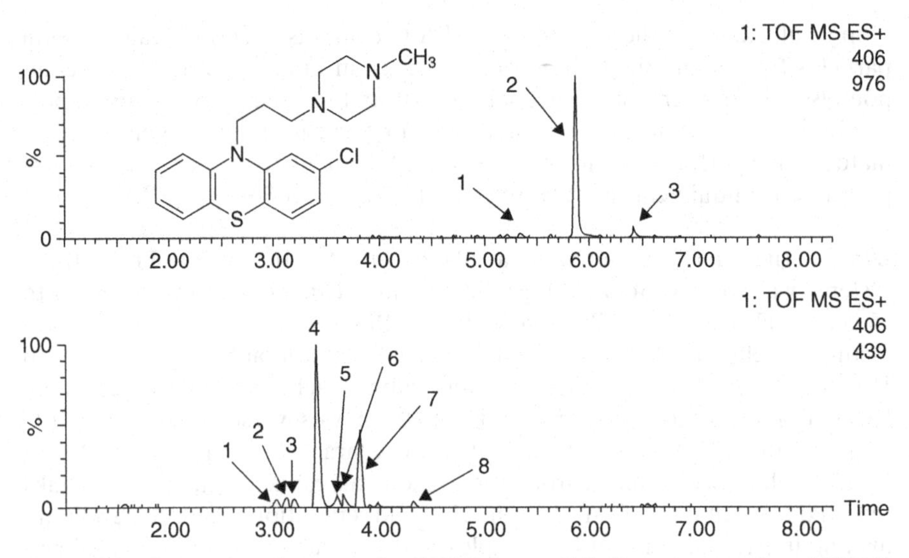

Figure 2.2 Comparison between the HPLC (upper LC-Chromatogram) and UPLC approaches (lower LC-chromatogram) for doubly hydroxylated metabolites of prochlorperazine (Castro-Perez et al., 2005) (Reprinted with permission from John Wiley & Sons, Inc.).

to realize its advantages in speed, resolution, and sensitivity. However, this requirement makes it unsuitable for most of the current mass analyzers used for metabolite identification.

Ionization Methods To date, electrospray ionization (ESI) and atmospheric pressure chemical ionization (APCI) are the two most popular API techniques to realize transformation of solution molecules from liquid phase to gas phase in LC-MS. These ionization techniques are easily interfaced to the chromatographic systems and are likely to yield parent molecular ions. In recent years, two other API techniques have been introduced: atmospheric pressure photoionization (APPI), as a useful alternative for nonpolar compounds at atmospheric pressure, and atmospheric pressure laser ionization (APLI), as a powerful addition to APPI.

ESI ESI is considered as the most universal and gentle API technique (soft ionization) to transform the molecule from liquid phase to gas phase. The use of electrospray as an ionization method was first explored by Dole et al. (Dole et al., 1968; Mack et al., 1970). The interest in ESI was reignited by the experiments carried out by Fenn and his colleagues in the 1980s, demonstrating the successful combination of ESI and mass spectrometry (Yamashita and Fenn, 1984a; Whitehouse et al., 1985). They also demonstrated ESI-MS application in the negative ion mode (Yamashita and Fenn, 1984b) and ESI ionization for

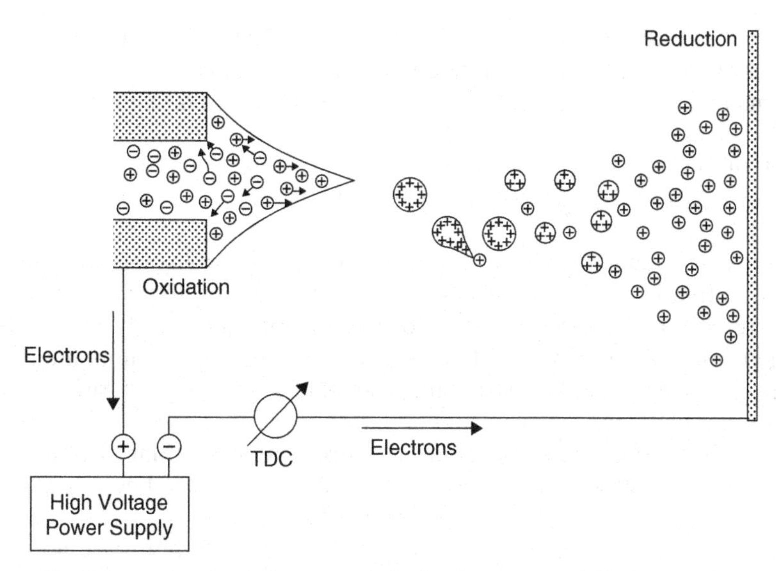

Figure 2.3 Schematic of an electrospray ionization source (TDC: total droplet current) (Kebarle and Tang, 1993) (Reprinted with permission from American Chemical Society).

large biomolecules (Fenn et al., 1989). The technique significantly advanced the applications of mass spectrometry, for which John Fenn received the 2002 Nobel Prize in Chemistry, shared with Koichi Tanaka (for matrix-assisted laser desorption/ionization [MALDI]) and Kurt Wüthrich (for NMR).

During ESI, the aqueous sample solution passes through a metallic capillary tube with a high electric field of potentially several kilovolts relative to the surrounding chamber wall and forms a fine spray of highly charged solution ion droplets where the analyte is ionized. The spray of droplets is further concentrated in a heating device within the source to increase the charging density and vaporized and introduced into the MS (Fig. 2.3; Kebarle and Tang, 1993).

The mechanism of the gas-phase ion formation from liquids is still in debate (Cole, 2000; de la Mora et al., 2000). Two leading theories are the charge residue model (CRM) and the ion evaporation model (IEM) (Kebarle, 2000). CRM assumes that the ions are formed when droplets eventually are ripped apart through Coulombic explosion when the charge repulsion of a droplet exceeds its surface tension because of solvent evaporation. IEM assumes that the ions are directly emitted from the droplets when the size of the charged droplets reduces to a certain radius (~10–20 nm) due to solvent evaporation and Coulomb droplet fissions. Recent studies by de la Mora and coworkers have suggested that small molecule ions are produced by IEM and large molecule ions, such as polyprotonated globular proteins, are produced by CRM

(Loscertales and de la Mora, 1995; de la Mora, 2000; Gamero-Castano and de la Mora, 2000a,b). Either positive ions $[M+H]^+$ or negative ions $[M-H]^-$ can be formed depending on the voltage polarity of the capillary probe:

Formation of positive ions: $M + H^+ \leftrightarrow [M+H]^+$;
Formation of negative ions: $M - H^+ \leftrightarrow [M-H]^-$.

Basic compounds tend to form positive ions while acid compounds tend to form negative ions. The charged species is usually a proton which increases or decreases the molecular weight of the parent molecule by 1Da. Other charged ions are possible, such as sodium, potassium, or ammonium ions. Depending on the chemical structure of an analyte, multiple charged molecular ions can be generated, which is optimal for the bioanalysis of macromolecules such as proteins. ESI is ideal for introduction of polar and thermally labile molecules into a mass spectrometer and therefore has been widely used for the analysis of polar and ionic compounds.

Despite its successful use in LC-MS, ESI has been found to be limited in its dynamic range: the response at higher analyte concentrations are independent of analyte concentration (nonlinearity), which may be caused by its inefficiency in converting all charges in the droplets to gas phase ions at higher analyte concentrations. Another significant limitation of ESI is the matrix effect where matrix components co-eluting with the analytes can often cause either ion suppression or enhancement and adversely affect analytical accuracy and reproducibility (Cappiello et al., 2008). A possible mechanism is a competition of co-eluted matrix components with the analytes for available charges and access to the droplet surface for gas-phase emission.

One development in ESI is nano-flow electrospray (nano-ESI) which was introduced in the 1990s primarily for protein identification and sequencing and is characterized by very small capillaries and low flow rate (~20nL/min) (Wilm and Mann, 1996). The use of nano-ESI in metabolite identification is relatively new. Biological samples are infused directly to nano-ESI or fractionated by HPLC first and then introduced to nano-ESI. Nano-ESI consumes low sample volume and can run longer acquisition times. It provides improved ionization efficiency, reduced MS detection noise, and higher sensitivity compared to conventional ESI. Meier and Blaschke used an ion trap mass spectrometer equipped with nano-ESI to investigate glucuronide conjugates (Meier and Blaschke, 2000). Due to its low flow rates and sample consumption, nano-ESI offers an opportunity to perform time-consuming and complicated multiple sequential fragmentation (MS^n) spectrometric analyses, which are very useful for structure elucidation of the glucuronide conjugates.

Two major limitations of traditional nano-ESI make it difficult to apply nano-ESI in metabolite identification, that is, difficulty in capillary preparation and capillary mounting/alignment in the ion source. As an alternative nanoflow approach, a silicon chip-based nano-ESI has been developed (Schultz et al., 2000; Van et al., 2002). In this new approach, the low-flow infusion is

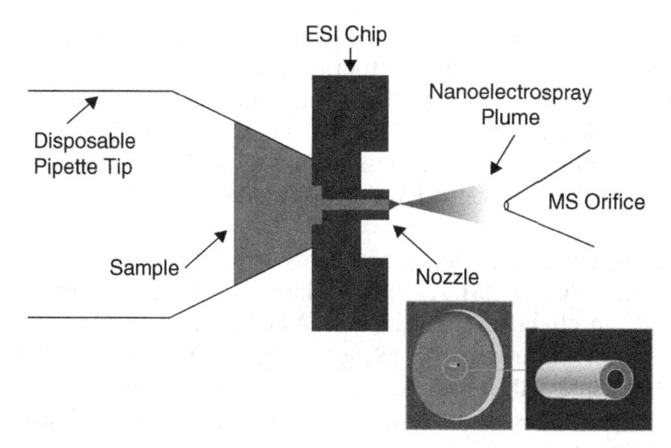

Figure 2.4 Diagram of the NanoMateTM chip-based nano-electrospray device. The close-up shows one of the nozzles (Hop, 2006) (Reprinted with permission from Bentham Science Publishers Ltd).

accomplished through one of the nano-ESI nozzles in a 10×10 array fabricated on a silicon substrate and each nozzle is only used once (Fig. 2.4) (Hop, 2006). The analyte is delivered to the chip through a pipette tip controlled by a robotic system. This new approach has been used in metabolite identification analyses and has demonstrated high speed of analysis, ease of use, and a lack of carryover problem, in addition to the advantages of traditional nano-ESI (Staack et al., 2005; Yu et al., 2005). This approach also significantly reduced variation in ionization efficiencies compared to traditional LC-MS and therefore has the potential for more accurate metabolite quantification where no authentic standard is available (Hop et al., 2005).

APCI APCI was initially introduced by Horning et al. in the mid-1970s as an interface between LC and MS (Horning et al., 1974). The ionization mechanism is slightly different from that of ESI. APCI requires the complete evaporation of solvents and analytes prior to ionization in the gas phase, while for ESI, ionization occurs in the liquid phase, followed by evaporation of solvent droplets.

In APCI, the aqueous solution is vaporized through a glass capillary by a combination of high temperature (450–550°C) and carrier gas and the vaporized solvents are then ionized by a corona discharge needle (Corona effect) to form protonated solvent ions (H_3O^+, $CH_3OH_2^+$, etc.) (Prakash et al., 2007). Ionization of analyte molecules is done through reaction between analyte molecules and ionized solvents as shown in the following equation: $[S+H]^+ + M \rightarrow [M+H]^+ + S$, where S is the solvent molecule and M is an analyte molecule. Because ionization takes place in the gas phase, APCI is less susceptible to matrix inferences and changes in mobile phase conditions compared to ESI. APCI is also less demanding for sample cleanup and can be used at higher flow rates of chromatography (>1 mL/min).

APCI is suitable for analysis of neutral compounds that are volatile and thermally inert. Compared to ESI, APCI has the advantage of ionizing less polar, low to medium molecular-mass compounds. However, it is not suitable for analytes with high molecular weight, like proteins.

APPI The ionization efficiency for analytes with little or no polarity is usually low with either APCI or ESI. APPI is a relatively new API technique that was introduced as an effective alternative to APCI and ESI for analysis of non-polar compounds as direct APPI (Syage et al., 2000; Syage and Evans, 2001) and dopant-assisted APPI (DA-APPI) (Robb et al., 2000).

The technique is based on a single-photon ionization, in which ionization is initiated from a photoionization lamp (typically a krypton lamp) rather than the corona discharge needle in the APCI ion source. The krypton lamp emits 10.2 eV vacuum–UV (VUV) photons which ionize species with ionization potentials (IPs) lower than the energy of the VUV photons (Fig. 2.5; Robb et al., 2000). This energy is generally higher than the IP of analytes because the IPs of most organic molecules are within the range of 7–10 eV, while the common solvents used in LC have higher IPs: water IP is 12.6 eV; methanol IP is 10.8 eV; acetonitrile IP is 12.2 eV (Van Berkel, 2003). Therefore, APPI

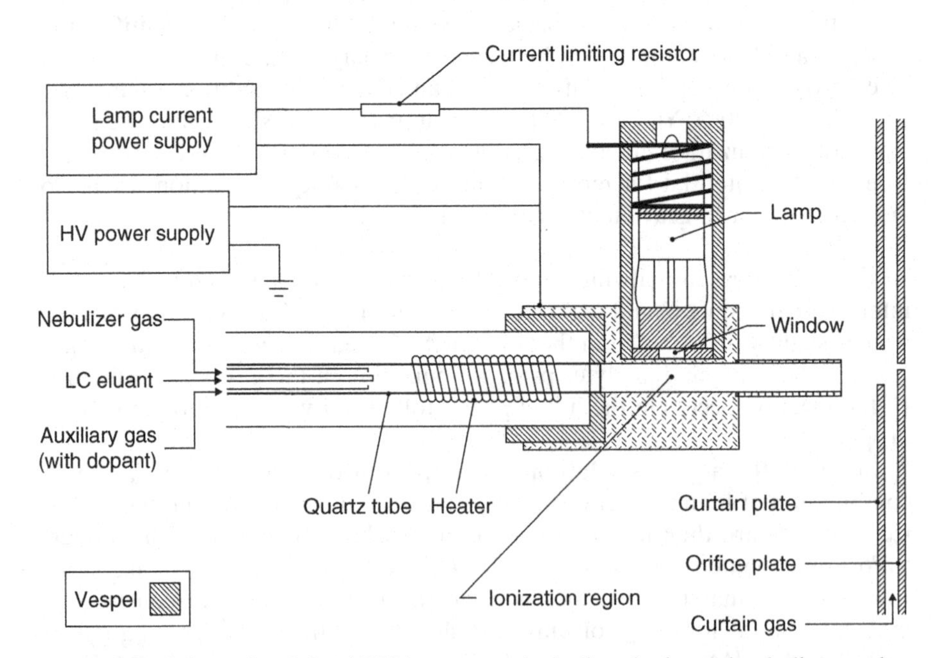

Figure 2.5 Schematic of the APPI ion source, including the heated nebulizer probe, photoionization lamp, and lamp mounting bracket (Robb et al., 2000) (Reprinted with permission from American Chemical Society).

will directly ionize the analytes. Some APPI applications involve the addition of a large amount of a dopant, an easily ionizable compound, to increase ionization efficiency, and thus are called DA-APPI. The dopant is added to the LC eluant or to the vapor generated from the eluant and forms photoions by photoionization, which then starts an ion–molecule reaction cascade with other compounds presented in the ionization region through charge exchange or photon transfer. Besides acetone and toluene, anisole was recently reported as an effective dopant for APPI (Kauppila et al., 2004).

The basic mechanism of photoionization is formation of the radical cation of the molecular species:

$$\mathrm{M\,(analyte\ molecule)} + h\upsilon \rightarrow \mathrm{M}^{\cdot+} + \mathrm{e}^{-}$$

However, the predominant ion produced through APPI is $[M+H]^+$. There are two possible mechanisms to produce $[M+H]^+$ from M:

$$\mathrm{M} + \mathrm{R}^{\cdot+} \rightarrow [\mathrm{M+H}]^{+} + [\mathrm{R-H}]^{\cdot}\ (\text{protonation by charge carrier } \mathrm{R}^{\cdot+})$$

$$\mathrm{M}^{\cdot+} + \mathrm{S} \rightarrow [\mathrm{M+H}]^{+} + [\mathrm{S-H}]^{\cdot}$$
(hydrogen atom abstraction from a protic molecule S)

The dominant mechanism for DA-APPI is believed to be the protonation mechanism, where the analyte M is ionized by the charge carrier $R^{\cdot+}$ formed by protoionization of the dopant (R), such as acetone or toluene (Robb et al., 2000). It is not clear which mechanism dominates in the ionization process for direct APPI (without a dopant). Studies by Syage suggested that the direct APPI mechanism is $M + h\upsilon \rightarrow M^{\cdot+} + e^-$ followed by hydrogen atom abstraction from protic solvent S, for example, CH_3OH, H_2O (Syage, 2004).

APPI has been reported to be more sensitive than the corona discharge APCI technique in some cases (Raffaelli and Saba, 2003), but its ionization process is directly dependent on the reactant ion composition which is influenced by the dopant, solvent, nebulizing gas and its impurities, and surrounding atmosphere (Van Berkel, 2003). Therefore, APPI needs to be further studied before broader implementation.

Fig. 2.6 shows the application of ESI, APCI, and APPI on a polarity versus molecular weight scale (Syage et al., 2008). Overall, the applications of ESI, APCI, and APPI are complementary.

APLI The most recent development in API is APLI, which was introduced by Constapel and coworkers as a promising addition to existing API techniques (Constapel et al., 2005). In APLI, the liquid phase is vaporized by a conventional APCI inlet (AP probe) without the corona needle, and ionization is performed through selective resonance-enhanced, multiphoton ionization schemes using a high-repetition-rate fixed-frequency excimer laser operating at 248 nm. Since the APLI mechanism generally operates primarily

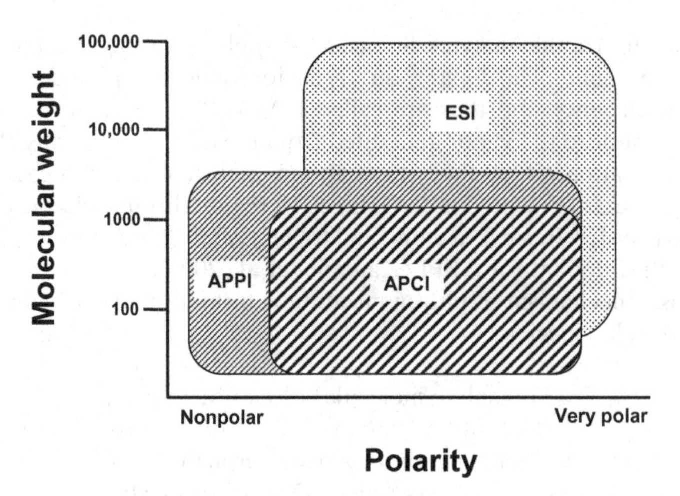

Figure 2.6 Ionization range of ESI, APCI, and APPI as a function of compound polarity and molecular weight (Syage et al., 2008) (Adapted with permission from 2008 Advanstar Communications Inc.).

on the analyte directly, it allows for efficient ionization even of nonpolar compounds such as polycyclic aromatic hydrocarbons (PAHs). APLI replaces the one-step VUV approach in APPI with a step-wise two-photon ionization, which can significantly enhance the selectivity of the ionization process. In addition, because the photon flux is drastically increased over that of APPI during an ionization event, the detection limit of APLI is very low, within the low fmol range.

Mass Analyzers A mass analyzer measures the mass-to-charge (m/z) ratio of ions via the interaction of charged particles with electric or magnetic fields. Although the fundamental operating principle is similar, the variety of instrument types are different in their capacities for metabolite profiling and are often complementary to each other. Therefore, it is very important to select the appropriate instrumentation to ensure quality and reliability of the analytical results and analytical efficiency.

Mass analyzers with different operating principles have been creatively combined to form a new generation of hybrid mass analyzers, such as triple quadrupole-linear ion trap (Q-trap), LTQ-Orbitrap, quadrupole time-of-flight (Q-TOF), and LTQ-Fourier transform ion cyclotron resonance mass analyzer (LTQ-FT), which are much more powerful and versatile than the mass analyzers used alone.

The most commonly used single and hybrid mass analyzers are summarized below.

Quadrupole Mass Analyzer The principle of quadropole mass analyzers was first described by Paul and Steinwedel in 1953 (Paul and Steinwedel,

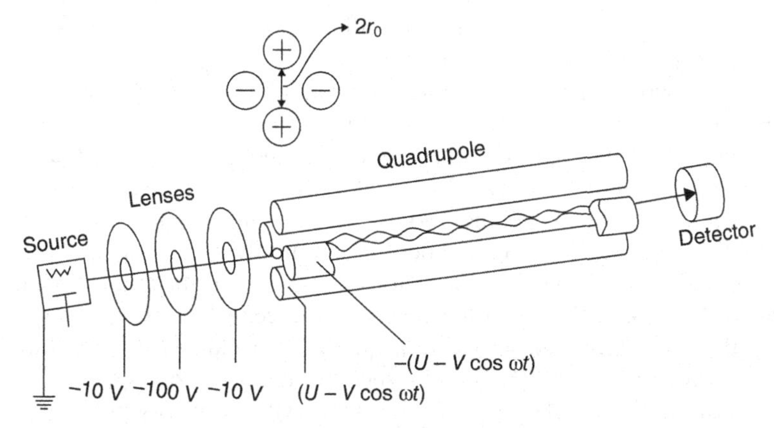

Figure 2.7 Schematic of a quadrupole mass analyzer (de Hoffmann and Stroobant, 2007) (Reprinted with permission from John Wiley & Sons, Inc.).

1953, 1960). Quadrupole mass analyzers consist of four parallel electrodes arranged in a radial array as shown in Fig. 2.7 (de Hoffmann and Stroobant, 2007), with the diagonally opposite rods connected electrically so that they have the same direct current (DC) and radio frequency (RF) potentials. The two pairs of electrodes have the same magnitude but opposite signs of potential at any given time. Ions entering the analyzer travel parallel to the electrodes. At a given RF/DC ratio, only ions with a certain *m/z* ratio can have a stable trajectory passing through the electrodes from one end to another. Other ions will collide with the rods and never reach the detector.

The relationship between ion motion and its *m/z* ratio in the quadrupole field can be described mathematically by the two parameters, a_u and q_u, of the Mathieu equation (Ioanoviciu, 1997; Westman-Brinkmalm and Brinkmalm, 2002):

$$a_u = a_x = -a_y = \frac{8zU}{mr_0^2\omega^2}$$

$$q_u = q_x = -q_y = \frac{4zV}{mr_0^2\omega^2}$$

where m is the ion mass, z is the ion charge, U is the DC voltage, V is the RF amplitude, ω is the RF frequency, and r_0 is half the distance between opposite electrodes. Only certain combinations of a and q give stable solutions to the Mathieu equation, which correspond to the ions passing through quadrupoles without touching the electrodes. Otherwise, the ion discharges itself against an electrode and will not be detected.

The quadrupole mass analyzers can perform both selected ion monitoring (SIM), in which only ions with a selected *m/z* ratio can go through the rods for detection while ions with higher or lower *m/z* will collide with the rods, and a full scan spectrum in which ions within an *m/z* range can pass through the rods and be detected by the analyzer.

There are two types of quadrupole mass analyzers: single quadrupole mass spectrometers (SQMS) and triple quadrupole mass spectrometers (TQMS). SQMS, although small in size and relatively inexpensive, are rarely used as a stand-alone instrument for metabolite identification because they can only provide molecular weight of metabolites and are not capable of elucidating metabolite structures. TQMS have been a commonly used mass spectrometer for quantitative analysis of known molecules and for structure elucidation. Similar to SQMS, TQMS also have an ion source, a lens (Q0) transporting ions to the mass analyzer, and Q1 (a quadrupole mass analyzer), but with addition of a collision cell (Q2) and a second mass analyzer (Q3).

This setup in the TQMS allows for ion selectivity by mass separation at two stages, which leads to a high detection sensitivity when used in multiple-reaction monitoring (MRM) mode. In MRM mode, Q1 is used as a mass filter for precursor ions (PIs), while Q3 is used as a mass filter for product ions after the desired PIs are fragmented by inert gas molecules (typically Ar or N_2) within Q2.

TQMS can also run product ion scans (single stage MS/MS or MS^2) to help with structure elucidation. In this case, the PIs are selected in Q1 and are transferred into Q2. The ions are fragmented in Q2, and product ions are then measured by scanning Q3.

TQMS perform well with PI scan and neutral loss (NL) scan modes. PI scans can be conducted to detect PIs which fragment to form a characteristic charged fragment. In this scan mode, Q1 is scanned across a defined mass range. Ions are passed into Q2 where they are fragmented. Q3 is fixed to transmit only the mass of the characteristic fragment. Therefore, only ions that are passed through Q1, and which fragment to form the characteristic charged fragment, will be detected. NL scan can be performed to detect PIs, which lose a characteristic neutral fragment. In this scan, Q1 is scanned across a defined mass range and ions are fragmented in Q2. Q3 is scanned over a mass range, offset by the neutral mass of the characteristic fragment. The above-mentioned multiple tandem mass spectrometric scan modes are illustrated in Fig. 2.8.

The detection sensitivity of TQMS is high when used in MRM mode. However, the detection sensitivity of TQMS decreases dramatically when high mass range is analyzed in a scanning mode. This is the major limitation of this system when used for unknown metabolite screening (Tolonen et al., 2009). More versatile and powerful mass analyzers, including the ion trap, the quadrupole linear ion trap, and the Q-TOF mass analyzers, have been developed and dominate the metabolite identification field as they can run fast full scan, multiple sequential fragmentations (MS^n), or exact mass measurement experiments.

Ion Trap The principle of quadrupole ion traps is the same as the quadrupole mass analyzers but they are set up in three dimensions rather than two, as

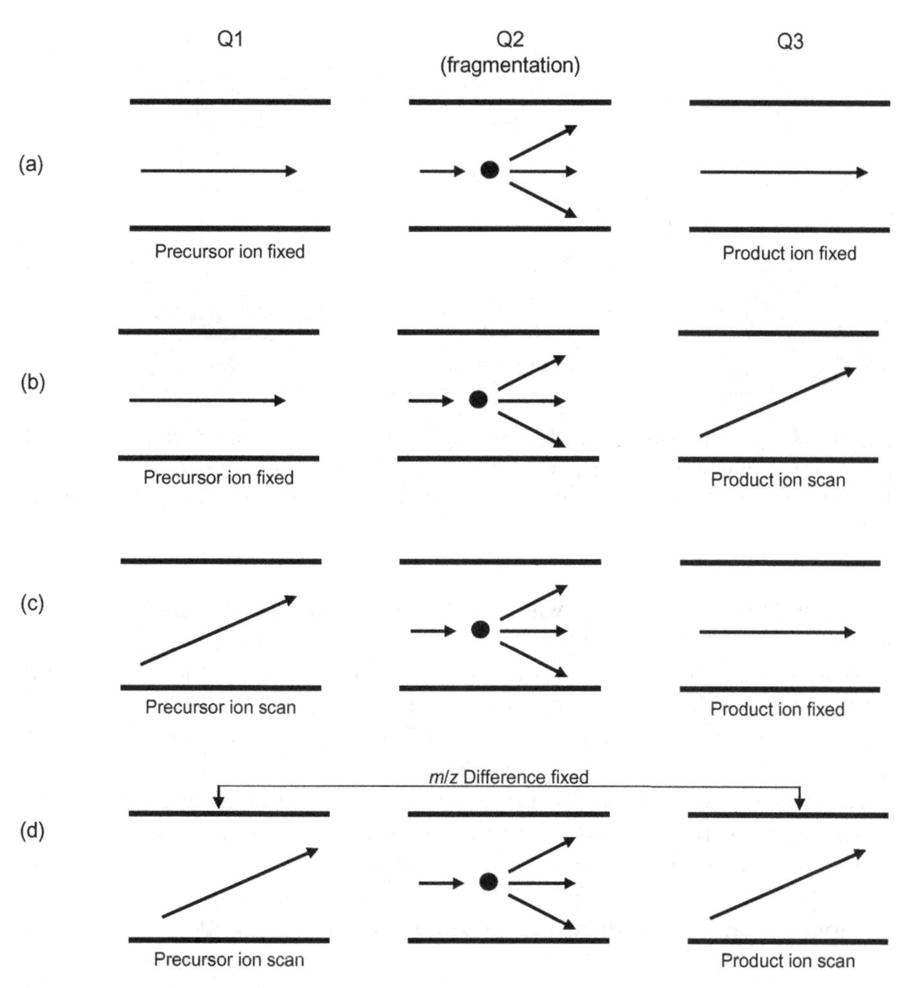

Figure 2.8 Schematic representation of various types of tandem mass spectrometric scan modes: (a) multiple-reaction monitoring; (b) product ion scan; (c) precursor ion scan; (d) neutral loss scan.

shown in Fig. 2.9 (Westman-Brinkmalm and Brinkmalm, 2002). A quadrupole ion trap has three electrodes: two end caps and one ring. Ions generated in the ion sources are guided to the ion trap through an opening in one of the end-cap electrodes. Within the quadrupole ion trap is a low pressure of helium (10^{-3} torr) to reduce the kinetic energy of the ions and focus them toward the center of the trap, which allows the ion traps to measure all ions retained in the trapping step and thus avoid sensitivity losses as well as to increase mass resolution during the full-scan mode (Kamel and Prakash, 2006). The end-cap electrodes are held at ground potential and only one RF potential is applied to the ring electrode in most commercial quadrupole ion traps. All ions above

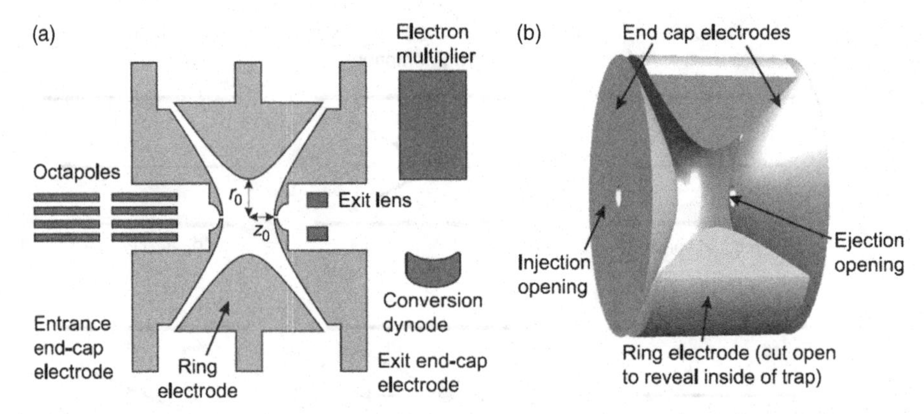

Figure 2.9 (a) A cross-section schematic of a quadrupole ion trap mass spectrometer; (b) A three-dimensional perspective view of the quadrupole ion trap (Westman-Brinkmalm and Brinkmalm, 2002) (Reprinted with permission from John Wiley & Sons, Inc.).

a certain *m/z* can be trapped when the RF amplitude is set to the low storage voltage.

A mass spectrum is acquired when ions are ejected from the ion trap and hit the external detector. There are different ways to eject ions and acquire a mass spectrum with quadrupole ion traps. One commonly used method is the mass selective instability scan, in which the ions with lower *m/z* become unstable when the RF voltage increases and are ejected. As the RF voltage continues to increase, heavier ions become unstable and are ejected as well. The ejected ions are detected by an electron multiplier or ion detector to generate a mass spectrum. The mass analysis equation for the mass selective instability method with the quadrupole ion traps is expressed as

$$m/z = \frac{8V}{q_z\left(r_0^2 + 2z_0^2\right)\omega^2}$$

where V is the RF potential, r_0 is the radius of the ring electrode, z_0 is the distance from the trap center to the end cap, and ω is the angular frequency (March, 1997; Ma and Chowdhury, 2008).

Because trapped ions oscillate with different frequencies corresponding to their *m/z*, a supplementary RF voltage can be added to the end-cap electrodes to selectively eject ions with a certain *m/z*. When the supplementary RF voltage of a certain frequency is added, ions with a certain *m/z* fall into resonance with the oscillating potential, and their oscillation amplitude increases and eventually ions are ejected. This technique is called resonant ejection. It can also be combined with an RF scan to eject ions with *m/z* ratios from low to high successively for detection.

When there is need for PI isolation for MS/MS, a mass selective stability method can be applied. Compared with mass selective instability ejection and resonant ejection, in which the trapping potential is defined only by an RF voltage, the mass selective stability method requires a DC voltage applied to the ring electrode. This will produce a narrow *m/z* stability range within the ion trap so that only the PI of interest is trapped while all other ions are forced out of the trap.

The theoretically infinite MS^n capacity is achieved by resonance excitation which effectively dissociates the isolated PI to product fragmentation ions. Similar to resonant ejection, resonant excitation is achieved with a supplemental RF voltage that induces additional translational energy into the ion without ejecting it out of the trap (resonant excitation voltage or tickle voltage) and allows for ion fragmentation within the trap. Resonant ejection can then be applied to the fragmented ions to generate an MS/MS spectrum. The isolation and fragmentation steps can be repeated a number or times to generate MS^n spectra and is only limited by the trapping efficiency of the instrument. Because fragmentation occurs in the same physical location as mass analysis, there is a one-third loss of the lower end of the possible mass scan range (low mass cutoff).

Compared with quadrupole mass analyzers, ion trap mass analyzers have improved sensitivity, specificity, and scan speed at full scan modes and are able to perform MS^n. MS^n provides more specific structural information compared to MS/MS by isolating and further fragmenting product ions. This approach narrows the potential sites of modification and provides a more complete assessment of the metabolic sites. All of these features make ion traps very useful in metabolite identification and structure elucidation. The ion traps are, however, incapable of performing PI and NL scans.

The main restriction of the traditional three-dimensional (3D) quadrupole ion traps is that there is a limit on the number of ions that can be trapped (about 10^5) before the space charge effects seriously alter performance. A relatively new variation ion trap called two-dimensional (2D) linear ion trap (LIT), traps and detects ions in similar ways as the quadrupole ion traps but with improved ion storage capacity and scan rate, and much higher detection sensitivity (Douglas et al., 2005). LIT is a geometrical variation of the traditional 3D quadrupole ion trap. It has four hyperbolic electrodes with two end caps. Ions in a LIT are confined radially by a two-dimensional RF field and axially by a stopping potential on both end cap electrodes. This setup allows for a significant increase in storage capacity and trapping efficiency with 10–100 times more ions than the 3D ion trap. Additionally, because ions enter into the LIT through one end cap and then are ejected from it, the LIT can also be coupled with two detectors, one on each side of the trap, to double the ion detection (Ma and Chowdhury, 2008). The LIT has been used in combination with TQMS (Q-Trap) or Orbitrap (LTQ-Orbitrap) for even more versatile and improved functionalities.

Triple Quadrupole-Linear Ion Trap (Q-Trap) The combination of triple quadrupole technology with linear ion traps yields the triple quadrupole-linear ion trap (Q-Trap) instrument, a highly flexible and powerful platform in which the last quadrupole of a TQMS is replaced with a linear ion trap. Q-Trap combines the ion trap MS^n capacity and high-sensitivity full scanning with high-quality fragmentation ion production of TQMS (no low mass cutoff) and can therefore perform a variety of bioanalytical applications for metabolite identification and simultaneous quantification. Q-Trap can perform NL or PI scans which the traditional ion traps cannot and performs MRM scanning with the same high sensitivity as equivalent TQMS. A particularly powerful application of Q-Trap is for data-dependent experiments in which ions that meet predetermined criteria of high intensity or expected m/z detected with ion trap high-sensitivity full scan mode, can automatically trigger MS/MS for structural characterization (Hopfgartner et al., 2003). Q-Trap has been widely used for rapid screening and characterization of drug metabolites (Hopfgartner et al., 2003; Xia et al., 2003; King and Fernandez-Metzler, 2006; Yao et al., 2008).

Orbitrap and LTQ-Orbitrap Orbitrap, TOF, and Fourier transform-ion cyclotron resonance mass analyzer (FTMS) (to be discussed later) are high-resolution mass analyzers. Exact mass measurement is very helpful for confirming proposed metabolites and determining or, at least, limiting the possible molecular formulae of unknown metabolite structures. The more accurate the mass measurement, the fewer choices are available for possible molecular structures.

Orbitrap was invented by Makarov and is the most recent addition to the mass analyzer family (Makarov, 1999, 2000). The Orbitrap is similar to a modified ion trap. Ions are trapped in a potential well and move in stable trajectories both around a central spindle electrode and in harmonic oscillation in the z direction as shown in Fig. 2.10 (Hu et al., 2005).

In contrast to ion trap, in which ions are ejected for external detection, the Orbitrap detects the ions by measuring their unique oscillation frequencies along the z-axis. Because the z-direction potential is completely quadratic, ion motion is independent in the z-direction and the ion m/z is related to ion oscillation frequency, f_z, as expressed below (Westman-Brinkmalm and Brinkmalm, 2008):

$$fz \propto \frac{1}{\sqrt{m/z}}$$

Orbitrap provides very good resolution and mass accuracy. In fact, it has been shown that the Orbitrap has better mass accuracy than all quadrupoles and TOF instruments and is just behind the Fourier transform ion cyclotron resonance instrument, at ~2 ppm with internal calibration (Hu et al., 2005; Yates et al., 2006).

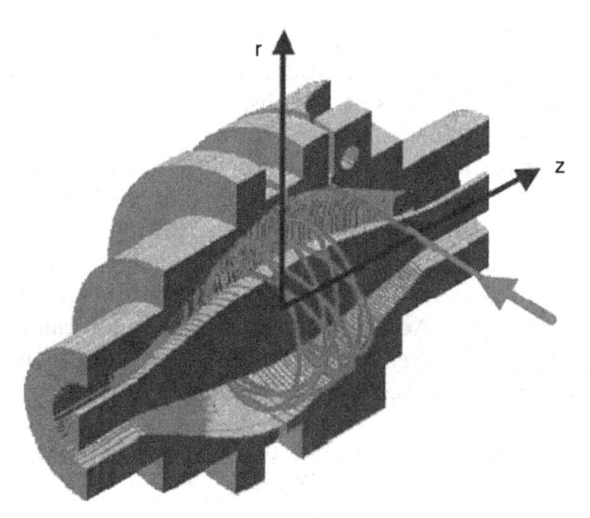

Figure 2.10 Cutaway view of the Orbitrap mass analyzer (Hu et al., 2005) (Reprinted with permission from John Wiley & Sons, Inc.).

Orbitrap alone cannot perform multistage mass spectrometry and therefore is of limited use in metabolite identification. It is often used in combination with a 2D linear ion trap (LTQ, one type of LIT), forming a powerful and versatile hybrid, high-resolution mass analyzer. In an LTQ-Orbitrap, ions formed by different ionization methods are stored and prepared by the LTQ utilizing AGC (automated gain control) and MS^n scanning. Desired ions exiting the LTQ can be further fragmented in a C-shaped ion trap (C-trap) or directly transferred to the Orbitrap for detection. More efficient high-energy collisional dissociation (HCD) fragmentation can be achieved in an additional octopole collision cell located behind the C-trap. Ions formed through HCD fragmentation are transferred back into the C-trap from where they are injected into the Orbitrap for detection.

The LTQ-Orbitrap is an effective instrument for metabolite identification, capable of high-sensitivity screening over a wide mass range and acquiring accurate mass data for both parent and fragment ions. An example of these strengths is provided by Ruan et al. who demonstrated the rapid detection and structural characterization of *in vitro* metabolites of indinavir using an LTQ-Orbitrap mass spectrometer to generate high-resolution mass spectral data with multiple techniques for postacquisition data mining. As a result, the study identified a total of 15 metabolites including two new indinavir metabolites in a rat liver S9 incubation sample (Ruan et al., 2008).

One limitation of the LTQ-Orbitrap is that it is not as effective as TOF instruments for fast chromatography applications or UPLC due to its relatively slow data acquisition rate (Tolonen et al., 2009).

Time-of-Flight (TOF) and Quadrupole Time-of-Flight (Q-TOF) TOF mass analyzers are suitable for fast and cost-efficient metabolite identification analyses. Its operating principle was first published in 1946 (Stephens, 1946). Ions are generated in the ion source and accelerated toward ground potential after entering the gas phase. After leaving the acceleration area, ions with different mass-to-charge ratios, but the same kinetic energy, drift at different speeds through a field-free zone in the TOF drift tube and reach the detector at different times (Ma et al., 2006). For ions with the same charge, high mass ions travel slower than low mass ions since ion velocity is inversely proportional to the square root of m/z. Accordingly, the m/z of an ion can be determined by measuring the time it takes to reach the detector after its formation, that is, the TOF. The time difference (t_{TOF}) can be expressed as:

$$t_{TOF} = \frac{L}{v} = L\sqrt{\frac{m}{2zUa}} \propto \sqrt{\frac{m}{z}}$$

where L is the ion drift length, v is the ion velocity after acceleration, m is the ion mass, z is the ion charge, and Ua is the accelerating electric potential difference.

Q-TOF, a hybrid mass analyzer that combines TOF instrumentation with the mass filter and collision cells of the TQMS, is a versatile instrument that can provide high sensitivity and high-resolution data with a wide mass range. As such, it is very useful for detection of known and unknown metabolites from a single run without pre-adjustment. The high mass resolution and mass accuracy of Q-TOF with its very high data acquisition rate ensures reliable, accurate, and fast metabolite identification and structure determination. Mortishire-Smith et al. used LC-QTOF with automated data analysis for metabolite identification of several structurally diverse compounds and demonstrated that this approach is able to indicate the presence of a wide range of metabolites with only a limited requirement for manual intervention (Mortishire-Smith et al., 2005). Similarly, Tiller et al. described a semi-automated high-throughput metabolite identification strategy on a Q-TOF system in support of optimization efforts in drug discovery (Tiller et al., 2008).

In addition, TOF is an ideal mass spectrometer to couple with UPLC due to its fast data acquisition speed and high mass resolution. Typical chromatographic peak widths in UPLC chromatograms are 1–3 s, which requires that the mass spectrometer must acquire data quickly enough to match the chromatographic output. High-resolution mass spectrometers such as FTMS or Orbitrap need a longer scan time to acquire high-resolution data, typically 1 s to achieve a resolution of 60,000 FWHM (full width at half maximum; please see note [b] of Table 2.1 for method to calculate resolution). As a result, few data points will be collected which might result in the loss of information. By contrast, TOF mass spectrometers only need about 0.1 s to complete a full scan, for example, from 1–1000 mass units, so it can acquire sufficient mass spectra over UPLC chromatographic peaks (Castro-Perez, 2007).

TABLE 2.1 Characteristics of Different Mass Analyzers[a]

Characteristics	Quadrupole Mass Analyzer	Ion Trap	Orbitrap	TOF	FTICR
Mass resolving power or resolution[b]	10^2–10^4	10^3–10^4	$>1.5 \times 10^5$[g,h]	10^3–10^4	10^5–10^6[j]
Mass accuracy[c]	100 ppm	50–100 ppm	2–5 ppm[g,h]	5–50 ppm	<1 ppm[j]
Mass range[d]	10^4	1.5×10^5	4000[i]	$>10^5$	$>10^4$
Linear dynamic range[e]	10^7	10^2–10^5	10^3–10^4[h]	10^2–10^6	10^2–10^5
Speed[f]	1–20 Hz	1–30 Hz	1 Hz at the resolution of 60,000[i]	10–10^4 Hz	0.001–10 Hz
Cost	Relatively low	Low to moderate	Moderate to high	Moderate to high	Moderate to high

[a] McLuckey and Wells (2001).
[b] Refers to $M/\Delta M_{1/2}$. M is the mass associated with the apex of a single well-resolved peak and $\Delta M_{1/2}$ is the full width of the peak at half-maximum peak height (FWHM).
[c] The ratio of the mass-to-charge (m/z) measurement error (i.e., the difference between the measured m/z and the true m/z) divided by the true m/z value and is usually stated in terms of parts per million.
[d] Refers to "m/z range." Only the upper limit of mass range is provided here, which means the ions with lower m/z values can also be analyzed.
[e] The range over which ion signal is linear with analyte concentration.
[f] The spectral generation rate in Hertz.
[g] Perry et al. (2008).
[h] Hu et al. (2005).
[i] Thermo Product Specifications.
[j] Brown et al. (2005).

Although the Q-TOF performs well in full scan mode, it is not very suitable for selected ion monitoring. Additionally, compared to instruments with ion traps, Q-TOF can run MS/MS product ion scans but does not have MS^n ($n > 2$) capacity.

Fourier Transform Ion Cyclotron Resonance Mass Analyzer (FTICR/MS or FTMS) The principle of ion cyclotron resonance (ICR) was developed in the early 1930s by Lawrence and colleagues (Lawrence and Livingston, 1932; Lawrence and Edlefsen, 1930) but was not used for MS until 1949 by Hipple et al. (Hipple et al., 1949; Sommer et al., 1951). ICR was combined with the Fourier Transform (FT) technique to create the FTICR/MS or FTMS by Comisarow and Marshall (1974). In 1985, the first report of FTICR coupled with external ion sources appeared (Kofel et al., 1985; McIver et al., 1985).

FTMS is one of the most sensitive mass analyzers with resolution ranging from 10^5–10^6 FWHM and mass accuracy of <1 ppm (Brown et al., 2005). With FTMS, source-generated ions are trapped in an analyzer cell which is contained in a high vacuum chamber and centered within a high-field superconducting magnet. Trapped ions are excited by varying RF excitation pulses from excitation plates and form an alternating current with frequency and intensity equal to the number of ions for a particular *m/z*. The stepped RF sweep simultaneously quantifies all detected ions to generate a frequency versus time spectrum, which is converted to a mass spectrum by FT.

Despite its advantages in high resolution and high sensitivity, FTMS is very expensive, is relatively difficult to use, and has a relatively low throughput (Sanders et al., 2006). In addition, FTMS is not compatible with standard HPLC columns and flow rates. Therefore, FTMS is rarely used on its own in metabolite identification. Several hybrid FTMS instruments have been developed to couple FTMS with quadrupole (i.e., quadrupole FTMS from Bruker Daltonics [Billerica, MA] and IonSpec [Lake Forest, CA, now Varian]) and linear ion trap (i.e., LTQ-FT from Thermo Scientific), which combine the excellent capacities of FT and the well-established features of quadrupole and ion trap, and eliminate the compatibility issues with HPLC.

The performance characteristics of hybrid FTMS are ideal for the types of complex mixtures encountered in high-throughput metabolite identification. The high resolving power of FTMS reduces the need for chromatographic separations prior to mass spectrometric analysis and its high mass accuracy indicates that metabolites can often be identified on the basis of their accurate mass alone. FTMS analyzers also allow for accurate mass measurements in MS^n, which simplifies structure elucidation and identification of complex metabolite structures.

Different types of mass analyzers have been covered so far. As a summary, the characteristics of different mass analyzers are included in Table 2.1 for comparative purposes.

2.4 STRATEGIES FOR METABOLITE IDENTIFICATION

A major challenge in metabolite identification is provided by low concentrations of metabolites and high abundance of endogenous materials in biological samples. Strategies for metabolite identification have grown significantly more sophisticated with advances in instrument technologies and application software. Optimizing these approaches requires significant knowledge and experience from researchers and an appropriate handling of the various types of analyses in order to generate the required outcome with high-quality data as well as providing insightful interpretation of those data.

A typical metabolite identification approach involves the following steps: (1) predict possible metabolites using *in silico* resources (as described below) or based on researcher's experience; (2) apply survey scans of the test and control samples; (3) compare ion chromatograms of the test and control samples to determine the *m/z* values of expected and unexpected metabolites; (4) conduct product ion scan and/or MS^n on the *m/z* of suspected metabolites and propose metabolite structures; (5) confirm metabolite structures using NMR, synthesis of authentic standards, or other chemical techniques combined with MS if needed (Anari and Baillie, 2005).

2.4.1 *In Silico* Prediction

A number of software programs have been developed to predict possible metabolites of a given compound, including METEOR, MetaDrug, MetaSite, and MDL Metabolite Database.

METEOR (LHASA Ltd., Leeds, UK) is a software based on "knowledge-based expert rules" to evaluate the likelihood of metabolic pathways and predict the metabolites likely to be formed from xenobiotics, including both Phase I (P450- and non-P450-mediated) and Phase II biotransformations (Anari and Baillie, 2005). The software contains a knowledge base of biotransformations and expert rules that are derived from historical drug metabolism literature. Users can also build their own biotransformations and expert rules from proprietary knowledge using a knowledge-based editor. In addition, METEOR contains a direct link to DEREK, which is toxicity prediction software, to produce an assessment of potential metabolite toxicity.

Another software that was developed recently, MetaDrug (GeneGo, Inc., St. Joseph, MI), also shows potential as a promising approach to predict drug metabolism based on a combined approach using rule-based metabolite prediction, quantitative structure–activity relationship models for important drug-metabolizing enzymes, and an extensive database of human protein-xenobiotic interactions (Ekins et al., 2005). It can be conveniently used to derive major Phase I and Phase II metabolic pathways, to determine the involvement of particular CYPs in biotransformation, and has been used for both human *in vitro* and *in vivo* drug metabolism studies.

Metabolite screening using MS techniques can be readily performed at lower nanogram levels but may not provide definitive structure elucidation. NMR generally requires microgram amounts and is usually quite definitive. Some metabolite structures are obvious from changes in mass, such as N- and S-oxidation, N-dealkylation, and many Phase II conjugations. However, other biotransformation pathways, including aromatic and aliphatic oxidations, require NMR for accurate regiochemical structure determination (Chen et al., 2007). Two *in silico* approaches show potential to resolve this issue.

MetaSite (Molecular Discovery Ltd., London, UK) is a computational program that uses the 3D structure of a compound to predict the sites of metabolic transformations related to CYP-mediated reactions (Cruciani et al., 2005). The program also predicts structure of the metabolites with a ranking derived from the site of the metabolism. Metasite could potentially become a powerful tool for accurate regiochemical structure determination. However, it is currently limited to predictions related to human CYPs only.

Another program, MDL Metabolite Database, is a comprehensive searchable database of published information on drug metabolism, pharmacokinetics, original metabolism literature, and new drug applications that is regularly updated with new entries. The MDL Metabolite Database can be searched graphically with exact, similar, or substructure options, or perform generic searches with keyword options to predict the metabolic pathways of a compound.

Currently, there are still significant gaps between *in silico* prediction and actual metabolite identification results. Therefore, *in silico* programs need further development for accurate prediction. *In silico* predictions cannot replace conventional metabolite identification studies and are most effective when used in combination with conventional methods, which will be described below.

2.4.2 MS Acquisition Strategies

Biotransformations can occur through both expected and novel or rarely encountered metabolic pathways. Expected biotransformations, such as hydroxylation and N-dealkylation, produce metabolites whose molecular weights can be readily predicted. However, metabolites formed through unexpected or multiple-step biotransformations are more difficult to project because their potential mass shifts from parent drugs are, by definition, less readily predictable. Comprehensive elucidation of metabolite profiles from a variety of biotransformations that could occur remains a difficult and time-consuming process, even with the wide use of LC-MS/MS and the many impactful advances in related technologies. Many different strategies have been developed. In general, an integrated MS acquisition strategy that combines multiple approaches is often required to effectively and efficiently detect metabolites from both expected and unexpected biotransformations in complex biologic matrices. The general strategy first focuses on identification of potential metabolites, followed by acquisition of product ion spectra of metabolites and structure determination.

Survey Scans A common method for metabolite screening is to conduct a full scan MS which analyzes the test and control samples over the full mass range. This nonselective approach ensures that most ionizable metabolites will generate a response in the acquired MS data. However, an obvious disadvantage of this approach is that a full scan will yield significant interfering data from nondrug-related components, such as endogenous materials in biological samples.

More specific detection of metabolites can be achieved by PI and NL scans. Both are conducted on a triple-quadrupole mass spectrometer and are important initial steps in metabolite identification. For PI scans, the analyst needs to know the fragmentation patterns of the parent compound and the PI scan then detects the likely metabolites by searching for an unaltered portion of the parent drug within the metabolite structure. Compared to PI scans, NL scans may not require any knowledge of the parent molecule. NL scans detect unknown metabolites by scanning characteristic neutral losses associated with unique structural features, and are widely used for detection of Phase II conjugates, including glucuronides (loss of 176), sulfates (loss of 80), and glutathione conjugates (loss of 129) (Liu and Hop, 2005). Since both PI and NL scans are quite effective in detecting metabolites with similar structural characteristics as the parent molecule, these scans greatly reduce the amount of data that need to be analyzed. PI and NL scans are more selective than the full scan mode; however, they may suffer significant loss of detection sensitivity with TQMS.

MRM, as an alternative to PI and NL scans for metabolite identification, maximally preserves the high detection sensitivity capacity of TQMS. Armed with predictions of possible metabolites and a knowledge of MS/MS fragmentation of the parent molecule, MRM can be applied to generate a list of targeted transitions for potential metabolites. While it is possible for MRM to miss some potential metabolites (unusual or less frequently occurring), this approach is a very powerful tool for metabolite screening, especially when detection sensitivity is important.

Data processing and interpretation are crucial, yet time-consuming processes and have been a major bottleneck in the overall scheme of metabolite identification. Several software packages have been developed to accelerate the process and reduce reliance on the analyst, such as Lightsight™ from Applied Biosystems/MDS Sciex (Foster City, CA), MetWorks™ from Thermo, and Metabolynx™ from Waters. These software tools can automatically run samples using LC-MS and also process acquired data. In general, the overall process starts with acquiring LC-chromatograms of control and treated samples. Then the LC-chromatograms of control samples are subtracted from that of the treated samples. This filtering helps to distinguish any new peaks in the sample chromatogram. The detected components are assigned as "expected metabolites" if their mass shift, relative to parent, matches predefined biotransformation pathways. Table 2.2 includes common Phase I and Phase II metabolic pathways and the mass shift associated with the metabolites (Baranczewski et al., 2006). Other detected components are assigned as "unexpected metabolites" if they do not fit into the specific biotransformation pathways. After potential metabolites are detected, the software can generate and run LC-MS/MS methods to acquire product ion spectra of the metabolites, which can lead to further structure elucidation. In addition to data filtration, the software also includes detection of isotope patterns for chlorine and bromine-containing compounds, which can aid in structure assignment.

Product Ion Scan Product ion scans generate fragmentation patterns of potential metabolites that provide key information for elucidation of their

TABLE 2.2 Common Metabolic Pathways and Mass Shift of Expected Metabolites Compared to Parent Compound (Baranczewski et al., 2006) (Reprinted with Permission from Institute of Pharmacology Polish Academy of Sciences)

Reaction		Phase of Metabolism/ Enzyme	Mass Shift Da (NL*, Parent Ion/Ion Mode)
Nitro reduction	$R–NO_2$ > $R–NH_2$	I/Amine oxidase	–30
N-, *O*- or *S*-demethylation	$R–NH–CH_3$ > $R–NH_2$	I/CYP	–14
N-, *O*- or *S*-dealkylation	R–NH-alkyl > $R–NH_2$	I/CYP	– depends on alkyl chain length
Dehydrogenation	$R–CH_2–OH$ > R–CHO	I/Dehydrogenase	–2
Hydroxylation	$R–CH_2$ > R–CH–OH Ar–H > AR–OH	I/CYP	+16
di-Hydroxylation		I/CYP	+32
Oxidation	$R_1–CH_2–R_2$ > $R_1–CO–R_2$	I/CYP	+14
N-oxidation	R–NH > R–N–OH	I/CYP and/or FMO	+16
Sulfoxidation	R–S–R > R–SO–R	I	+16
Aldehyde oxidation	R–CHO > R–COOH	I/Alcohol dehydrogenase	+16
Alcohol oxidation	$R–CH_2–OH$ > R–COOH	I/Alcohol dehydrogenase	+16
Oxidation of CH_3-group to carboxylic acid	$R–CH_3$ > R–COOH	I/CYP	+30
Epoxide hydroxylation	R–CH(O)–R > R–CH(OH)–CH(OH)–R	I/Epoxide hydratase	+18
Epoxide formation and hydroxylation		I/Epoxide hydratase & CYP	+34 (+18, +16)
Sulfation aromatic	Ar–OH > $Ar–O–SO_3H$	II/Sulfotransterase	+80 (Precursor m/z 97/–)
Sulfation allphatic	R–OH > $R–O–SO_3H$		

Glucuronidation	R–OH > R–O-GlcA	II/UDP-transferase	+176 (NL 176/+ or –)
Carbamoyl-glucuronide	primary & secondary amines	II/UDP-transferase	+220 (NL 176/+)
Glycosylation hexose (Glc)		II/GDP-transferase	+162
Glutathione conjugation	R–CH=CH_2 > R–CH_2–CH_2–SG	II/Glutathione transferase	+307 (305) Aliphatic (NL 129/+), Aromatic (NL 273/+)
N-Acetylcysteines mercapturic acid	R–CH_2–CH_2-CysOAc		+163 (NL 129/+)
Glutathione conjugation	Epoxide + GSH-H_2O > GS-parent		+305
Acetylated GSH			+347 (305 + 42)
Glutathione conjugation	Epoxide + GSH > R–CHOH–HCSG		+323 (129)
GSH	R–CH_2–CH_2–CysGly		178 (176)
GSH	R–CH_2-CH_2–CysGlu		250 (248)
Cysteine conjugation	R–CH=CH_2 > R–CH_2–CH_2-Cys		+121 (119)
Mercapturic acid (from GSH conj.)			+161
Reduction of NO_2	R–CH_2–NO_2 > R–CH_2-SG	II/GSH transferase	+160
Gly-conjugation	R–COOH > R–CO-Gly	I or II	+57
Ala-conjugation	R–COOH > R–CO-Ala	I or II	+71
Methylation	R–OH > R–O–CH_3	I/Methyl transferase	+14
Acetylation (1°, 2° amines)	R–NH_2 > R–NH–CO–CH_3	I/*N*-acetyltransferase	+42
Phosphorylation	R–OH > R–O–PO_3H		+79 (Precursor m/z 63/– Precursor m/z 79/–)

structures. During product ion scans, the ions of interest are selectively isolated and are further fragmented, either within the same space (e.g., ion trap) or a different space (e.g., TQMS, Q-TOF) depending on the instrument, to provide a series of fragment ion masses that correspond to part of the unknown metabolite. The general approach is to compare the product ion spectra of metabolites with those of the parent molecule. As discussed previously, many instruments are capable of product ion scans, including TQMS, ion trap, Q-TOF, and FTMS. Also as mentioned before, the first stage of a product ion scan is usually MS/MS or MS^2. Consecutive fragmentation of the product ions generated from the first stage is referred to as MS^n.

MS/MS and multistage MS^n can be collected "on-the-fly" in a single analysis, using data-dependent acquisitions (or information-dependent acquisition [IDA]) where the user defines parameters such as mass range and an arbitrary intensity threshold for a survey scan. The survey scans can be any type of the scans discussed previously, such as full scan, PI scan, or NL scan. When a survey scan is running, once the detection of ions exceeds the preset threshold, the IDA software automatically triggers different data acquisition, such as product ion scans for the specific ion. This approach can generate survey scan LC-chromatograms and product ion spectra in a single LC-MS run, thus maximizing the dimension of data that are collected with minimal sample injections. One drawback of this approach is that there is a risk of missing some components that are outside of the preset parameters. This approach may also potentially generate a large amount of data unrelated to the parent compound. Users can also predefine a list of masses of interest for MS/MS and MS^n, but this is obviously subject to the user's prior experience and knowledge and may also miss some metabolites or lead to incomplete lists of metabolites as a result of user's biases. Anari et al. evaluated an approach to address this problem through the integration of *in silico* metabolite prediction with IDA (Anari et al., 2004). This approach used a list-dependent LC-MS^n data acquisition protocol formulated using knowledge-based predictions of metabolic pathways derived from a commercial database, the MDL metabolism database. Using indinavir as an example, a substructure similarity search in the MDL metabolism database indicated potential metabolic pathways, including two hydrolytic, two N-dealkylation, three N-glucuronidation, one N-methylation, and several aromatic and aliphatic oxidation pathways. This information was integrated with data-dependent LC-MS^n analysis using an ion trap mass spectrometer. In an incubation with human hepatic S9 fraction, a total of 18 metabolites of indinavir were identified with a single LC/MS^n run, which is a significant saving in instrument use and operator time.

Although there are many sophisticated techniques to acquire product ion spectra for structure elucidation, the interpretation of these data has been mostly manual and has relied heavily on the intervention of an expert. Some software programs have been developed recently to aid the interpretation process, including Mass Frontier from Thermo and ACD from ACD/

Labs (Toronto, Ontario, Canada). Both programs can provide information of some product ions from a given chemical structure.

Exact Mass Measurement A few high-resolution mass spectrometers, including Orbitrap, TOF, and FTMS, are capable of exact mass measurement, which greatly enhances the capability of structure determination. Exact mass measurement is especially useful for differentiating biotransformations which lead to different molecular formulae but with the same nominal mass changes. For example, metabolites formed by a combination of hydroxylation and dehydrogenation have the same nominal mass weight gain of 14 units as metabolites formed by methylation. These metabolites can be easily differentiated due to their different accurate masses. The use of accurate mass for metabolite identification has been exemplified by Liu and Hop (2005). Following incubation in hepatocytes of a parent compound which had a methoxy substituent ($R–CH_2–O–CH_3$), a metabolite with the same nominal *m/z* value and product ion spectra as the parent compound appeared. Using Q-TOF accurate mass measurement, the metabolite was identified as a carboxylic acid (R–COOH) resulting from O-demethylation and subsequent oxidation of the alcohol, which was recognized as its molecular ion was 0.0363 Da lower than that of the parent compound.

A rapid screening approach, MS^E, has been described in several publications that allows for simultaneous acquisition of molecular ion and fragmentation ion data for all detected analytes within one single experiment (Castro-Perez, 2007; Tiller et al., 2008; Tolonen et al., 2009). The MS^E approach works by alternating low (MS) and elevated (MS^E) collision energy settings without MS/MS PI selection, using a hybrid quadrupole orthogonal TOF mass spectrometer. The MS scan collects accurate mass information of the intact metabolites and the MS^E scan acquires accurate mass information of fragmented ions of the metabolites. Data between MS and MS^E scans can be linked by their retention times and analyzed together using software tools such as MetaboLynx MS^E. Because there is no PI selection, it is essential to have good chromatographic separation so that the fragmented ions are derived predominantly from the analytes of interest. This approach is suitable for samples of relatively simple matrices such as microsomal and hepatocyte incubations but may not work well for bile samples or other complex matrices (Tiller et al., 2008).

Mass Defect Filter (MDF) An MDF technique has been developed to detect drug metabolites via post-acquisition processing of high-resolution LC-MS data (Zhang et al., 2003). Mass defect is the difference between the exact atomic weight and the nominal atomic weight of an element. The exact mass of all other elements is scaled relative to carbon-12 which has an exact mass of 12 Da, and therefore, all other elements have their unique mass defect. For example, the mass defect for hydrogen is 0.007825 Da and for oxygen is –0.005085 Da. Accordingly, each molecule has its own mass defect, and this value will change as a result of metabolism. For Phase I biotransformations, the change in mass defect should fall within ±40 mDa with the exception of major dealkylations which would lead to larger, negative shifts within 180 mDa (Castro-Perez, 2007). The change in mass defect for most Phase II metabolites should fall within ±90 mDa considering glutathione adducts and other biotransformations. Because of the requirement for an accurate mass measurement, the MDF has to be used in combination with Orbitrap, TOF instruments, or FTMS. The MDF is a featured software incorporated into the Metabolynx software from Waters and Xcalibur from Thermo Scientific. Based on these principles, the MDF allows users to set criteria beforehand so that possible metabolites can be differentiated from endogenous components. This approach greatly simplifies the total ion chromatogram (TIC), as data exceeding the expected change in mass defects are removed. The MDF technique is also very helpful for identification of unknown metabolites from unexpected biotransformations. However, metabolites with substantial structure modification (e.g., cleavage or dealkylation metabolites) may have greater mass defect changes than preset criteria and therefore could be potentially missed after MDF filtering. In addition, MDF-processed ion chromatograms still contain false positive peaks from endogenous components. Therefore, the MDF approach needs to be combined with other survey scan methods for complete screening of possible metabolites.

Due to its significant benefit, the MDF approach has been applied to metabolite identification in various complex biological samples (Zhu et al., 2006; Mortishire-Smith et al., 2009; Rousu et al., 2009; Zhang et al., 2009). One example is the detection of omeprazole metabolites in plasma. The TIC of omeprazole metabolites, without MDF filtering, displayed significant endogenous interferences, without apparent metabolite peaks (Fig. 2.11a). However, after MDF processing, the TIC exhibited all metabolite peaks and minimal endogenous peaks (Fig. 2.11b) (Zhu et al., 2006). Therefore, MDF processing could significantly minimize the data mining process and streamline further analysis.

LC-Radioactivity Detector-MS Radioactive labeling of a parent compound with ^{14}C or ^{3}H isotopes is widely used to aid in metabolite identification in drug development. The biological samples containing radiolabeled compound and its metabolites are usually analyzed by LC coupled with both a mass spectrometer and a flow scintillation analyzer (FSA), which can generate mass spectra and radiochromatograms simultaneously. This is a very powerful approach because the detection of radioactivity locates the metabolites in a chromatogram while the mass spectra provide structural information on the metabolites. Common online radiochemical flow detectors use liquid cells where liquid scintillant is added to the HPLC effluent passing through the cell, also called homogenous detection (Egnash and Ramanathan, 2002). The advantage with this method is its high throughput, while a major disadvantage is its relatively low detection sensitivity. Using eluent fraction collection and

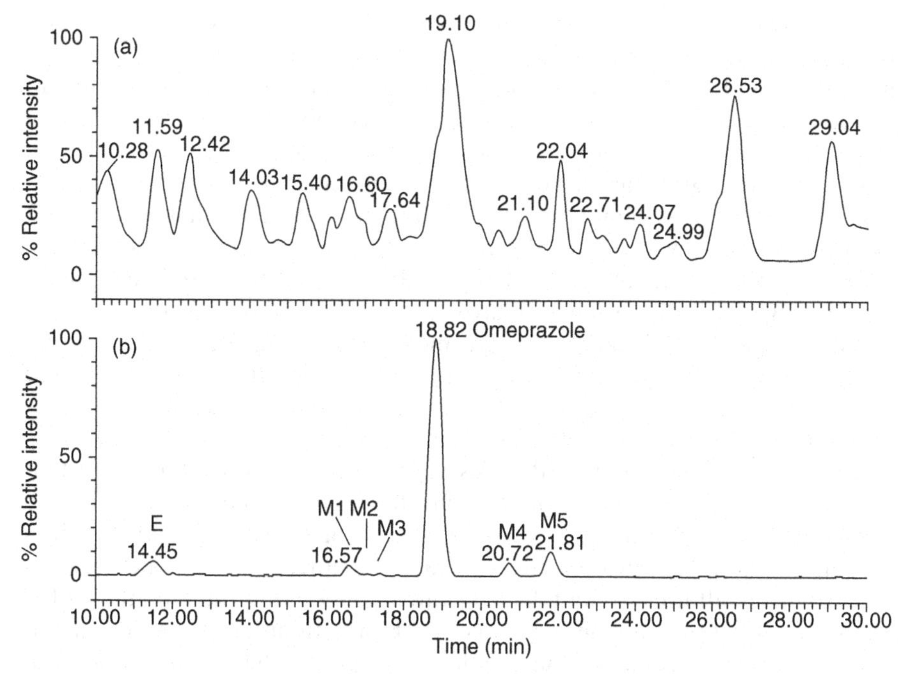

Figure 2.11 LC-chromatographic profiles of omeprazole metabolism samples analyzed by Q-TOF LC-MS (Zhu et al., 2006). (a) TIC profile of plasma sample without MDF processing; (b) TIC profiles of plasma samples obtained after MDF processing (Reprinted with permission from American Society for Pharmacology and Experimental Therapeutics).

offline scintillant counting can significantly improve counting efficiency and therefore increase the detection sensitivity; however, it is also more labor-intensive and time-consuming and is not suitable for volatile metabolites. Radioactive labeling has also often been used for metabolite quantification, which will be further discussed in Section 2.5.

2.4.3 Other Chemical Techniques Combined with MS

Some of the older, more established chemistry techniques are still useful in metabolite identification and structure elucidation when applied in combination with MS. Analysis of MS/MS data prior to and post chemical intervention can provide very important structural information for metabolites. These techniques include chemical derivatization, hydrogen/deuterium (H/D) exchange, stable isotope labeling, and hydrolysis. H/D exchange and isotope labeling will be discussed in further detail in later chapters.

Chemical derivatization has been used in GC-MS for analysis of nonvolatile compounds. As it can be a very labor-intensive process, it is considered a major drawback of GC-MS. Chemical derivatizaion is, however, still useful for characterization of unexpected metabolites and quantification of unknown metabolites in LC-MS. Chemical derivatization can be used before instrumental analysis to stabilize labile molecules, increase lipophilicity and extraction yield, improve separation from endogenous materials, and enhance ionization efficiency and fragmentation. For example, Shen and colleagues used chemical oxidation on a crude sample of pioglitazone metabolites with Jones reagent at microscale followed by LC-MS/MS analysis (Shen et al., 2003). Oxidation helped identify the site of hydroxylation at the terminal ethyl group and therefore distinguished the terminal hydroxyl from ω-1 hydroxyl of two isomeric metabolites with identical protonated molecules at *m/z* 373. Schaefer et al. used a selective acetylation strategy (selective acetylation of hydroxyl and amine groups under different conditions) to determine the positions of glucuronidation of carvedilol (Schaefer et al., 1992).

Chemical and enzymatic hydrolysis of metabolites can play an important role in metabolite identification. Specifically, β-glucuronidase and sulfatase can be used for determination of the presence of Phase II metabolites in a biomatrix. These enzymes selectively cleave the ether or ester bond between glucuronic or sulfuric acid and the aglycone of the Phase II metabolite. HPLC analyses of the same biological samples, with or without enzymatic treatment using β-glucuronidase or sulfatase, helped to identify Phase II metabolites of nabumetone (Nobilis et al., 2004).

Overall, the above strategies for metabolite identification typically are well suited for identification of all possible metabolites but may be not sufficient for complete structure characterization and quantification of the identified metabolites. NMR is generally needed for ultimate structural proof, either through stand-alone NMR or tandem LC-NMR/LC-MS-NMR.

2.5 METABOLITE QUANTITATION

Assessment of metabolite abundance is important in determining the major metabolic pathways of parent compounds and comparing metabolite exposure in different species including human. A well-established method for metabolite quantitation in biological samples is to employ LC-MS (or LC-MS/MS) detection with authentic standards, which are used to establish standard calibration curves. The main advantage of LC-MS (or LC-MS/MS) is its high sensitivity and specificity that allows for quantification of drug metabolites at very low concentrations in complex biological samples. UV and fluorescence detection have also been used for quantification when standards are available. In early drug discovery, authentic standards are typically not available. In this case, MS and UV can still be used to get an approximate estimate of the relative abundance of drug metabolites to the parent, with the assumption that MS and UV responses of the metabolites are similar to those of the parent.

However, this approach obviously has the potential risk of providing misleading estimates of concentrations.

If radioactively labeled compounds are available, LC-radioactivity detection can be used for quantitation of metabolites from *in vitro* and *in vivo* samples generated with radiolabeled parent compounds (so called hot samples). In addition, LC-MS, in combination with radioactive detection, is a good alternative approach for quantification of metabolites present in cold samples (generated with non-radiolabeled parent compounds). With this approach, the radiolabeled metabolites from *in vitro* or *in vivo* samples serve as pseudo-standards for the metabolites and their concentrations are quantified by radioactive detectors. The amount of the metabolites from cold biological samples can then be determined by comparing their MS responses to those of the corresponding metabolite pseudo-standards. This approach was demonstrated by Yu et al. using ^{14}C-labeled metabolites from dog hepatocytes and rat bile samples as the metabolite standards to quantify metabolites in cold human plasma (Yu et al., 2007).

As discussed previously, FSA is quite limited for detecting or quantifying metabolites of low abundance because of its low sensitivity. Eluent fraction collection, followed by an offline liquid scintillation spectrometry (LSS), offers much higher detection sensitivity than FSA. However, since the samples have to be completely dry, volatile metabolites may be lost during drying. In addition, offline analysis is very time-consuming and labor-intensive and is therefore not suitable for high-throughput analysis. A newly developed microplate scintillation counter (TopCount™, PerkinElmer, Waltham, MA), which collects fractions from HPLC eluent into 96-well plates, contains an yttrium silicate-based solid scintillator for offline scintillation counting. Other than eliminating the laborious step of adding scintillation cocktail to the eluent fractions, microplate scintillation counting also significantly shortens radioactivity analysis time by counting up to 12 wells at a time, which is an order of magnitude faster than traditional LSS. However, because microplate scintillation counting uses offline counting with completely dry samples, it still has relatively low throughput and runs the risk of losing volatile compounds and nonpolar compounds (which can be adsorbed to the surface of the plate). Another novel method that combines LC-MS with accurate radioisotope counting (ARC) appears to overcome the limitations of other radioactive detection approaches described above (Nassar et al., 2003, 2004). The LC-ARC system uses online stop flow scintillation technology to increase the counting time which improves the detection sensitivity by ~20 times over conventional FSA. The simple interface of LC-ARC with MS also facilitates the online acquisition of MS data and eliminates the need for fraction collection and sample preparation. The system also uses predefined threshold values for the radioactive signal to stop the flow, which can shorten analysis time.

Another technology used with radioactive compounds is accelerator mass spectrometry (AMS), which has extremely high sensitivity compared to LC-ARC, microplate scintillation counter, or LSS. AMS has been used to detect very small quantities of isotopes and predominately measures ^{14}C (Garner, 2000). Because of its high sensitivity, AMS can accurately measure metabolite levels in biological samples from a study where the radioactive dose administered to human is equivalent to the level of radiation naturally occurring in the adult human body. With this very low level of radioactivity, the regulatory requirements needed to administer the radioactive dose are significantly less stringent and allows the use of radiolabel earlier in drug discovery and development. However, this approach requires preparation of ^{14}C-labeled compounds and sample fraction collection for offline analysis, which makes the process very labor-intensive and slow throughput. In addition, the application of AMS technology is somewhat limited by the high costs and availability of the instruments.

When neither standard nor radiolabeled compounds are available, alternative metabolite quantitation approaches can be considered which include chemiluminescent nitrogen detection (CLND), inductively coupled plasma mass spectrometry (ICP-MS), and NMR, all of which can give a response independent of the compound structure. CLND produces a response proportional to the number of nitrogen atoms present in the compound structure and therefore can be used to quantify a metabolite as long as the number of nitrogen atoms in the structure is known (Corens et al., 2004). A main limitation of CLND is low selectivity due to interferences from endogenous substances containing numerous nitrogen atoms. This limitation may be minimized by appropriate sample cleanup or good chromatographic separation. Another limitation of CLND is that the responses for compounds containing N–N and N=N bonds are less reliable (Taylor et al., 2002). ICP-MS is another technique with decent sensitivity and selectivity. It works by atomization and ionization of certain elements, including metals, Br, Cl, I, P, and S, and therefore is suitable for quantification of metabolites of certain types containing the abovementioned elements (Gammelgaard et al., 2008). Similar to CLND, ICP-MS only requires knowledge of the number of atoms of the element of interest. NMR is commonly used for ultimate structure determination but in principle can also be used for metabolite quantification because its signal intensity is proportional to the number of resonating nuclei. ^{1}H NMR could, in theory, be used for quantitation, but its application is limited for crude samples due to interference from endogenous components. A commonly used NMR technique is ^{19}F-NMR which can selectively detect and quantify fluorine containing metabolites since ^{19}F is not naturally present in endogenous materials in biofluids (Prakash et al., 2007).

2.6 CONCLUSIONS AND FUTURE TRENDS

Metabolite identification is a critical component of modern drug discovery and development. LC-MS has been, and will continue to be, the mainstay for identification, structural characterization, and quantification of drug metabolites. Many technical advances have been made in recent years to improve the sensitivity, selectivity, accuracy, and efficiency of LC-MS. Introduction of API methods including ESI, APCI, APPI, and APLI greatly improves the efficient ionization of various analytes with different properties. Development of high-resolution mass analyzers (TOF, Orbitrap, and FTMS) and hybrid mass analyzers, such as Q-trap, LTQ-Orbitrap, Q-TOF, and LTQ-FTMS, allows for more powerful, versatile, and efficient metabolite identification. The recently introduced LC-MS-NMR technique provides ultimate structure characterization of metabolites. ^{19}F-NMR, ICP-MS, and CLND technologies also make it possible to quantify metabolites without reference standards or radiolabeled parent compounds. Many software programs have been developed to assist with data-dependent analysis, data interpretation, and to improve analytical automation.

Even with all the significant technical advances, there are still opportunities to further improve metabolite identification research, as the process can still be labor-intensive and time-consuming. One key opportunity for improvement is to focus on enhancing sample preparation and introduction into the analytical instrument. The rapid development of microfluidics holds the promise of integrating all steps of metabolite identification on one microchip, and therefore providing complete analysis cycles including sample preparation, chemical reactions, sample separation, detection, and data processing on one single microfabricated device (Weigl et al., 2003; Clayton, 2005). This approach is very appealing because the "lab-on-a-chip" technology will realize a more integrated automation and therefore greatly improve analytical efficiency, increase throughput, and reduce overall time and resource. Another area of interest is to improve column technology for ultrafast sample separation to achieve high throughput analysis, including higher temperature, higher pressure, smaller particle sizes, faster flow rates, and shorter columns. In addition, manual data interpretation has become a bottleneck for high-throughput analysis because of the enormous amount of data that are generated. Therefore, software will become more sophisticated and more user-friendly for prediction and identification of both expected and unexpected metabolites. Other emerging technologies, such as ion mobility mass spectrometry and MALDI imaging, also have the potential to play important roles in metabolite identification and determination of the distribution of parent drug and its metabolites in tissues. Ion mobility MS is a new technology for ion separation at atmospheric pressure that can be used as a pre-processing tool to reduce background ions by isolating ions of interest and simplifying spectra of complex mixtures by dividing the mixture into a series of simpler subsets of ions (Guevremont, 2004). MALDI imaging is capable of localization of test article or spatial identification of metabolites which may be critical to the understanding of distribution of parent drug and its metabolites in individual tissues and organs and the relevance of the information to clinical safety (Drexler et al., 2007).

As a final note, there are advantages and disadvantages associated with different strategies for metabolite identification and quantitation, and many of them provide complementary information. In addition, there is an impressive array of technology currently available to scientists. Therefore, it is important for scientists to design creative and thoughtful strategies for metabolite identification that combine knowledge of the molecules of interest (chemistry, physicochemical properties), prior experiences (metabolic pathways both common and unusual), with available instrumentation and resources.

ACKNOWLEDGMENT

The author would like to thank Dr. Donald Tweedie for reviewing and providing valuable suggestions.

REFERENCES

Anari MR, Baillie TA. 2005. Bridging cheminformatic metabolite prediction and tandem mass spectrometry (Review). *Drug Discov Today* 10:711–717.

Anari MR, Sanchez RI, Bakhtiar R, Franklin RB, Baillie TA. 2004. Integration of knowledge-based metabolic predictions with liquid chromatography data-dependent tandem mass spectrometry for drug metabolism studies: Application to studies on the biotransformation of indinavir. *Anal Chem* 76:823–832.

Baillie TA, Cayen MN, Fouda H, Gerson RJ, Green JD, Grossman SJ. et al. 2002. Drug metabolites in safety testing. *Toxicol Appl Pharmacol* 182:188–196.

Baranczewski P, Stanczak A, Kautiainen A, Sandin P, Edlund PO. 2006. Introduction to early in vitro identification of metabolites of new chemical entities in drug discovery and development (Review). *Pharmacol Rep* 58:341–352.

Biddlecombe RA, Pleasance S. 1999. Automated protein precipitation by filtration in the 96-well format. *J Chromatogr B Biomed Sci Appl* 734:257–265.

Brown SC, Kruppa G, Dasseux JL. 2005. Metabolomics applications of FT-ICR mass spectrometry. *Mass Spectrom Rev* 24:223–231.

Cappiello A, Famiglini G, Palma P, Pierini E, Termopoli V, Trufelli H. 2008. Overcoming matrix effects in liquid chromatography-mass spectrometry. *Anal Chem* 80:9343–9348.

Castro-Perez J, Plumb R, Granger JH, Beattie I, Joncour K, Wright A. 2005. Increasing throughput and information content for in vitro drug metabolism experiments using ultra-performance liquid chromatography coupled to a quadrupole time-of-flight mass spectrometer. *Rapid Commun Mass Spectrom* 19:843–848.

Castro-Perez JM. 2007. Current and future trends in the application of HPLC-MS to metabolite-identification studies (Review). *Drug Discov Today* 12:249–256.

Chen Y, Monshouwer M, Fitch WL. 2007. Analytical tools and approaches for metabolite identification in early drug discovery (Review). *Pharm Res* 24:248–257.

Christie WW. 1998. Gas chromatography-mass spectrometry methods for structural analysis of fatty acids. *Lipids* 33:343–353.

Clarke NJ, Rindgen D, Korfmacher WA, Cox KA. 2001. Systematic LC/MS metabolite identification in drug discovery (Review). *Anal Chem* 73:430A–439A.

Clayton J. 2005. Go with the microflow. *Nat Methods* 2:621–627.

Cole RB. 2000. Some tenets pertaining to electrospray ionization mass spectrometry. *J Mass Spectrom* 35:763–772.

Comisarow MB, Marshall AG. 1974. Fourier transform ion cyclotron resonance spectroscopy. *Chem Physics Lett* 25:282–283.

Constapel M, Schellentrager M, Schmitz OJ, Gab S, Brockmann KJ, Giese R, et al. 2005. Atmospheric-pressure laser ionization: A novel ionization method for liquid chromatography/mass spectrometry. *Rapid Commun Mass Spectrom* 19:326–336.

Corens D, Carpentier M, Schroven M, Meerpoel L. 2004. Liquid chromatography-mass spectrometry with chemiluminescent nitrogen detection for on-line quantitative analysis of compound collections: Advantages and limitations. *J Chromatogr A* 1056:67–75.

Cruciani G, Carosati E, De BB, Ethirajulu K, Mackie C, Howe T, et al. 2005. MetaSite: Understanding metabolism in human cytochromes from the perspective of the chemist. *J Med Chem* 48:6970–6979.

de Hoffmann E., Stroobant V. 2007. *Mass Spectrometry Principles and Applications*, 3rd ed. Chichester, UK: John Wiley & Sons.

de la Mora JF. 2000. Electrospray ionization of large multiply charged species proceeds via Dole's charged residue mechanism. *Anal Chim Acta* 406:93–104.

de la Mora JF, Van Berkel GJ, Enke CG, Cole RB, Martinez-Sanchez M, Fenn JB. 2000. Electrochemical processes in electrospray ionization mass spectrometry. *J Mass Spectrom* 35:939–952.

Dole M, Mack LL, Hines RL, Chemistry DO, Mobley RC, Ferguson LD, et al. 1968. Molecular beams of macroions. *J Chem Phys* 49:2240–2249.

Douglas DJ, Frank AJ, Mao D. 2005. Linear ion traps in mass spectrometry. *Mass Spectrom Rev* 24:1–29.

Drexler DM, Garrett TJ, Cantone JL, Diters RW, Mitroka JG, Prieto Conaway MC, et al. 2007. Utility of imaging mass spectrometry (IMS) by matrix-assisted laser desorption ionization (MALDI) on an ion trap mass spectrometer in the analysis of drugs and metabolites in biological tissues. *J Pharmacol Toxicol Methods* 55:279–288.

Egnash LA, Ramanathan R. 2002. Comparison of heterogeneous and homogeneous radioactivity flow detectors for simultaneous profiling and LC-MS/MS characterization of metabolites. *J Pharm Biomed Anal* 27:271–284.

Ekins S, Andreyev S, Ryabov A, Kirillov E, Rakhmatulin EA, Bugrim A, et al. 2005. Computational prediction of human drug metabolism (Review). *Expert Opin Drug Metab Toxicol* 1:303–324.

Evans DC, Watt AP, Nicoll-Griffith DA, Baillie TA. 2004. Drug-protein adducts: An industry perspective on minimizing the potential for drug bioactivation in drug discovery and development. *Chem Res Toxicol* 17:3–16.

Fedeniuk RW, Shand PJ. 1998. Theory and methodology of antibiotic extraction from biomatrices. *J Chromatogr A* 812:3–15.

Fenn JB, Mann M, Meng CK, Wong SF, Whitehouse CM. 1989. Electrospray ionization for mass spectrometry of large biomolecules. *Science* 246:64–71.

Fura A, Shu YZ, Zhu M, Hanson RL, Roongta V, Humphreys WG. 2004. Discovering drugs through biological transformation: Role of pharmacologically active metabolites in drug discovery. *J Med Chem* 47:4339–4351.

Gad SC. 2003. Active drug metabolites in drug development. *Curr Opin Pharmacol* 3:98–100.

Gamero-Castano M, de la Mora JF. 2000a. Kinetics of small ion evaporation from the charge and mass distribution of multiply charged clusters in electrosprays. *J Mass Spectrom* 35:790–803.

Gamero-Castano M, de la Mora JF. 2000b. Mechanisms of electrospray ionization of singly and multiply charged salt clusters. *Anal Chim Acta* 406:67–91.

Gammelgaard B, Hansen HR, Sturup S, Moller C. 2008. The use of inductively coupled plasma mass spectrometry as a detector in drug metabolism studies. *Expert Opin Drug Metab Toxicol* 4:1187–1207.

Garner RC. 2000. Accelerator mass spectrometry in pharmaceutical research and development—A new ultrasensitive analytical method for isotope measurement. *Curr Drug Metab* 1:205–213.

Guengerich FP. 2001. Common and uncommon cytochrome P450 reactions related to metabolism and chemical toxicity. *Chem Res Toxicol* 14:611–650.

Guevremont R. 2004. High-field asymmetric waveform ion mobility spectrometry: A new tool for mass spectrometry. *J Chromatogr A* 1058:3–19.

Hipple JA, Sommer H, Thomas HA. 1949. A precise method of determining the Faraday by magnetic resonance. *Phys Rev* 76:1877–1878.

Hoh E, Lehotay SJ, Pangallo KC, Mastovska K, Ngo HL, Reddy CM, et al. 2009. Simultaneous quantitation of multiple classes of organohalogen compounds in fish oils with direct sample introduction comprehensive two-dimensional gas chromatography and time-of-flight mass spectrometry. *J Agric Food Chem* 57:2653–2660.

Hop CE. 2006. Use of nano-electrospray for metabolite identification and quantitative absorption, distribution, metabolism and excretion studies (Review). *Curr Drug Metab* 7:557–563.

Hop CE, Chen Y, Yu, LJ. 2005. Uniformity of ionization response of structurally diverse analytes using a chip-based nanoelectrospray ionization source. *Rapid Commun Mass Spectrom* 19:3139–3142.

Hop CE, Tiller PR, Romanyshyn L. 2002. In vitro metabolite identification using fast gradient high performance liquid chromatography combined with tandem mass spectrometry. *Rapid Commun Mass Spectrom* 16:212–219.

Hopfgartner G, Husser C, Zell M. 2003. Rapid screening and characterization of drug metabolites using a new quadrupole-linear ion trap mass spectrometer. *J Mass Spectrom* 38:138–150.

Horning EC, Carroll DI, Dzidic I, Haegele KD, Horning MG, Stillwell RN. 1974. Atmospheric pressure ionization (API) mass spectrometry. Solvent-mediated ionization of samples introduced in solution and in a liquid chromatograph effluent stream. *J Chromatogr Sci* 12:725–729.

Hu Q, Noll RJ, Li H, Makarov A, Hardman M, Graham CR. 2005. The Orbitrap: A new mass spectrometer. *J Mass Spectrom* 40:430–443.

Ioanoviciu D. 1997. Ion trajectories in quadrupole mass filters for a and q parameters near the stability region tip. *Rapid Commun Mass Spectrom* 11:1383–1386.

Kalgutkar AS, Gardner I, Obach RS, Shaffer CL, Callegari E, Henne KR, et al. 2005. A comprehensive listing of bioactivation pathways of organic functional groups. *Curr Drug Metab* 6:161–225.

Kamel A, Prakash C. 2006. High performance liquid chromatography/atmospheric pressure ionization/tandem mass spectrometry (HPLC/API/MS/MS) in drug metabolism and toxicology (Review). *Curr Drug Metab* 7:837–852.

Kauppila TJ, Kostiainen R, Bruins AP. 2004. Anisole, a new dopant for atmospheric pressure photoionization mass spectrometry of low proton affinity, low ionization energy compounds. *Rapid Commun Mass Spectrom* 18:808–815.

Kebarle P. 2000. A brief overview of the present status of the mechanisms involved in electrospray mass spectrometry. *J Mass Spectrom* 35:804–817.

Kebarle P, Tang L. 1993. From ions in solution to ions in the gas phase—The mechanism of electrospray mass spectrometry. *Anal Chem* 65:972A–986A.

King R, Fernandez-Metzler C. 2006. The use of Qtrap technology in drug metabolism (Review). *Curr Drug Metab* 7:541–545.

Kofel P, Allemann M, Kellerhals H, Wanczek KP. 1985. External generation of ions in ICR spectrometry. *Int J Mass Spectrom Ion Process* 65:97–103.

Kostiainen R, Kotiaho T, Kuuranne T, Auriola S. 2003. Liquid chromatography/atmospheric pressure ionization-mass spectrometry in drug metabolism studies (Review). *J Mass Spectrom* 38:357–372.

Lawrence EO, Edlefsen NE. 1930. On the production of high speed protons. *Science* 72:376–377.

Lawrence EO, Livingston MS. 1932. The production of high speed light ions without the use of high voltages. *Phys Rev* 40:19–35.

Lee MS, Kerns EH. 1999. LC/MS applications in drug development. *Mass Spectrom Rev* 18:187–279.

Liu DQ, Hop CE. 2005. Strategies for characterization of drug metabolites using liquid chromatography-tandem mass spectrometry in conjunction with chemical derivatization and on-line H/D exchange approaches (Review). *J Pharm Biomed Anal* 37:1–18.

Loscertales IG, de la Mora JF. 1995. Experiments on the kinetics of field evaporation of small ions from droplets. *J Chem Phys* 103:5041–5060.

Ma S, Chowdhury SK. 2008. Application of liquid chromatography/mass spectrometry for metabolite identification. In *Drug Metabolism in Drug Design and Development*, ed. Zhang D, Zhu M, Humphreys WG, pp. 319–367. Hoboken, NJ: John Wiley & Sons.

Ma S, Chowdhury SK, Alton KB. 2006. Application of mass spectrometry for metabolite identification (Review). *Curr Drug Metab* 7:503–523.

Mack LL, Kralik P, Rheude A, Dole M. 1970. Molecular beams of macroions: II. *J Chem Phys* 52:4977–4986.

Makarov A. 1999. Mass spectrometer. U.S. Patent 5886346.

Makarov A. 2000. Electrostatic axially harmonic orbital trapping: A high-performance technique of mass analysis. *Anal Chem* 72:1156–1162.

March RE. 1997. An introduction to quadrupole ion trap mass spectrometry. *J Mass Spectrom* 32:351–369.

McIver J, Hunter RL, Bowers WD. 1985. Coupling a quadrupole mass spectrometer and a Fourier transform mass spectrometer. *Int J Mass Spectrom Ion Process* 64:67–77.

McLuckey SA, Wells JM. 2001. Mass analysis at the advent of the 21st century. *Chem Rev* 101:571–606.

Meier H, Blaschke G. 2000. Capillary electrophoresis-mass spectrometry, liquid chromatography-mass spectrometry and nanoelectrospray-mass spectrometry of praziquantel metabolites. *J Chromatogr B Biomed Sci Appl* 748:221–231.

Mortishire-Smith RJ, O'Connor D, Castro-Perez JM, Kirby J. 2005. Accelerated throughput metabolic route screening in early drug discovery using high-resolution liquid chromatography/quadrupole time-of-flight mass spectrometry and automated data analysis. *Rapid Commun Mass Spectrom* 19:2659–2670.

Mortishire-Smith RJ, Castro-Perez JM, Yu K, Shockcor JP, Goshawk J, Hartshorn MJ, et al. 2009. Generic dealkylation: A tool for increasing the hit-rate of metabolite rationalization, and automatic customization of mass defect filters. *Rapid Commun Mass Spectrom* 23:939–948.

Mullett WM. 2007. Determination of drugs in biological fluids by direct injection of samples for liquid-chromatographic analysis. *J Biochem Biophys Methods* 70:263–273.

Nassar AE, Bjorge SM, Lee DY. 2003. On-line liquid chromatography-accurate radioisotope counting coupled with a radioactivity detector and mass spectrometer for metabolite identification in drug discovery and development. *Anal Chem* 75:785–790.

Nassar AE, Parmentier Y, Martinet M, Lee DY. 2004. Liquid chromatography-accurate radioisotope counting and microplate scintillation counter technologies in drug metabolism studies. *J Chromatogr Sci* 42:348–353.

Nobilis M, Holcapek M, Kolarova L, Kopecky J, Kunes M, Svoboda Z, et al. 2004. Identification and determination of phase II nabumetone metabolites by high-performance liquid chromatography with photodiode array and mass spectrometric detection. *J Chromatogr A* 1031:229–236.

Obach RS. 2003. Drug-drug interactions: An important negative attribute in drugs. *Drugs Today (Barc)* 39:301–338.

O'Connor D. 2002. Automated sample preparation and LC-MS for high-throughput ADME quantification. *Curr Opin Drug Discov Devel* 5:52–58.

Pasikanti KK, Ho PC, Chan EC. 2008. Gas chromatography/mass spectrometry in metabolic profiling of biological fluids (Review). *J Chromatogr B: Analyt Technol Biomed Life Sci* 871:202–211.

Paul W, Steinwedel H. 1953. Ein Neues Massenspektrometer Ohne Magnetfeld. *Z Naturforsch* 8a, 448–450.

Paul W, Steinwedel H. 1960. Apparatus for separating charged particles of different specific charges. U.S. Patent 2939952.

Pelkonen O, Turpeinen M, Uusitalo J, Rautio A, Raunio H. 2005. Prediction of drug metabolism and interactions on the basis of in vitro investigations. *Basic Clin Pharmacol Toxicol* 96:167–175.

Perry RH, Cooks RG, Noll RJ. 2008. Orbitrap mass spectrometry: Instrumentation, ion motion and applications. *Mass Spectrom Rev* 27:661–699.

Polson C, Sarkar P, Incledon B, Raguvaran V, Grant R. 2003. Optimization of protein precipitation based upon effectiveness of protein removal and ionization effect in liquid chromatography-tandem mass spectrometry. *J Chromatogr B Analyt Technol Biomed Life Sci* 785:263–275.

Prakash C, Shaffer CL, Nedderman A. 2007. Analytical strategies for identifying drug metabolites (Review). *Mass Spectrom Rev* 26:340–369.

Raffaelli A, Saba A. 2003. Atmospheric pressure photoionization mass spectrometry. *Mass Spectrom Rev* 22:318–331.

Robb DB, Covey TR, Bruins AP. 2000. Atmospheric pressure photoionization: An ionization method for liquid chromatography-mass spectrometry. *Anal Chem* 72:3653–3659.

Rousu T, Pelkonen O, Tolonen A. 2009. Rapid detection and characterization of reactive drug metabolites in vitro using several isotope-labeled trapping agents and ultra-performance liquid chromatography/time-of-flight mass spectrometry. *Rapid Commun Mass Spectrom* 23:843–855.

Ruan Q, Peterman S, Szewc MA, Ma L, Cui D, Humphreys WG, et al. 2008. An integrated method for metabolite detection and identification using a linear ion trap/orbitrap mass spectrometer and multiple data processing techniques: Application to indinavir metabolite detection. *J Mass Spectrom* 43:251–261.

Rubiolo P, Liberto E, Sgorbini B, Russo R, Veuthey JL, Bicchi C. 2008. Fast-GC-conventional quadrupole mass spectrometry in essential oil analysis. *J Sep Sci* 31:1074–1084.

Rufer CE, Glatt H, Kulling SE. 2006. Structural elucidation of hydroxylated metabolites of the isoflavan equol by gas chromatography-mass spectrometry and high-performance liquid chromatography-mass spectrometry. *Drug Metab Dispos* 34:51–60.

Sanders M, Shipkova PA, Zhang H, Warrack BM. 2006. Utility of the hybrid LTQ-FTMS for drug metabolism applications (Review). *Curr Drug Metab* 7: 547–555.

Schaefer WH, Goalwin A, Dixon F, Hwang B, Killmer L, Kuo G. 1992. Structural determination of glucuronide conjugates and a carbamoyl glucuronide conjugate of carvedilol: Use of acetylation reactions as an aid to determine positions of glucuronidation. *Biol Mass Spectrom* 21:179–188.

Schultz GA, Corso TN, Prosser SJ, Zhang S. 2000. A fully integrated monolithic microchip electrospray device for mass spectrometry. *Anal Chem* 72:4058–4063.

Shen Z, Reed JR, Creighton M, Liu DQ, Tang YS, Hora DF, et al. 2003. Identification of novel metabolites of pioglitazone in rat and dog. *Xenobiotica* 33:499–509.

Sommer H, Thomas HA, Hipple JA. 1951. The measurement of em by cyclotron resonance. *Phys Rev* 82:697–702.

Souverain S, Rudaz S, Veuthey JL. 2004. Restricted access materials and large particle supports for on-line sample preparation: An attractive approach for biological fluids analysis. *J Chromatogr B Analyt Technol Biomed Life Sci* 801:141–156.

Staack RF, Varesio E, Hopfgartner G. 2005. The combination of liquid chromatography/tandem mass spectrometry and chip-based infusion for improved screening and characterization of drug metabolites. *Rapid Commun Mass Spectrom* 19: 618–626.

Stephens W. 1946. A pulsed mass spectrometer with time dispersion. *Phys Rev* 69:691.

Syage JA. 2004. Mechanism of [M + H]+ formation in photoionization mass spectrometry. *J Am Soc Mass Spectrom* 15:1521–1533.

Syage JA, Evans MD. 2001. Photoionization mass spectrometry as a powerful new tool for drug discovery. *Spectroscopy* 16(11):14, 16, 18, 20–21.

Syage JA, Evans MD, Hanold KA. 2000. Photoionization mass spectrometry. *Am Lab* 32(24):24–29.

Syage JA, Short LC, Cai S-S. 2008. APPI: The second source for LC-MS. http://chromatographyonline.findanalytichem.com/lcgc/article/articleDetail.jsp?id=504702&sk=&date=&pageID=3 [Online].

Taylor EW, Jia W, Bush M, Dollinger GD. 2002. Accelerating the drug optimization process: Identification, structure elucidation, and quantification of in vivo metabolites using stable isotopes with LC/MSn and the chemiluminescent nitrogen detector. *Anal Chem* 74:3232–3238.

Tiller PR, Yu S, Castro-Perez J, Fillgrove KL, Baillie TA. 2008. High-throughput, accurate mass liquid chromatography/tandem mass spectrometry on a quadrupole time-of-flight system as a "first-line" approach for metabolite identification studies. *Rapid Commun Mass Spectrom* 22:1053–1061.

Tolonen A, Turpeinen M, Pelkonen O. 2009. Liquid chromatography-mass spectrometry in in vitro drug metabolite screening. *Drug Discov Today* 14:120–133.

Van PC, Zhang S, Henion J. 2002. Characterization of a fully automated nanoelectrospray system with mass spectrometric detection for proteomic analyses. *J Biomol Tech* 13:72–84.

Van Berkel GJ. 2003. An overview of some recent developments in ionization methods for mass spectrometry. *Eur J Mass Spectrom* 9:539–562.

Vizcaino E, Arellano L, Fernandez P, Grimalt JO. 2009. Analysis of whole congener mixtures of polybromodiphenyl ethers by gas chromatography-mass spectrometry in both environmental and biological samples at femtogram levels. *J Chromatogr* 1216:5045–5051.

Watt AP, Morrison D, Locker KL, Evans DC. 2000. Higher throughput bioanalysis by automation of a protein precipitation assay using a 96-well format with detection by LC-MS/MS. *Anal Chem* 72:979–984.

Watt AP, Mortishire-Smith RJ, Gerhard U, Thomas SR. 2003. Metabolite identification in drug discovery (Review). *Curr Opin Drug Discov Devel* 6:57–65.

Weigl BH, Bardell RL, Cabrera CR. 2003. Lab-on-a-chip for drug development. *Adv Drug Deliv Rev* 55:349–377.

Westman-Brinkmalm A, Brinkmalm G. 2002. Mass spectrometry instrument. In *Mass Spectrometry and Hyphenated Techniques in Neuropeptide Research*, ed. Silberring J, Ekman R, 47–105. New York: John Wiley & Sons.

Westman-Brinkmalm A, Brinkmalm G. 2008. A mass spectrometer's building blocks. In *Mass Spectrometry: Instrumentation, Interpretation, and Applications*, ed. Ekman R, Silberring J, Westman-Brinkmalm A, Agnieszka K, 15–87. Hoboken, NJ: John Wiley & Sons.

Whitehouse CM, Dreyer RN, Yamashita M, Fenn JB. 1985. Electrospray interface for liquid chromatographs and mass spectrometers. *Anal Chem* 57:675–679.

Wilm M, Mann M. 1996. Analytical properties of the nanoelectrospray ion source. *Anal Chem* 68: 1–8.

Xia YQ, Miller JD, Bakhtiar R, Franklin RB, Liu, DQ. 2003. Use of a quadrupole linear ion trap mass spectrometer in metabolite identification and bioanalysis. *Rapid Commun Mass Spectrom* 17:1137–1145.

Xu RN, Fan L, Rieser MJ, El-Shourbagy TA. 2007. Recent advances in high-throughput quantitative bioanalysis by LC-MS/MS. *J Pharm Biomed Anal* 44:342–355.

Yamashita M, Fenn JB. 1984a. Electrospray ion source. Another variation on the free-jet theme. *J Phys Chem* 88:4451–4459.

Yamashita M, Fenn JB. 1984b. Negative ion production with the electrospray ion source. *J Phys Chem* 88:4671–4675.

Yao M, Ma L, Humphreys WG, Zhu M. 2008. Rapid screening and characterization of drug metabolites using a multiple ion monitoring-dependent MS/MS acquisition method on a hybrid triple quadrupole-linear ion trap mass spectrometer. *J Mass Spectrom* 43:1364–1375.

Yates JR, Cociorva D, Liao L, Zabrouskov V. 2006. Performance of a linear ion trap-orbitrap hybrid for peptide analysis. *Anal Chem* 78:493–500.

Yu C, Chen CL, Gorycki FL, Neiss TG. 2007. A rapid method for quantitatively estimating metabolites in human plasma in the absence of synthetic standards using a combination of liquid chromatography/mass spectrometry and radiometric detection. *Rapid Commun Mass Spectrom* 21:497–502.

Yu LJ, Chen Y, Deninno MP, O'Connell TN, Hop CE. 2005. Identification of a novel glutathione adduct of diclofenac, 4′-hydroxy-2′-glutathion-deschloro-diclofenac, upon incubation with human liver microsomes. *Drug Metab Dispos* 33:484–488.

Zhang H, Zhang D, Ray K. 2003. A software filter to remove interference ions from drug metabolites in accurate mass liquid chromatography/mass spectrometric analyses. *J Mass Spectrom* 38:1110–1112.

Zhang H, Zhang D, Ray K, Zhu M. 2009. Mass defect filter technique and its applications to drug metabolite identification by high-resolution mass spectrometry. *J Mass Spectrom* 44:999–1016.

Zhu M, Ma L, Zhang D, Ray K, Zhao W, Humphreys WG, et al. 2006. Detection and characterization of metabolites in biological matrices using mass defect filtering of liquid chromatography/high resolution mass spectrometry data. *Drug Metab Dispos* 34:1722–1733.

CHAPTER 3

TOOLS OF CHOICE FOR ACCELERATING METABOLITE IDENTIFICATION: MASS SPECTROMETRY TECHNOLOGY DRIVES METABOLITE IDENTIFICATION STUDIES FORWARD

ALA F. NASSAR
Chemistry Department, Brandeis University, Waltham, MA

The best part of being a scientist is to share the experience with other scientists.

Acquisition of detailed knowledge of how a drug candidate is metabolized as early as possible in the drug development process is crucial for a pharmaceutical company. It can save time and precious resources, and maximize return-on-investment. The final selection of a successful drug candidate relies enormously on the metabolism studies that are performed *in vitro* and *in vivo*. Absorption, distribution, metabolism, excretion, and toxicology (ADMET) studies are widely used in drug discovery and development to help obtain the optimal balance of properties necessary to convert lead compounds into drugs that are safe and effective for human use. Drug discovery efforts have been aimed at identifying and addressing metabolism issues at the earliest possible stage, by developing and applying innovative liquid chromatography–mass spectrometry (LC-MS)-based techniques and instrumentation, which are both faster and more accurate. In the area of drug metabolism, for example, revolutionary changes have been achieved by the combination of LC-MS with innovative instrumentation such as triple quadrupoles, ion traps, orbitrap, and time-of-flight (TOF) MS. Such new approaches are demonstrating considerable potential to improve the overall safety profile of drug candidates throughout the drug discovery and development process. These emerging techniques stream-

Biotransformation and Metabolite Elucidation of Xenobiotics, Edited by Ala F. Nassar

line and accelerate the process by eliminating potentially harmful candidates earlier and improving the safety of new drugs. In turn, most ADMET studies have come to rely on LC-MS for the analysis of an ever-increasing workload of potential candidates. This chapter provides a discussion on the important tools and techniques that are used in supporting metabolite characterization.

3.1 INTRODUCTION

MS and nuclear magnetic resonance (NMR) are critical to the success of such ADMET studies. NMR spectroscopic techniques are used to confirm and elucidate metabolite identification in drug metabolism studies. Liquid chromatography (LC)-NMR is a good choice for these studies, but MS has advantages over NMR with respect to sensitivity, smaller sample size, and greater speed. LC-MS-NMR has become a commercially available technique and is used in the late discovery stages to confirm and characterize metabolites. LC-MS is an analytical technique that still shows room for development; already, significant improvements have been made in sensitivity and resolution. It is probably the most powerful technique currently available for pharmaceutical analysis, and has significantly accelerated the drug discovery and development process. LC-MS has become the dominant technique for performing almost all of the analyses involved in ADMET studies, and is likely to remain the principal tool for such studies (Thompson, 2000; Nassar, 2003; Nassar and Adam, 2003; Roberts, 2003; Kassel, 2004; Nassar and Talaat, 2004; Lin et al., 2003; Taylor et al., 2002). International Union of Pure and Applied Chemistry (IUPAC) MS terms and definitions are presented at the end of this chapter (see Abbreviations and Glossary).

The major aim of LC-MS is the application of its analytical power to create straightforward, sensitive, fast, and reliable data. Improved hardware and software for LC-MS have led to greater sensitivity, greater ease-of-use, and improved postanalysis of data (Kubinyi, 1977; Bakke et al., 1995; Lesko et al., 2000; Meyboom et al., 2000; Thompson, 2000; Greene, 2002; Lasser et al., 2002; Taylor et al., 2002; Tiller and Romanyshyn, 2002; Yang et al., 2002; Jemal et al., 2003; Kostiainen et al., 2003; Lin et al., 2003; Nassar, 2003; Nassar and Adam, 2003; Plumb et al., 2003; Roberts, 2003; Kassel, 2004; Nassar and Talaat, 2004; Nassar and Lopez-Anaya, 2004; Balimane et al., 2005; Castro-Perez et al., 2005a; Johnson and Plumb, 2005; Leclercq et al., 2005). Techniques such as electron impact, chemical ionization, atmospheric pressure chemical ionization (APCI), fast-atom bombardment, thermospray, gas chromatography-MS, and electrospray ionization (ESI) are used in ADMET studies. Metabolite characterization ion trap (MS^n—this refers to multistage MS/MS experiments where n is the number of product ion stages [progeny ions]) and Quadrupole time-of-flight (QTOF)-MS are widely used MS^n because it allows the relatively rapid construction of fragmentation maps, and a higher degree of specificity than with other methods. QTOF-MS has advantages for metabolite identification over MS^n or triple quadrupole MS, including fast mass spectral acquisition speed with high full-scan sensitivity, enhanced mass resolution, and accurate mass measurement capabilities which allow for the determination of elemental composition. Exact mass measurement is a valuable tool for solving structure elucidation problems by helping to confirm elemental composition, and is also invaluable in eliminating false positives and determining nontrivial metabolites. Accurate mass measurements are routine experiments performed by modern mass spectrometers. The accuracy of a measurement refers to the degree of conformity of a measured quantity to its actual true value. Accurate mass measurements may not be sufficient if mixtures are analyzed or if the isotopic fine structure of compounds needs to be evaluated. In such cases, high resolving power is needed; this can be easily achieved by using Orbitrap or Fourier transform ion cyclotron resonance (FTICR)-MS. Also, chromatography or comprehensive chromatography (GC-GC, LC-LC), or multidimensional chromatography (LC-GC) can help to enhance the efficiency.

Data-dependent scans on QTOF or MS^n provide a highly useful tool. Increasing sample complexity, sample volume restrictions and throughput requirements necessitate that the maximum amount of useful information is extracted from a single experiment. Data-directed analysis (DDA) or data-dependent scans enable intelligent MS and MS/MS acquisitions to be performed automatically on multiple co-eluting components. DDA is able to make intelligent decisions about which ions to select for MS/MS using its inherent high resolution, exact mass measurement capability, and full mass range. One shortcoming, however, is that due to limited sensitivity, DDA often overlooks minor metabolites that could be toxic or active in nature. MS/MS or MS^n fragmentation data contain a tremendous variety of information. Unfortunately, this wealth of information is contained within a mosaic of rearrangments, fragmentations, ion physics, and gas phase chemistry reactions, which we do not yet fully understand. MS/MS fragmentation data cannot solve all our problems without an expert scientist to analyze and interpret this information, and use it to identify and characterize metabolites. While this analysis does not provide certain, "yes-or-no" answers, it does aid in directing the researcher as to which further studies will be needed. In this chapter, one demonstrably successful analytical strategy for metabolite characterization and identifying reactive metabolites by LC-MS is presented.

3.2 CRITERIA FOR LC-MS METHODS

The growing realization of the importance of ADMET properties early in the drug discovery process has led to a dramatic increase in the numbers of compounds requiring screening. Because of the resulting necessity for speed, errors, as well as false positives and/or false negatives, will always be a risk factor with any screening procedure in the early stages of drug discovery. While discovery scientists seek to improve their turnaround time, it is vital to maintain or even improve the quality of chromatographic resolution of the metabolites produced in screening efforts. The high-throughput methods needed to achieve the required speed would not be possible without the innovative enabling technologies of computing, automation, new sample preparation technologies, and highly sensitive and selective detection systems. To ensure the integrity of the data, the screening procedure should meet criteria such as relevance, effectiveness, speed, robustness, accuracy, and reproducibility (White, 2000).

3.3 MATRIX EFFECTS

The analysis of compounds in complex biological matrices, such as blood, plasma, bile, urine, and feces samples, is probably the largest application of LC-MS/MS. LC-MS has become the principal technique used in quantitative bioanalysis due to a combination of factors, such as cost, ease, and speed of performing selected reaction monitoring (SRM), and a wide dynamic range and good sensitivity due to favorable signal-to-noise ratio (S/N) (Jemal and Bergum, 1992; Jemal, 2000; Nassar et al., 2001a,b; Zhang and Henion, 2001; Jemal and Ouyang, 2003; Guevremont, 2004; Castro-Perez et al., 2005b; Kapron et al., 2005). It is interesting to note that it was initially thought that the invention of MS/MS would enable the analysis of matrices without separation, but suppression effects have prevented this. The presence of endogenous components in the matrix can suppress the analyte response, probably by competition for ESI droplet surface and hence ionization. These effects can cause differences in response between standards and samples in matrices, leading to difficulties in quantitative analysis and compound identification. This emphasizes the importance of the chromatographic step in the analysis, where good separation can reduce or eliminate these effects. Because of the high degree of selectivity routinely provided by SRM, bioanalytical method development time for quantitative determinations of one or several analytes has been reduced to a few days or less. Although the SRM approach demonstrates excellent sensitivity, selectivity, and efficiency, one drawback is that it fails to produce the qualitative information required to support the recognition and structural elucidation of metabolites that could be present in the samples. Metabolite identification and structure elucidation requires development of a method that is separate and distinct from the quantitation of the known components present, which requires additional time, effort, and expertise. During spectral scanning, the duty cycle of a quadrupole mass spectrometer is such that only a small fraction of the total time is spent monitoring any one ion. To obtain an optimum S/N, a quadrupole analyzer must allow a limited number of selected ions to pass. Because most ions are filtered out, much of the qualitative information is lost. A potential strategy to deal with this situation is the use of LC-TOF-MS to generate data that will simultaneously provide qualitative and quantitative information about drug candidate metabolism and disposition (Jemal and Bergum, 1992; Jemal, 2000; Zhang and Henion, 2001; Castro-Perez et al., 2005a).

3.4 TOOLS OF CHOICE FOR METABOLITE CHARACTERIZATION

There is no one magic tool that can solve all problems of metabolite identification. A combination of the tools listed below need to be used; each tool provides specific information and addresses specific problems. It is one's responsibility as a good scientist to know the breadth of applicability of a technique, what it is good for—and what it is NOT good for—and to use what is needed when it is needed. Do not fall into the trap—if one's favorite tool is a hammer, then everything looks like a nail—of attempting to solve all problems with one's favorite technique. Below is a general list of tools that can be used to identify and characterize metabolites. It is not necessary to use all of these tools for each analysis; for example, LC-NMR can be saved for use in the late stage of drug discovery.

List of Tools Which Can Be Used for Metabolite Elucidation
In silico
Full scan LC-MS positive and negative modes
Multistage MS (MS^n) positive and negative modes
N-rule
RDBE
Isotopic patterns
Exact mass measurements
HD experiments
Stable isotope
LC-MS-NMR
Radiolabeled compounds

Metabolite identification is crucial to the drug discovery and development process, as it can be used to investigate the Phase I metabolites likely to be formed *in vivo*, the differences between species in drug metabolism, the major circulating metabolites of an administered drug, and the Phase I and Phase II pathways of metabolism and pharmacologically active or toxic metabolites, and also to help determine the effects of metabolizing enzyme inhibition and/or induction. The ability to produce this information early in the discovery phase is becoming increasingly important as a basis for judging whether a drug candidate merits further development. Table 3.1 shows a list of a wide variety of Phase 1 and Phase 2 biotransformations, together with the mass changes from the parent drug (modified from Mortishire et al., 2005). This table could also be used retroactively to find the potential reactions responsible for certain metabolites.

TABLE 3.1 Common Biotransformation Reactions

Metabolic Reaction	Monoisotopic Mass Change
Oxidative debromination	–61.9156
Tert-butyl dealkylation	–56.0624
Hydrolysis of nitrate esters	–44.9851
Decarboxylation	–43.9898
Isopropyl dealkylation	–42.0468
Propyl ketone to acid	–40.0675
Tert-butyl to alcohol	–40.0675
Reductive dechlorination	–33.9611
Hydroxymethylene loss	–30.0106
Nitro reduction	–29.9742
Propyl ether to acid	–28.0675
Deethylation	–28.0312
Decarboxylation	–27.9949
Ethyl ketone to acid	–26.0519
Isopropyl to alcohol	–26.0519
Alcohols dehydration	–18.0105
Dehydration of oximes	–18.0105
Reductive defluorination	–17.9906
Oxidative dechlorination	–17.9662
Sulfoxide to thioether	–15.9949
Thioureas to ureas	–15.9772
Ethyl ether to acid	–14.0519
Demethylation	–14.0157
Tert-butyl to acid	–12.0726
Methyl ketone to acid	–12.0363
Ethyl to alcohol	–12.0363
Two sequential desaturation	–4.0314
Hydroxylation and dehydration	–2.0157
Primary alcohols to aldehyde	–2.0157
Secondary alcohols to ketone	–2.0157
Desaturation	–2.0157
1,4-Dihydropyridines to pyridines	–2.0157
Oxidative defluorination	–1.9957
Oxidative deamination to ketone	–1.0316
Demethylation and methylene to ketone	–0.0365
2-Ethoxyl to acid	–0.0363
Oxidative deamination to alcohol	0.984
Isopropyl to acid	1.943
Demethylation and hydroxylation	1.9792
Ketone to alcohol	2.0157
Methylene to ketone	13.9792
Hydroxylation and desaturation	13.9792
Alkene to epoxide	13.9792
(O, N, S) methylation	14.0157
Ethyl to carboxylic acid	15.9586

TABLE 3.1 (*Continued*)

Metabolic Reaction	Monoisotopic Mass Change
Hydroxylation	15.9949
Secondary amine to hydroxylamine	15.9949
Tertiary amine to N-oxide	15.9949
Thioether to sulfoxide, sulfoxide to sulfone	15.9949
Aromatic ring to arene oxide	15.9949
Demethylation and two hydroxylation	17.9741
Hydration, hydrolysis (internal)	18.0106
Hydrolysis of aromatic nitriles	18.0106
Hydroxylation and ketone formation	29.9741
Quinone formation	29.9741
Demethylation to carboxylic acid	29.9742
Hydroxylation and methylation	30.0105
Two hydroxylation	31.9898
Thioether to sulfone	31.9898
Alkenes to dihydrodiol	34.0054
Acetylation	42.0106
Three hydroxylation	47.9847
Aromatic thiols to sulfonic acids	47.9847
Glycine conjugation	57.0215
Sulfate conjugation	79.9568
Hydroxylation and sulfation	95.9517
Cysteine conjugation	103.0092
Taurine conjugation	107.0041
S-cysteine conjugation	119.0041
Decarboxylation and glucuronidation	148.0372
N-acetylcysteine conjugation	161.0147
Glucuronide conjugation	176.0321
Two sulfate conjugation	191.9035
Hydroxylation + glucuronide	192.027
GSH conjugation	289.0732
Desaturation + S-GSH conjugation	303.0525
S-GSH conjugation	305.0682
Epoxidation + S-GSH conjugation	321.0631
Two glucuronide conjugation	352.0642

Metabolite identification enables early detection of potential metabolic liabilities or issues, provides a metabolism perspective to guide synthesis efforts with the aim of either blocking or enhancing metabolism so as to optimize the pharmacokinetic and safety profiles of newly synthesized drug candidates, and assists the prediction of the metabolic pathway(s) of potential drug candidates for development (Thompson, 2000; Kantharaj et al., 2003; Nassar et al., 2004a,b). Experience has demonstrated that the best combination for

TABLE 3.2 Techniques That Help in the Search for, and Identification of, Metabolites in Drug Metabolism

Stage of Compound	Technique	Comments
Very early stage of drug discovery	*In silico as well as in vitro screening*	— Useful during synthesis to help select/eliminate compounds
Ranking compounds for drug discovery	*In silico as well as in vitro screening* LC-MS and LC-MS/MS techniques	— Useful during synthesis efforts to adjust metabolism — Helps in Identifying simple/major metabolites, for example, dealkylations, and conjugations such as glucuronide — Predicts likely metabolites formed *in vivo*
Selected candidate in late drug discovery	*In silico as well as in vitro screening* LC-MS and LC-MS/MS techniques Online HD exchange, ion trap, QTOF and Oribtrap	— Aids in determination of metabolic differences between species — Used to identify potential pharmacologically active or pharmacologically toxic metabolites
Nominate compound for clinical trials	Same as selected candidate in the late drug discovery. Also, LC-MS-NMR and LC-Radioactive-MS	— Used to determine the percentage of metabolite formed *in vitro* or *in vivo* — Aids synthesis of metabolites for toxicology testing — Aids in comparison of human pathways — Aids identification of drug–drug interactions

accomplishing rapid and accurate metabolite identification involves a robotic system for sample preparation and in silico software to predict and find possible metabolites as well as to predict hypothetical metabolite chemical structure (Obach, 1997, 1999; Greene, 2002). This can be combined with the use of LC-MS to determine exact mass measurements (accurate mass) for sample analysis and LC-MS-NMR, and online hydrogen–deuterium (HD) exchange for further metabolite structure confirmation and elucidation. In the following sections, the relative merits of current and potential strategies for dealing with metabolite characterization in various stages of drug discovery and development are examined and illustrated with examples. Techniques, tools and approaches are suggested for each of these stages, as summarized in Table 3.2.

3.4.1 Use of LC-MS to Identify and Characterize Metabolites

LC-MS is used with various strategies in drug discovery for identifying compounds and/or their metabolites. One area in which the technique is employed is in confirming the structure of a known compound. A second area of application, and most relevant to this discussion, is the identification of unknown metabolites of drug candidates.

3.5 STRATEGIES FOR IDENTIFYING UNKNOWN METABOLITES

When using full-scan mass spectra to search for metabolites, one should search for the most intense ion with singly charged ions, or doubly charged ions at 0.5 mass to charge ratio (m/z), adducts, multiply-charged ions, and/or dimers in the full-scan spectrum.

3.5.1 Nitrogen Rule

According to the nitrogen rule, if a compound contains either no nitrogen atoms or an even number, its molecular ion will be at an even mass number, while an odd number of nitrogen atoms will produce an odd mass number for the molecular ion. This rule, applied to the molecular ion, determines whether the unknown agent has an even or odd number of nitrogen atoms. A check of the isotopic peak of the molecular ions serves to confirm patterns, being aware of interferences from other ions. It is worth mentioning that the nitrogen rule for MS is not a rule per se but a principle. It is true for unit masses, but not for accurate mass measurements.

3.5.2 Ring Double-Bond-Equivalents (RDBE) (Kind et al., 2009)

RDBE or double bond equivalents (DBE) are calculated from valence values of elements contained in a formula and should tell the number of bonds—or rings. The formula generators report an RDBE range. By applying the RDBE rule, one can determine or confirm the total number of rings plus double bonds for a compound containing carbon, hydrogen, nitrogen, or oxygen, as well as elements with the same valences as these. The value of DBE can be calculated according to Equation 1.1:

$$\mathrm{RDBE} = \mathrm{C} - \frac{1}{2}[\mathrm{H}+\mathrm{F}+\mathrm{Cl}+\mathrm{Br}+\mathrm{I}] + \frac{1}{2}[\mathrm{N}+\mathrm{P}] + 1 \qquad \text{(Eq. 1.1)}$$

Next, one would perform accurate mass measurement to confirm/determine possible molecular formulae, and then compare the product ion MS/MS or MS^n spectra of the parent with the metabolites. Having identified molecular ions for possible metabolites, MS/MS or MS^n should be performed, and comparisons made between the product ion MS/MS or MS^n spectra of the parent and metabolites. Given the structure of the parent drug, with its corresponding fragmentation, elucidation of metabolite structure is greatly facilitated. The specific fragment ion that creates a shift in the *m/z* value leads to identification of the site of the modification on the molecule.

Once drug candidates reach the late drug discovery/candidate selection phase, LC-MS/MS, QTOF, Orbitrap (high resolution and exact mass measurement), ion trap (MS^n), and HD exchange come into prominence in order to determine metabolic differences between species and identify potential pharmacologically active or toxic metabolites (Table 3.2). Exact mass analysis serves to confirm the elemental composition of metabolites. The exact mass (accurate mass) shift between a drug candidate and its metabolites can be used to predict the elemental composition of those metabolites. Knowledge of the molecular formulae of unknown metabolites is one of the tools that can be used to identify the metabolites and then propose their structure using additional MS data and tools.

3.6 ONLINE HD-LC-MS

Substantial amounts of work appear in the literature in recent years on the study of backbone amide HD exchange studies of protein folding, based upon the pH dependence of the rates of HD exchange of amide protons (Bai et al., 1993; Krishna et al., 2004; Englander 2006). Half-lives for exchange can be greater than 1 h at pH 2 and pH 3 at 0°C for unprotected amides. This slow rate of exchange permits conducting exchange experiments on intact proteins, then cleaving those proteins with acid-active peptidases and determining the regions of the original protein accessible to exchange.

The utility of deuterium exchange, however, is not limited to large molecules. Identifying and confirming the structures of small molecules can benefit considerably from determining the numbers (and nature) of exchangeable hydrogen atoms in the molecule. Nassar et al. (2003) have reported on the general utility of deuterium exchange LC-MS experiments for structure elucidation and impurity identification. HD exchange occurs in solution in the presence of exchangeable (labile) hydrogen atoms in a molecule. One advantage of the HD exchange method is that, with LC-MS/MS, it offers an easy estimation of the number of labile hydrogen atoms in such groups as -SH, -OH, -NH, $-NH_2$, and -COOH(Bakke et al., 1995). This number is useful in comparing metabolite structure with that of the parent drug to determine the presence or absence of the above groups. HD-exchange experiments are also valuable for structural elucidation and interpretation of MS/MS fragmentation processes (Nassar, 2003). A method for metabolite identification in drug discovery and development utilizing online HD exchange and a tandem QTOF mass spectrometer coupled with LC (LC-QTOF-MS) recently has been developed (Bakkeet al., 1995). This method apparently works very well for the identification of metabolites produced by dehydrogenation, oxidation, glucuronidation and dealkylation. It provides discrimination between N- or S-oxide formation and mono-hydroxylation; also, it becomes easy to identify conjugations such as quaternary amine glucuronide versus primary or secondary glucuronides using this technique. The generic method is simple, easy, fast, sensitive, robust, and reliable, as well as achieving enhanced throughput; these factors facilitate rapid characterization of metabolites *in vitro* or *in vivo*. This time saving, combined with the particular benefits of QTOF or Orbitrap, such as higher resolution, makes this technique a valuable tool for structure elucidation. This method also works well with low-resolution instruments. Derivatization can be useful to stabilize unstable metabolites, improve chromatographic properties for highly polar compounds and reduce volatility for volatile metabolites. Furthermore, it can be useful for characterizing chirality and site of metabolism. Many methods of derivatization have been reported that can be used to identify most functional groups. These techniques have the limitation of time-consuming sample preparation. Derivatization of metabolites in a sample demands either isolation of a specific metabolite before derivatization, or derivatization of the entire sample, which will create more complex ions in MS. The speed and reduction in preparation time and effort involved with HD-LC-MS, and the fact that LC-MS-NMR has become more widely available and sensitive, has pushed derivatization methods out of favor.

As an example of how this process works, five metabolites of nimodipine formed in human liver microsomes; all were identified and characterized by LC-QTOF (Nassar, 2003). Figure 3.1 shows the chemical structure of nimodipine with the labile H exchanges as indicated by asterisks; the arrows indicate the positions of the possible metabolites as predicted using Pallas software. Table 3.3 shows the exchange of labile hydrogens in nimodipine and its metabolites formed *in vitro*. The labile hydrogens ranged from none to two, which provides a significant means of identification, particularly with dehydrogenation metabolites, suggesting that the dehydrogenation took place on the pyridine moiety, resulting in a loss of one labile hydrogen. For example, in

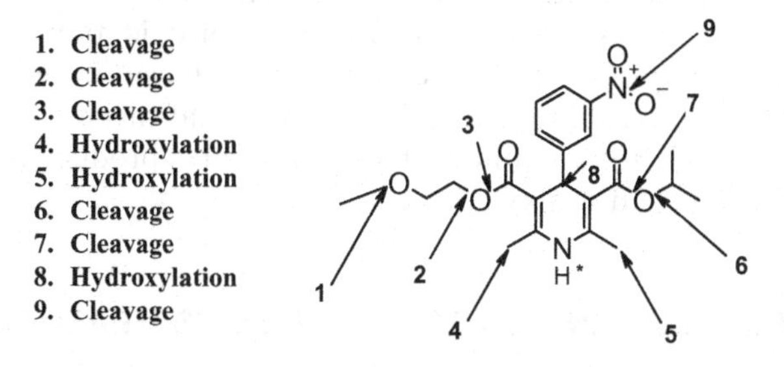

Figure 3.1 The chemical structure of nimodipine with the labile H exchange as indicated by asterisks. The arrows indicate the positions of the possible metabolites as predicted using Pallas software.

TABLE 3.3 Exchange of Labile Hydrogens in Nimodipine and Metabolites Formed *In Vitro*

Compound	Labile Hydrogen	Mass $[M_H + H]^+$, m/z	Mass $[M_D + D]^+$, m/z	
			Predicted	Measured
Nimodipine	1	419	421	421
M-1	1	359	361	361
M-2	2	435	438	438
M-3	1	403	405	405
M-4	2	405	408	408
M-5	0	417	418	418

M_H, the molecular weight in H_2O.
M_D, the molecular weight in D_2O.

Fig. 3.2a, the full-scan mass spectrum of M-3 revealed a protonated molecular ion $[M_H + H]^+$ at m/z 403, 16 amu lower than nimodipine, suggesting that this metabolite was a cleaved product. The product-ion spectrum of $[M_H + H]^+$ at m/z 403 showed the fragment ions of m/z 361, 343, 317, and 301. When H_2O was replaced with D_2O in the mobile phase, the full-scan mass spectrum of M-3 revealed a molecular ion $[M_D + D]^+$ at m/z 405 (Fig. 3.2b), 2 mu higher than $[M_H + H]^+$ M-3, indicating the presence of one exchangeable hydrogen atom in M-3. The product-ion spectrum of m/z 405 showed the fragment ions of m/z 363, 344, 319, and 302. Nimodipine has one exchangeable hydrogen atom, with the cleavage producing another exchangeable hydrogen atom. This eliminates the possibility that M-3 is due to direct cleavage of nimodipine. It is possible that M-3 can be formed from the dehydrogenation and cleavage of nimodipine. The proposed TOF-MS/MS fragmentation for M-3 in D_2O and H_2O is shown in Fig. 3.2.

The proposed metabolic pathways of nimodipine in hepatic microsomal incubations in H_2O and D_2O are shown in Figs. 3.3 and 3.4. Nimodipine is metabolized by means of dehydrogenation, demethylation of the methoxy group, cleavage of the ester groups by hydrolysis or oxidation and hydroxylation of methyl groups. HD exchange provided significant information for all five metabolite identifications. These results show that this method should be particularly desirable for identification of metabolites produced by dehydrogenation, oxidation, and dealkylation.

3.7 HIGH RESOLUTION AND ACCURATE MASS MEASUREMENTS

High resolution, accurate mass spectral analysis is an essential tool for structural elucidation in the fields of chemistry and biochemistry (Tyler et al., 1996; Kind and Fiehn, 2006). Mass resolution is the ability to separate two narrow mass spectral peaks. High resolution implies the ability to make confident and

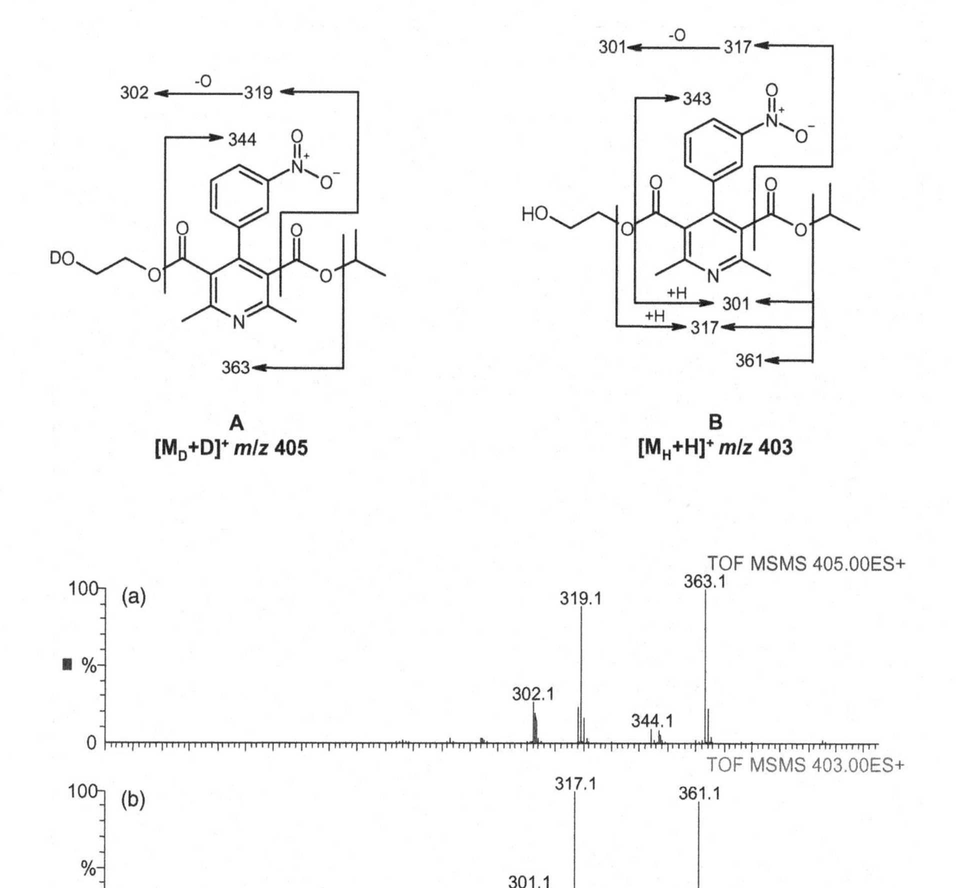

Figure 3.2 QTOF MS/MS spectra of Metabolite M-3 in (a) D_2O and (b) H_2O.

reliable measurements of small differences. High-resolution analysis allows isobaric compounds to be separated from one another or from chemical noise. For the identification of unknowns, it is desirable to perform the highest mass accuracy measurements possible in order to eliminate false positives. The ideal instrument for this kind of application must produce high resolution full-scan accurate mass MS and MS^n data without the need for internal calibration or frequent recalibration. The need for high resolution is emphasized for peptide analysis due to greater numbers of possible matches and frequent appearance of isobaric compounds. For a mixture of two isobaric peptides, it is necessary to have much higher resolution than that used for small molecules. Peptides mostly consist of C, H, O, N, and S and therefore, isobaric peptides may differ by only few millimass units. High mass resolution detection allows baseline

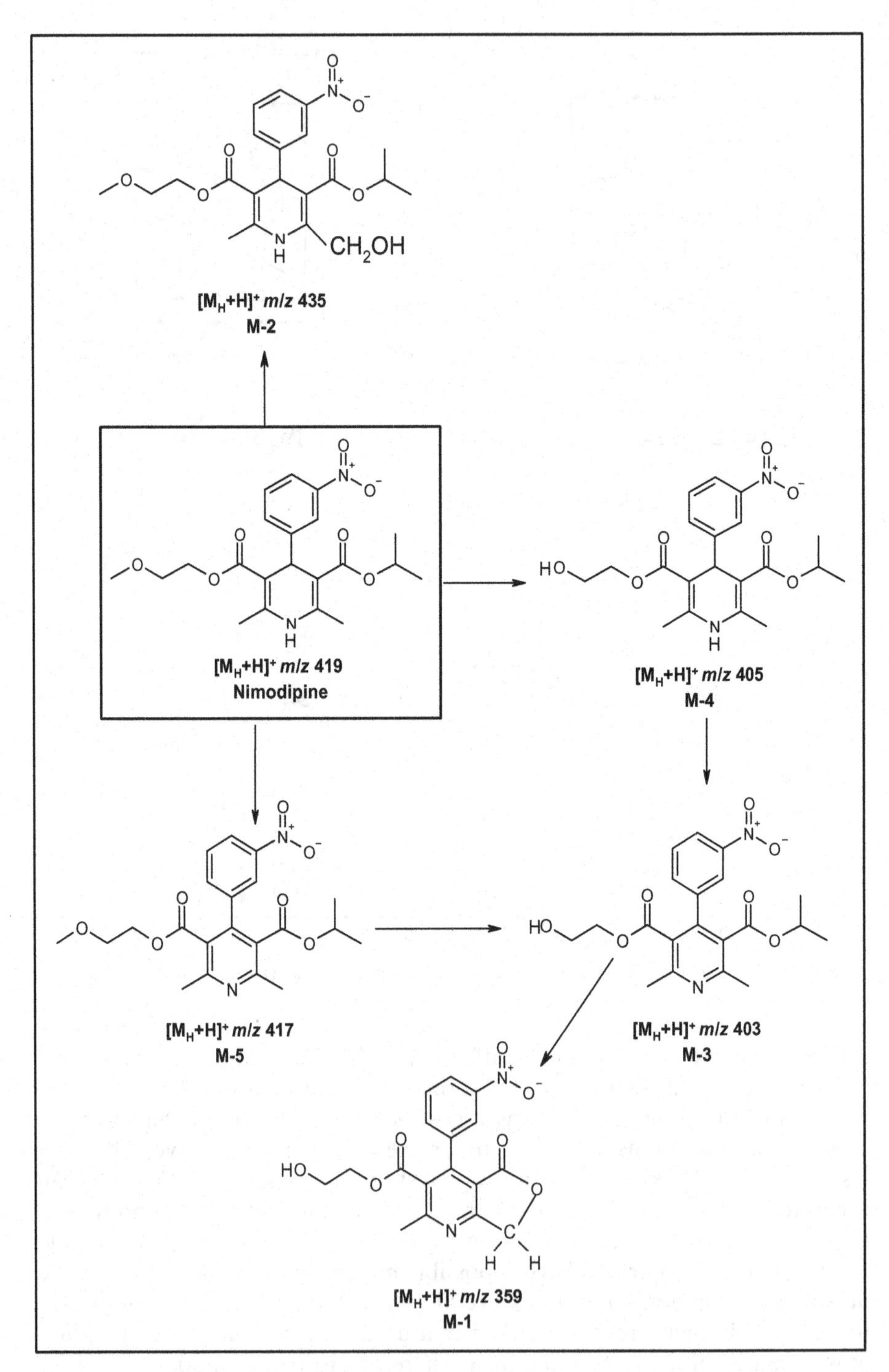

Figure 3.3 Proposed metabolic pathways of nimodipine in hepatic microsomal incubations in H_2O.

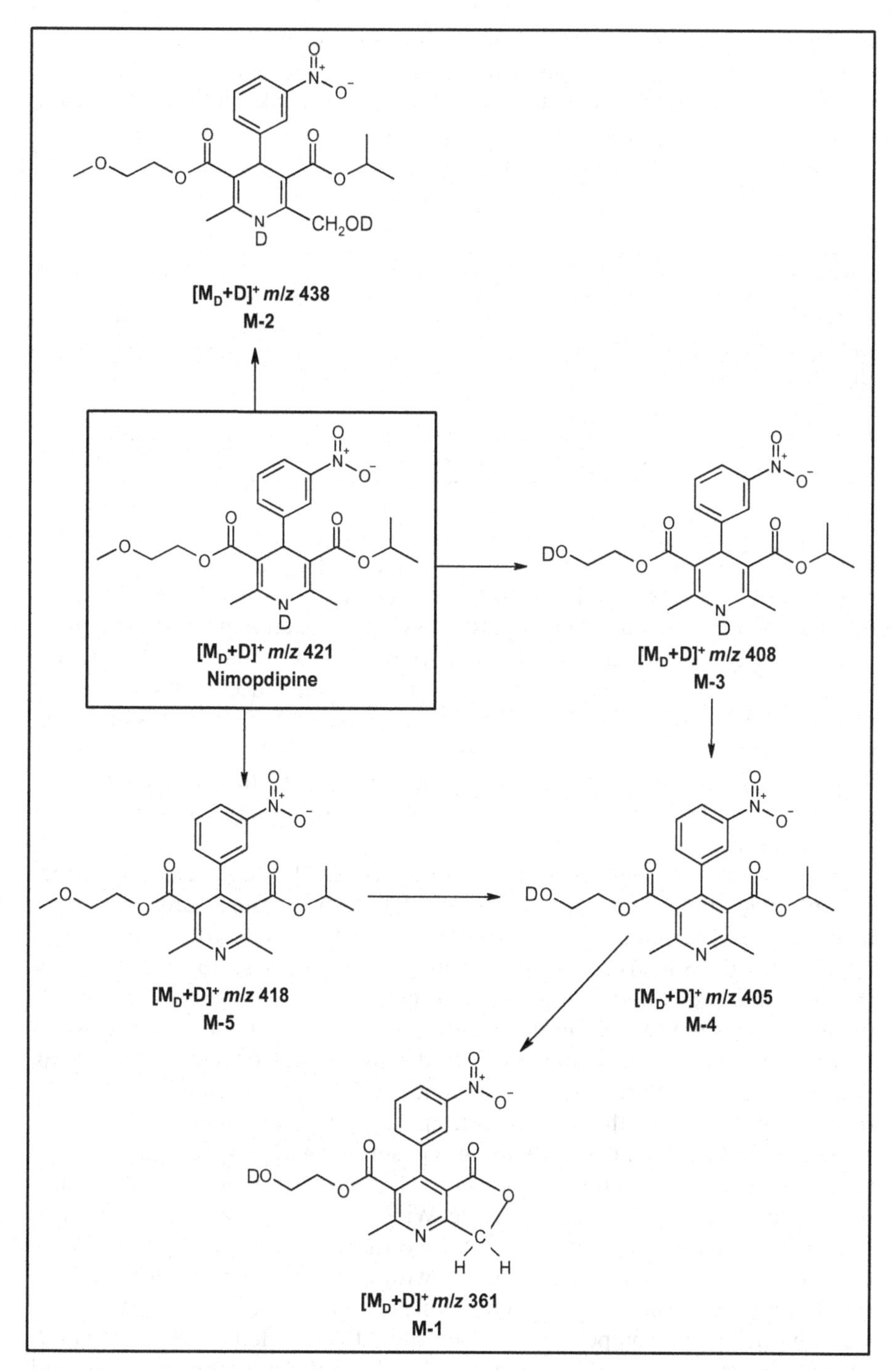

Figure 3.4 Proposed metabolic pathways of nimodipine in hepatic microsomal incubations in D_2O.

separation of the isobaric peaks which ensures the exact determination of the peak centroid for accurate mass calculation. Mass resolving power is defined as m/dm; where m designates the mass and dm the peak width necessary for separation at mass m. The resolving power and mass accuracy of several commonly used instruments are shown below:

Instrument	Resolving Power (FWHM)	Mass Accuracy (ppm)
FT-ICR-MS	1,000,000	0.1–1
FT-Orbitrap	100,000	0.5–1
High-Res-TOF-MS	60,000	3–5
Q-TOF	10,000	3–5
Triple Quad	10,000	3–5
Linear IonTrap	10,000	50–200

High-resolution MS is a beneficial tool when looking at relatively complex mixtures of peptides/proteins without any LC separation. FT-MS offers the highest possible resolution among all mass spectrometers with its maximum peak capacity exceeding that of conventional HPLC separation by greater than 200-fold. Also, FT-MS is known for its high mass accuracy due to how mass/charge is detected. The mass to charge ratio (m/z) of any ion is inversely proportional to its ion cyclotron frequency (higher m/z ~ lower frequency). Since we can measure frequency very accurately, and the superconducting magnet is very homogeneous, we can indirectly calculate the mass of a given ion very accurately.

Accurate mass measurements were until recently the nearly exclusive province of magnetic sector instruments. Making such measurements presented difficulties and required careful attention to detail. Reflectron TOF instruments and FT-MS instruments have challenged magnetic sector dominance of this area and facilitated a somewhat easier route to making these measurements. Accurate mass measurement refers to the ability to confidently assign a precise relative molecular mass (four decimal places or more) to a signal. Accurate mass measurements can be made at low resolution, given a stable, reproducible scan by the instrument, appropriate accurate mass reference standards, and the confidence that one is observing a mass spectral peak composed of a single elemental composition. High mass accuracy provides confident assignment of empirical formulae. Wide mass range allows detection of small molecules as well as peptides. Structural elucidation is supported by a combination of accurate mass analysis with compound characteristics (e.g., RDB or nitrogen rule). Use of accurate mass information facilitates the identification of known compounds and determination of elemental composition for unknowns. High mass accuracy is not sufficient if mixtures are analyzed or if the isotopic fine structure of compounds needs to be evaluated. In such a case, high resolving power is needed; such power can be achieved by chromatography or can be obtained by MS itself (such as Orbitrap or FT-MS). The instrument manufacturers often imply that accurate mass measurements (and therefore elemental composition determinations) are a single-measurement solution to many structural problems. This is an oversimplification. A recent report, stemming from a metabolomics application, presents data that high mass measurements, even at sub-1 ppm accuracy, are insufficient to "identify" metabolites. Isotopic abundance pattern analysis is a necessary additional step to eliminate false elemental composition candidates.

3.7.1 LC-MS-NMR

For determination of the structure of metabolites, NMR spectroscopy has become one of the most commonly used techniques. The sensitivity of NMR has increased significantly with the introduction of modern high field strength (500, 600, and 800 MHz) NMR spectrometers. For the purposes of drug metabolism, the most common accessible nuclei are ^{1}H, ^{19}F, ^{13}C, ^{31}P, and ^{15}N. When using NMR, sample concentrations must be between 1 μg and 10 mg, with analysis times between 5 min and 1 h. Several operational modes are available for LC-MS-NMR such as isocratic or gradient elution, continuous flow, stop flow, time-sliced stop flow, peak collection into capillary loops for post-chromatographic analysis, and automatic detection of chromatographic peaks with triggered NMR acquisition. Given considerations for sample concentration and chromatographic resolution, any of these techniques are available.

The online HD-LC-MS method has the ability to rapidly provide characterization of metabolites, such as those formed by dealkylation from *in vitro* or *in vivo* samples, and can distinguish between *N*- or *S*-oxide versus monohydroxyl and quaternary amine versus primary or secondary glucuronide. Despite these advantages, NMR is still necessary for identifying the regiochemistry of aromatic oxidation, determining the site of aliphatic oxidation where fragmentation pathways are unavailable or inconclusive, and to locate groups such as OH, epoxide, and sulfate by comparing NMR spectra of the parent with the metabolite. When seeking to identify unknown metabolites, NMR and MS are clearly complementary methods, and confirmation of a definite metabolite structure often requires data from both methods. There are strong benefits from the high quality and dual nature of information gained from a single run when using LC-MS-NMR; as the instrumentation and techniques involved undergo continued evolution, efficiency and sensitivity in terms of the chromatographic properties are bringing LC-MS-NMR well on the way to becoming a mainstream approach.

3.7.2 Stable Isotope Labeling

To improve the reliability of metabolite identification by full-scan MS, stable isotopic labeling (^{2}H, ^{13}C, ^{15}N, ^{18}O, ^{34}S) has been used. Also, the use of stable isotope-labeled drugs allows safe experiments in humans, which provides great advantages over the use of radiolabels. In this type of study, a mixture of known amounts of labeled and nonlabeled drug is used for the metabolic experiments and analyzed by LC-MS. The identification of a metabolite should meet the following criteria: (1) two peaks with identical shapes and retention times must be recorded in the ion chromatograms; (2) the mass difference of the two peaks must be the same as the mass difference between the labeled and the unlabeled parent drug; and (3) the relative abundance ratio of the peaks must be the same as the concentration ratio of the labeled and the unlabeled parent drug. Labeling may not be necessary in those cases where the compound includes one or more Cl or Br atoms, which will show abundant "mass + 2" isotopes with known abundance ratios in the mass spectra. Also, the metabolite standards must be synthesized, and in some cases, chemical synthesis of metabolites may be difficult. Alternatively, enzymatic synthesis of metabolites has been found to be an easy and rapid way to produce a few milligrams of desired metabolites for structure confirmation.

3.8 "ALL-IN-ONE" RADIOACTIVITY-DETECTOR, STOP-FLOW, AND DYNAMIC-FLOW FOR METABOLITE IDENTFICATION

Many of the studies of drug metabolism such as absorption, bioavailability, distribution, biotransformation, excretion, metabolite identification, and other pharmacokinetic research use radiolabeled drugs, with the radioactive isotopes ^{14}C or tritium (^{3}H) most commonly used for the labeling of a given drug. Radioactive labeling is a good fit with HPLC separation, as it allows high-resolution, quantitative detection of unknown metabolites, and real-time monitoring by connecting the HPLC-radioactivity detector outlet to MS. The use of these detector interfaces aids the generation of data for structural elucidation of metabolites and biotransformation pathways for an administered drug. The detection sensitivity of both radioactivity and MS is the key to successfully identify and characterize metabolites (Nassar et al., 2003, 2004c).

Regulatory policy dictates that exposure to administered radioactivity be held as low as possible in most studies, which demands much greater sensitivity of the radioactivity detector to be able to detect metabolites. There has been great progress in the emerging science of detecting trace amounts of radiolabeled or non-radiolabeled drugs and their metabolites. The availability of these technologies should have a dramatic impact on drug discovery and development for metabolite profiling studies. It has been reported that a microplate scintillation counter combined with capillary LC can be used to enhance sensitivity by eluent fractionation and subsequent offline counting. The limitations with this method are that the sample must be completely dry before counting, that any volatile compounds are likely to be lost, and that there is the potential for apolar compounds to adsorb on the surface of the plate. In addition, this technique has the limitations of time-consuming sample

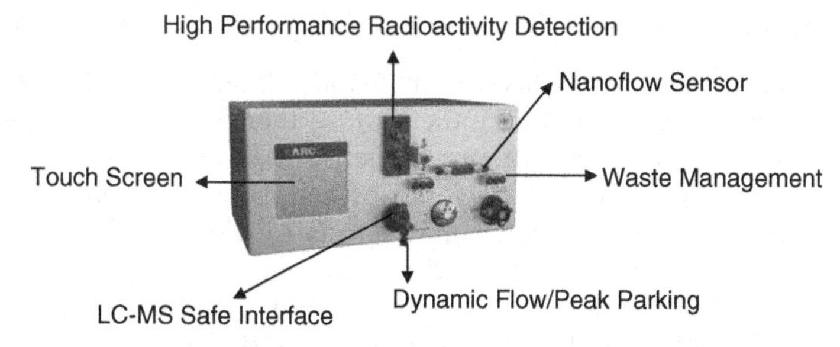

Figure 3.5 Hardware schematic diagram of the LC-ARC system.

preparation, high analysis costs, and the inability to elucidate metabolite structure.

A new development for metabolite identification in drug discovery and development is a detection method which couples online stop-flow and dynamic-flow with a radioactivity detector and mass spectrometer. Figure 3.5 shows the hardware schematic diagram of the v.ARC system. This system is simple and requires no custom-made hardware. The v.ARC system uses ARC's specially designed cells, and is operated under the ARC Data System which controls the entire radio-LC system, including LC and the radioactivity detector. When interfaced with LC-MS, the v.ARC system enhances detection of low-level radioactive peaks while increasing the flexibility and productivity of MS testing. This system has the major benefit of enhancing the sensitivity of radioisotope measurement for metabolite identification in drug metabolism studies. Another advantage to this system is the easy interface with the mass spectrometer, which allows acquisition of mass spectrometric data online. This system dramatically improves the sensitivity for ^{14}C peaks by up to 10-fold over conventional flow-through detection methods, and eliminates the need for a fraction collector and time-consuming sample preparation. In addition, the system produces accurate column recovery, quantification of low-level radioactivity, and consistently high resolution throughout the run. These factors give the combination of radioactivity-detector, stop-flow, and dynamic-flow and MS great potential as a powerful tool for improving the sensitivity of radioisotope measurement in metabolite identification studies; it also highlights the impressive progress that has been made in the technology of radioisotope counting in drug metabolism.

When online dynamic-flow coupled with a radioactivity detector and mass spectrometer is used, the total run time remains similar to conventional radio-LC. A technique has been developed to collect the peaks online and then infuse them to the MS, which allows acquisition of a higher order of multistage fragmentation for both major and minor metabolites. The multistage MS fragmentation pattern obtained for the metabolites enables determination of the sites of metabolism.

3.8.1 Online Peak Collection and Multistage Mass Experiments

It is important to note that this system can also be used to collect the peaks of interest online (peak parking), then infuse them for extended periods of time at flow rates as low as 1 μL/min while maintaining the column pressure. The peaks of interest can be triggered by radioactivity signals, UV signals or specified retention time, allowing analysts to use direct infusion with a low flow rate, with only one run and a small sample size. The peak parking feature allows sustained analytical signal input (infusion fashion), which allows any or all of the following experiments to be done as desired: optimizing mass spectrometric condition; tuning for any individual metabolite, tuning for both positive and negative polarities, and optimizing collision energy. Also, it allows the operator to conduct multistage mass experiments, automatically acquire ion mass data, perform MS^n, compare isotopic mass spectra, and perform neutral loss tests. This method provides the capability of identifying the structures of unknown metabolites or impurities, again requiring a limited sample amount and a single run. This is a significant improvement over an offline fraction collector, which may lose volatile compounds during the fraction collection process. Because this method retains these compounds for analysis, it greatly expands the ability to characterize and identify metabolites of a given compound, which in turn is of significant benefit to analysts.

For purposes of evaluating this system, *in vitro* human liver microsomal (HLM) incubations were performed with only [^{14}C]-dextromethorphan (Fig. 3.6), a semisynthetic narcotic cough-suppressing ingredient in a variety of over-the-counter cold and cough medications. The online separation and identification of dextromethorphan metabolites did not require intensive sample preparation, concentration, or fraction collection. Following incubation of [^{14}C]-dextromethorphan with human liver microsomes for 60 min in the presence of nicotinamide adenine dinucleotide phosphate (NADPH), three metabolites were detected: M-1, M-2, and M-3, along with the parent drug. The retention times of dextromethorphan and these metabolites were

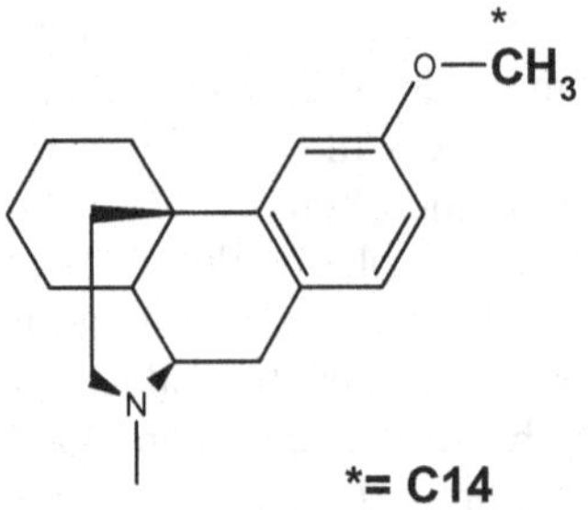

O—$\overset{*}{C}H_3$

N

*= C14

Figure 3.6 Chemical structure of [^{14}C]dextromethorphan, the test compound used for this study.

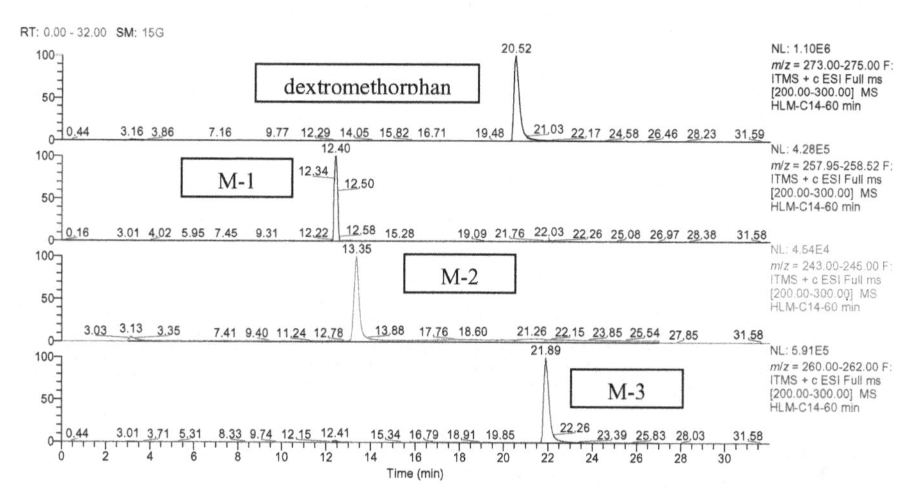

Figure 3.7 HPLC-MS chromatograms of [^{14}C]dextromethorphan following incubation with pooled human liver microsomes for 60 min in the presence of NADPH showing M-1, M-2, M-3, and nonmetabolized dextromethorphan.

between 12 and 23 min with excellent separation efficiency. Metabolites M-1, M-2, and M-3 have retention times of 12.4, 13.3, and 22.9 min, respectively. Figure 3.7 shows the LC-MS chromatograms of dextromethorphan following incubation with HLM for 60 min in the presence of NADPH. The LC-radio chromatogram shows that M-1 and M-2 did not have radioactive peaks, while M-3 and the parent dextromethorphan did. The full-scan mass spectra for M-1, M-2, and M-3 revealed protonated molecular ions $[M + H]^+$ at *m/z* 258, 244, and 260, suggesting that M-1 lost one methyl that contained C14, M-2 lost two methyl groups (one of which contained C14), and M-3 lost one methyl group.

The structures of dextromethorphan and metabolites were elucidated by LTQ-MS/MS analysis. These peaks were collected online and then infused at flow rate 10 μL/min; MS^2, MS^3, and MS^4 were performed with sufficient time to examine the MS data and decide on the next fragment ions to be used for metabolite characterizations. Separate runs were performed for MS^2, MS^3, and MS^4, each without collecting the peaks; the results were similar. Figure 3.8 represents LC-MS spectra of nonmetabolized [^{14}C]dextromethorphan $[M + H]^+$ at *m/z* 274: (a) MS^2, (b) MS^3, and (c) MS^4. The fragment ions are *m/z* 217, 215, 201, 175, 149, 123, 121, and 91. Comparison of the fragmentation ions for M-1 with the parent drug fragmentation suggests that *N*-dealkylation took place on the dextromethorphan molecule. Figure 3.9 shows proposed metabolic pathways of dextromethorphan in hepatic microsomal incubations. Mass spectrometric analysis showed the presence of dextromethorphan metabolites formed by *N*- and *O*-dealkylation, correlating with previously published results.

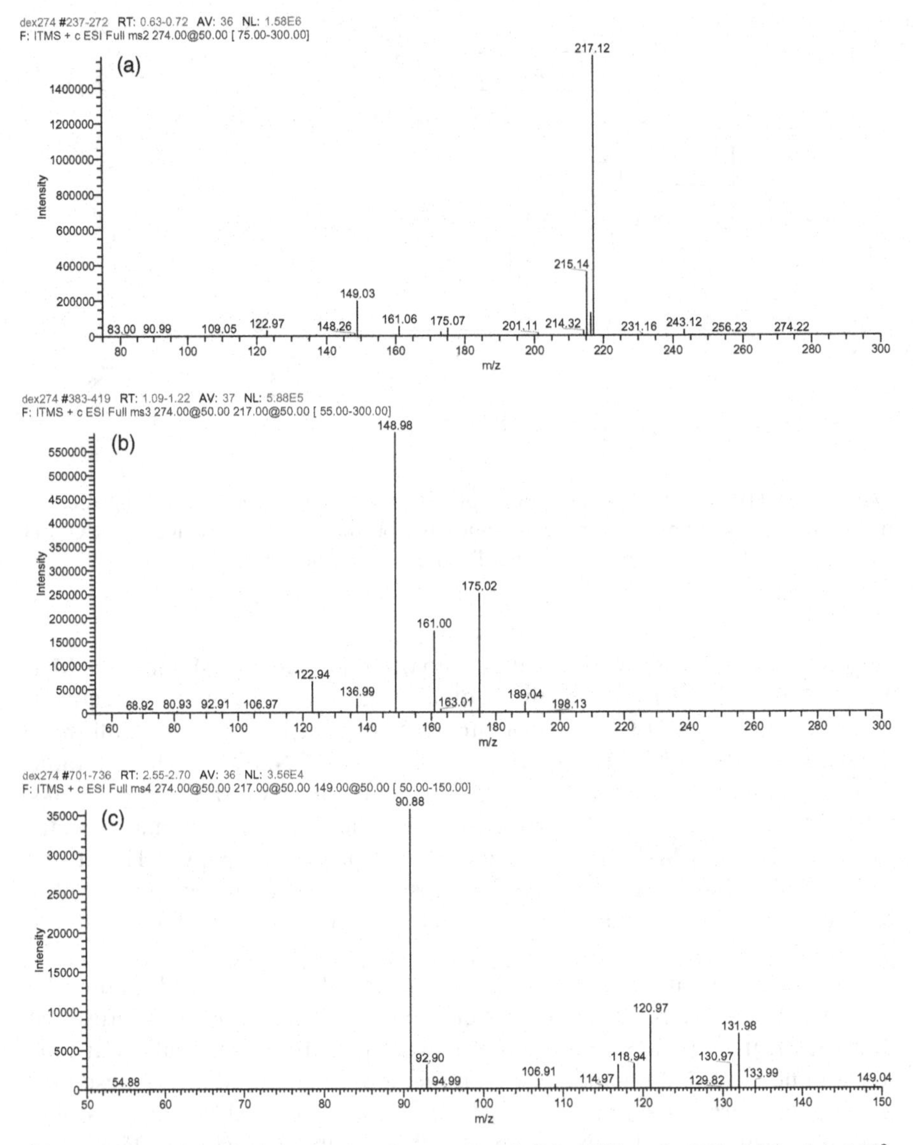

Figure 3.8 Representative LC-MS spectra of $[M+H]^+$ *m/z* 274, M-1, (a) MS^2, (b) MS^3, and (c) MS^4 following incubation of $[^{14}C]$dextromethorphan with human liver microsomes.

One of the major benefits of this method is that it is up to 10 times more sensitive in detecting ^{14}C peaks than commercially available flow-through radioactivity detectors. This method enhances the resolution of radiochromatograms and is able to measure volatile metabolites. Another advantage to this system is the easy interface with the mass spectrometer, which allows

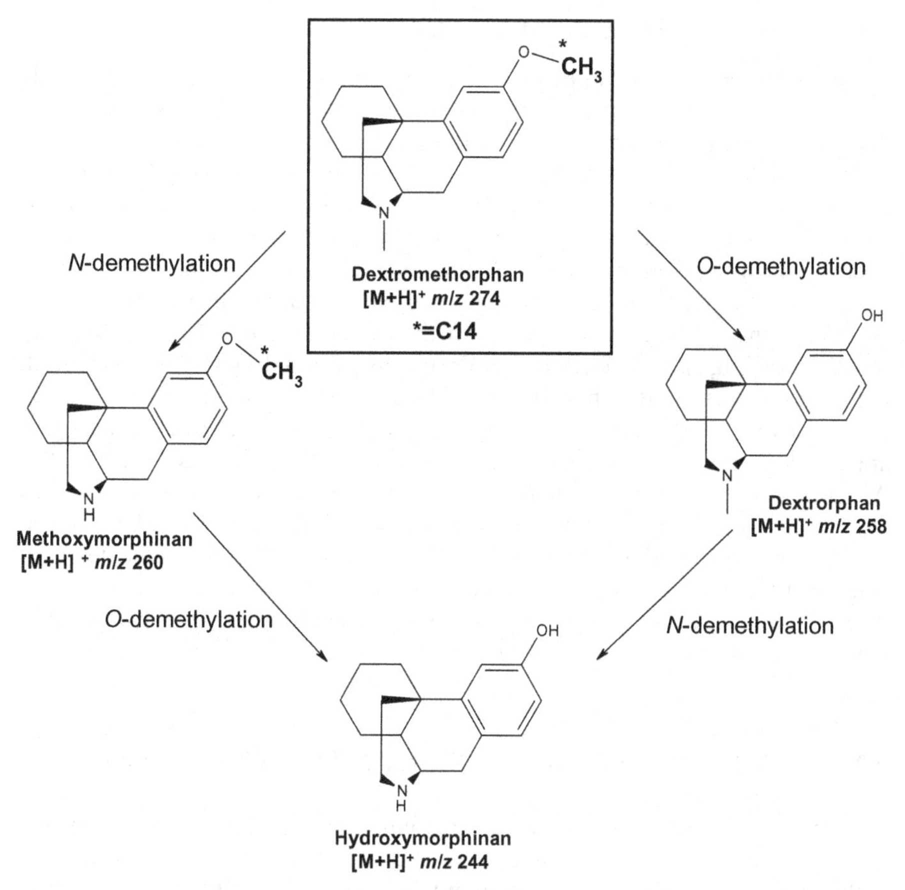

Figure 3.9 Proposed metabolic pathways of dextromethorphan in human hepatic microsomal incubations based on LC-ARC-MS/MS data.

acquisition of mass spectrometric data online. The method gives accurate column recovery and quantification of low-level radioactivity and high resolution throughout the run. An important safety benefit is that by using this method, injection size has been reduced, thereby decreasing potential exposure to radioactivity and reducing the amount of radioactive wastes. This smaller injection size, and the capability to perform many tests on a single run, also reduces both time and expenses significantly. Furthermore, it is easier because it reduces manual operations. This study showed that impressive progress has been made in the technology of radioisotope counting and metabolite characterization in drug metabolism studies using LC-ARC. The overall results suggest that the combination of LC-ARC dynamic-flow with radioactivity detection and MS has great potential as a powerful tool for radioisotope measurement in metabolite identification studies during drug discovery and development.

3.8.2 Metabolic Activation Studies by MS

Adverse drug reactions can be classified into type A (predictable) and type B (unpredictable or idiosyncratic) reactions. Type A reactions are dose-dependent and predictable based on the pharmacology of the drug; it is possible to reverse the reactions by reducing the dosage or, if necessary, discontinuing the drug altogether. Type B reactions cannot be predicted on the basis of the pharmacology of the drug. Type B reactions include anaphylaxis, blood dyscrasias, hepatoxicity, and skin reactions, and may result from a combination of genetic predisposition and environmental factors. Such unacceptable consequences can result from the occurrence of idiosyncratic drug reactions during late-stage clinical trials or even after a drug has been released. It is this uncertainty that has driven researchers to develop improved means for predicting the potential for a drug to cause such reactions. Currently, it appears that most idiosyncratic drug reactions are due to reactive metabolites. A typical cause of toxicity is the formation of electrophilic reactive metabolites that bind to nucleophilic groups present in vital cellular proteins and/or nucleic acids. Significant evidence from previously published reports indicates that chemically reactive metabolites may be responsible for serious forms of toxicity, including cellular necrosis, mutagenesis, carcinogenesis, teratogenesis, hypersensitivity, and blood disorders.

3.9 STRATEGIES TO SCREEN FOR REACTIVE METABOLITES

LC-MS used to screen for reactive electrophiles generated from *in vitro* experiments in several tissues (e.g., liver microsomes) have been designed to react with glutathione (GSH), with subsequent analysis of the GSH adducts (Nassar et al., 2004c; Nassar, 2009). There are several types of reactive metabolites (epoxides, arene oxides, quinones, quinone imines, quinone methides, imino-quinone methides, nitroso derivatives, nitrenium ions, nitro reduction products, nitro radicals, iminium ions, Michael acceptors, *S*-oxides, aliphatic aldehydes, or hydrolysis/acetylation products) which can be trapped in the presence of either GSH or an equimolar mixture of *N*-acetyl-cysteine and *N*-acetyl-lysine. This method is relatively simple and rapid, and has been implemented for high-throughput screening to identify reactive metabolites generated by bioactivation with Phase I and Phase II enzymes. Since thousands of compounds must be screened in a typical pharmaceutical industry setting, it is almost impossible, and highly expensive, to have all of them radiolabeled. Using radiolabeled reagents such as GSH and potassium cyanide is easier and more cost-effective for obtaining an accurate measurement of adduct formation. The reactive intermediates may be formed in other subcellular fractions such as liver S9, mitochondria, and cytosol (e.g., valproic acid forms reactive metabolites in subcellular mitochondria), and thus, screening for reactive intermediates in these fractions will minimize false negatives. As an example of such a screening procedure, the author compared *in vitro* cellular and subcellular models for identifying drug candidates with the potential to produce reactive metabolites. For purposes of evaluating these models, six known compounds currently in the clinic but with dose limitations due to hepatotoxicity were selected. Labetalol, acetaminophen, niacin, iproniazid, 8-hydroxyquinoline, and isoniazid were incubated separately with human liver mitochondria, S9, microsomes, cytosol, and hepatocytes. LC-QTOF-MS methods were used to elucidate metabolic profiles of selected compounds. The Phase II metabolites were in good agreement with previously published results; although without supplementation of reagents in subcellular fractions, Phase II metabolites were not detected. Hepatocytes also formed GSH conjugates, but the relative abundance was lower, indicating that the best model to identify potential reactive metabolites may involve the subcelluar fractions. The results from incubation of acetaminophen and labetalol, respectively, in subcellular fractions are described elsewhere. The results demonstrated that for acetaminophen, when reactive intermediates are produced, they can be trapped by GSH, while in the case of labetalol, no reactive intermediates were produced, and subsequently, no GSH adducts were detected.

3.10 SUMMARY

There is no doubt as to the value of LC-MS in drug metabolism research, as it has proved to be a sensitive and specific technique, which gives it important roles in ADMET studies. This chapter has demonstrated that LC-MS can provide a rapid, accurate, and relatively easy-to-use approach for metabolism studies without compromising the quality of the assays. Although LC-MS has improved the efficiency of metabolite analysis, it is still a daunting challenge to identify all of the metabolites generated by a particular drug candidate. Despite the high resolution MS provides, neutral-loss and precursor ion scans may miss some metabolites with unexpected fragmentation. Also, the current MS techniques may not provide the sensitivity required to detect trace quantities of metabolites in complex biological matrices. Future implementation plans can be applied in the following areas:

- Modern MS instrumentations can greatly assist the analyst in providing high-quality data in a more rapid and time-efficient manner. Techniques employing QTOF, Orbitrap, micro and capillary separation techniques with nanospray MS are emerging and gaining wider use. As the sensitivity and resolution of MS continue to improve, the technique will further enhance metabolite characterization. For example, another recent innovation, the hybrid two-dimensional quadrupole ion trap-fourier transform ion cyclotron resonance mass spectrometer, offers good potential for metabolite identification with high sensitivity and resolution.

- Good progress has been made and will continue in the automation of sample preparation and data handling.
- Advances in column technology and LC-MS have made chiral separation routine; more assays now have direct coupling of various chiral columns with an atmospheric pressure ion source without compromising the detector sensitivity or the LC resolving power.
- It is not always possible to determine the site of metabolic reaction in a drug molecule with MS. LC-MS-NMR provides unambiguous structural characterization of metabolites. Although the sensitivity of NMR is currently not sufficient for the analysis of metabolites in trace quantities, continuous improvements are being made.
- Online HD-LC-MS has become more widely used, because it is much faster and requires less preparation than NMR for metabolite confirmation, and further progress is anticipated.
- An important consideration is that any screening process for idiosyncratic drug reactions will produce false results, both positive and negative, and a goal must be to reduce or eliminate these, which will in turn lead to improved drug safety. More studies are required to improve our knowledge of, and hence the techniques used for, screening with the use of radiolabeled or stable isotope reagents, which would give an accurate measurement of the amount of reactive intermediate. Reactive intermediate formation should be investigated in subcellular fractions, which will aid in reducing false results.

REFERENCES

Bai Y, Milne JS, Mayne L, Englander SW. 1993. Primary structure effects on peptide group hydrogen exchange. *Proteins: Structure Function Genet* 17:75–86.

Bakke OM, Manocchia M, de Abajo F, Kaitin KI, Lasagna L. 1995. Drug safety discontinuations in the United Kingdom, the United States and Spain from 1974 through 1993: A regulatory perspective. *Clin Pharmacol Ther* 58(1):108–117.

Balimane PV, Pace E, Chong S, Zhu M, Jemal M, Pelt CK. 2005. A novel high-throughput automated chip-based nanoelectrospray tandem mass spectrometric method for PAMPA sample analysis. *J Pharm Biomed Anal* 39(1-2):8–16.

Castro-Perez J, Plumb R, Granger JH, Beattie I, Joncour K, Wright A. 2005a. Increasing throughput and information content for in vitro drug metabolism experiments using ultra-performance liquid chromatography coupled to a quadrupole time-of-flight mass spectrometer. *Rapid Commun Mass Spectrom* 19(6):843–848.

Castro-Perez J, Plumb R, Liang L, Yang E. 2005b. A high-throughput liquid chromatography/tandem mass spectrometry method for screening glutathione conjugates using exact mass neutral loss acquisition. *Rapid Commun Mass Spectrom* 19(6):798–804.

Englander SW. 2006. Hydrogen exchange and mass spectrometry: A historical perspective. *J Am Soc Mass Spectrom* 17:1481–1489.

Greene N. 2002. Computer systems for the prediction of toxicity: An update. *Adv Drug Deliv Rev* 54(3):417–431.

Guevremont R. 2004. High-field asymmetric waveform ion mobility spectrometry: A new tool for mass spectrometry. *J Chromatogr A* 1058(1-2):3–19.

Jemal M. 2000. High-throughput quantitative bioanalysis by LC/MS/MS. *Biomed Chromatogr* 14(6):422–429.

Jemal M, Bergum J. 1992. Effect of the amount of internal standard on the precision of an analytical method. *J Clin Pharmacol* 32(7):676–677.

Jemal M, Ouyang Z. 2003. Enhanced resolution triple-quadrupole mass spectrometry for fast quantitative bioanalysis using liquid chromatography/tandem mass spectrometry: Investigations of parameters that affect ruggedness. *Rapid Commun Mass Spectrom* 17(1):24–38.

Jemal M, Ouyang Z, Zhao W, Zhu M, Wu WW. 2003. A strategy for metabolite identification using triple-quadrupole mass spectrometry with enhanced resolution and accurate mass capability. *Rapid Commun Mass Spectrom* 17(24):2732–2740.

Johnson KA, Plumb R. 2005. Investigating the human metabolism of acetaminophen using UPLC and exact mass oa-TOF MS. *J Pharm Biomed Anal* 39(3-4):805–810.

Kantharaj E, Tuytelaars A, Proost PE, Ongel Z, Van Assouw HP, Gilissen RA. 2003. Simultaneous measurement of drug metabolic stability and identification of metabolites using ion-trap mass spectrometry. *Rapid Commun Mass Spectrom* 17(23):2661–2668.

Kapron JT, Jemal M, Duncan G, Kolakowski B, Purves R. 2005. Removal of metabolite interference during liquid chromatography/tandem mass spectrometry using high-field asymmetric waveform ion mobility spectrometry. *Rapid Commun Mass Spectrom* 19(14):1979–1983.

Kassel DB. 2004. Applications of high-throughput ADME in drug discovery. *Curr Opin Chem Biol* 8(3):339–345.

Kind T, Fiehn O. 2006. Metabolomic database annotations via query of elemental compositions: Mass accuracy is insufficient even at less than 1 ppm. *BMC Bioinformatics* 7:234–246.

Kind T, Wohlgemuth G, Lee DY, Lu L, Palazoglu M, Shahbaz S, Fiehn O. 2009. FiehnLib: Mass spectral and retention index libraries for metabolomics based on quadrupole and time-of-flight gas chromatography/mass spectrometry. *Anal Chem* 81:10038–10048.

Kostiainen R, Kotiaho T, Kuuranne T, Auriola S. 2003. Liquid chromatography/atmospheric pressure ionization-mass spectrometry in drug metabolism studies. *J Mass Spectrom* 38(4):357–372.

Krishna MMG, Hoang L, Lin Y, Englander SW. 2004. Hydrogen exchange methods to study protein folding. *Methods* 34:51.

Kubinyi H. 1977. Quantitative structure-activity relationships: 7. The bilinear model, a new model for nonlinear dependence of biological activity on hydrophobic character. *J Med Chem* 20(5):625–629.

Lasser KE, Allen PD, Woolhandler SJ, Himmelstein DU, Wolfe SM, Bor DH. 2002. Timing of new black box warnings and withdrawals for prescription medications. *J Am Med Assoc* 287(17):2215–2220.

Leclercq L, Delatour C, Hoes I, Brunelle F, Labrique X, Castro-Perez J. 2005. Use of a five-channel multiplexed electrospray quadrupole time-of-flight hybrid mass spectrometer for metabolite identification. *Rapid Commun Mass Spectrom* 19(12):1611–1618.

Lesko LJ, Rowland M, Peck CC, Blaschke TF. 2000. Optimizing the science of drug development: Opportunities for better candidate selection and accelerated evaluation in humans. *Pharm Res* 17(11):1335–1344.

Lin J, Sahakian DC, de Morais SM, Xu JJ, Polzer RJ, Winter SM. 2003. The role of absorption, distribution, metabolism, excretion and toxicity in drug discovery. *Curr Top Med Chem* 3(10):1125–1154.

Meyboom RH, Lindquist M, Egberts AC. 2000. An ABC of drug-related problems. *Drug Saf* 22(6):415–423.

Mortishire RJ, O'Connor D, Castro-Perez JM, Kirby J. 2005. Accelerated throughput metabolic route screening in early drug discovery using high-resolution liquid chromatography/quadrupole time-of-fligh mass spectrometry and automated data analysis. *Rapid Commun Mass Spectrom* 19:2659–2670.

Nassar AE. 2003. Online hydrogen-deuterium exchange and a tandem quadrupole time-of-flight mass spectrometer coupled with liquid chromatography for metabolite identification in drug metabolism. *J Chromatogr Sci* 41(8):398–404.

Nassar A-EF. 2009. Minimizing the potential for drug bioactivation of drug candidates to success in clinical development. In *Drug Metabolism Handbook: Concepts and Applications*, ed. Nassar A-EF, Hollenberg P, Scatina J, 283–306. Hoboken, NJ: John Wiley & Sons.

Nassar AE, Adam P. 2003. Metabolite characterization in drug discovery utilizing robotic liquid-handling, quadrupole time-of-flight mass spectrometry and in silico prediction. *Curr Drug Metab* 4(4):259–271.

Nassar AE, Bjorge SM, Lee DY. 2003. On-line liquid chromatography-accurate radioisotope counting coupled with a radioactivity detector and mass spectrometer for metabolite identification in drug discovery and development. *Anal Chem* 75(4):785–790.

Nassar AE, Lopez-Anaya A. 2004. Strategies for dealing with reactive intermediates in drug discovery and development. *Curr Opin Drug Discovery Dev* 7(1):126–136.

Nassar AE, Talaat R. 2004. Strategies for dealing with metabolite elucidation in drug discovery and development. *Drug Discov Today* 9(7):317–327.

Nassar AE, Varshney N, Getek T. 2001a. Quantitative analysis of hydrocortisone in human urine using a high-performance liquid chromatographic-tandem mass spectrometric-atmospheric-pressure chemical ionization method. *Anal Lett* 34 (12):2047–2062.

Nassar AE, Varshney N, Getek T, Cheng L. 2001b. Quantitative analysis of hydrocortisone in human urine using a high-performance liquid chromatographic-tandem mass spectrometric-atmospheric-pressure chemical ionization method. *J Chromatogr Sci* 39(2):59–64.

Nassar AE, Kamel AM, Clarimont C. 2004a. Improving the decision-making process in structural modification of drug candidates: Reducing toxicity. *Drug Discov Today* 9(24):1055–1064.

Nassar AE, Kamel AM, Clarimont C. 2004b. Improving the decision-making process in the structural modification of drug candidates: Enhancing metabolic stability. *Drug Discov Today* 9(23):1020–1028.

Nassar AE, Parmentier Y, Martinet M, Lee DY. 2004c. Liquid chromatography-accurate radioisotope counting and microplate scintillation counter technologies in drug metabolism studies. *J Chromatogr Sci* 42(7):348–353.

Obach RS. 1997. Nonspecific binding to microsomes: Impact on scale-up of in vitro intrinsic clearance to hepatic clearance as assessed through examination of warfarin, imipramine, and propranolol. *Drug Metab Dispos* 25(12):1359–1369.

Obach RS. 1999. Prediction of human clearance of twenty-nine drugs from hepatic microsomal intrinsic clearance data: An examination of in vitro half-life approach and nonspecific binding to microsomes. *Drug Metab Dispos* 27(11):1350–1359.

Plumb RS, Stumpf CL, Granger JH, Castro-Perez J, Haselden JN, Dear GJ. 2003. Use of liquid chromatography/time-of-flight mass spectrometry and multivariate statistical analysis shows promise for the detection of drug metabolites in biological fluids. *Rapid Commun Mass Spectrom* 17(23):2632–2638.

Roberts SA. 2003. Drug metabolism and pharmacokinetics in drug discovery. *Curr Opin Drug Discovery Dev* 6(1):66–80.

Taylor EW, Jia W, Bush M, Dollinger GD. 2002. Accelerating the drug optimization process: Identification, structure elucidation, and quantification of in vivo metabolites using stable isotopes with LC/MS^n and the chemiluminescent nitrogen detector. *Anal Chem* 74(13):3232–3238.

Thompson TN. 2000. Early ADME in support of drug discovery: The role of metabolic stability studies. *Curr Drug Metab* 1(3):215–241.

Tiller PR, Romanyshyn LA. 2002. Liquid chromatography/tandem mass spectrometric quantification with metabolite screening as a strategy to enhance the early drug discovery process. *Rapid Commun Mass Spectrom* 16(12):1225–1231.

Tyler AN, Clayton E, Green BN. 1996. Exact mass measurement of polar organic molecules at low resolution using electrospray ionization and a quadrupole mass spectrometer. *Anal Chem* 68:3561–3569.

White RE. 2000. High-throughput screening in drug metabolism and pharmacokinetic support of drug discovery. *Annu Rev Pharmacol Toxicol* 40 133–157.

Yang L, Amad M, Winnik WM, Schoen AE, Schweingruber H, Mylchreest I, Rudewicz PJ. 2002. Investigation of an enhanced resolution triple quadrupole mass spectrometer for high-throughput liquid chromatography/tandem mass spectrometry assays. *Rapid Commun Mass Spectrom* 16(21):2060–2066.

Zhang H, Henion J. 2001. Comparison between liquid chromatography-time of-flight mass spectrometry and selected reaction monitoring liquid chromatography-mass spectrometry for quantitative determination of idoxifene in human plasma. *J Chromatogr B Biomed Sci Appl* 757(1):151–159.

ABBREVIATIONS AND GLOSSARY

Accelerator Mass Spectrometry A mass spectrometry technique in which atoms extracted from a sample are ionized, accelerated to MeV energies and separated according to their momentum, charge, and energy.

Accurate Mass An experimentally determined mass that can be used to determine a unique elemental formula. For ions less than 200 μ, a measurement with 5 ppm accuracy is sufficient to determine the elemental composition.

Appearance Energy The minimum energy that must be imparted to an atom, molecule, or molecular species in order to produce a specified ion X+. The term "appearance potential" is not recommended.

Array Collector The array collector (detector) is made up of a number of ion collection elements or devices, arranged in a line or grid where each element is an individual detector.

Atmospheric Pressure Chemical Ionization A type of *chemical ionization* that takes place at atmospheric pressure as opposed to the reduced pressure which is normally used for the ionization region of *chemical ionization*. See also *chemical ionization*.

Atmospheric Pressure Ionization This term refers to any ionization processes whereby ionized species are caused to be in the gas phase at atmospheric pressure, including *atmospheric pressure chemical ionization*, *atmospheric pressure photo-ionization*, and *electrospray ionization*.

Atmospheric Pressure Photo-Ionization This term refers to *photo-ionization* and subsequent *atmospheric pressure chemical ionization* that takes place at atmospheric pressure as opposed to the reduced pressure which is normally used for the ionization region of *photo-ionization*. See also *atmospheric pressure chemical ionization* and *photo-ionization*.

Average Mass The mass of an ion or molecule calculated using the average mass of each element weighted for its natural isotopic abundance.

Base Peak The peak in a mass spectrum that has the greatest intensity. This term may be applied to the spectra of pure substances or mixtures.

Channeltron A horn-shaped continuous dynode particle multiplier. The ion strikes the inner surface of the device and induces the production of secondary electrons that in turn impinge on the inner surfaces to produce more secondary electrons. This avalanche effect produces an increase in signal in the final measured current pulse.

Chemical Ionization Describes the process whereby new ionized species are formed when gaseous molecules interact with ions. The process may involve transfer of an electron, a proton, or other charged species between the reactants. When a positive ion results from chemical ionization, the term may be used without qualification. When a negative ion results, the term *negative ion chemical ionization* should be used. Note that this term is not synonymous with *chemi-ionization* or with negative chemical ionization.

Chemi-Ionization A process by means of which gaseous molecules are ionized when they interact with other internally excited gaseous molecules or molecular species. See also *associative ionization*. Note that this term is not synonymous with *chemical ionization*.

Collisional Excitation An ion/neutral species process wherein there is an increase in the reactant ion's internal energy at the expense of the translational energy of either (or both) of the reactant species.

Collisionally Activated Dissociation (CAD) An ion/neutral species process wherein excitation of a projectile ion of high translational energy is brought about by the same mechanism as in collision-induced dissociation.

Collision-Induced Dissociation (CID) An ion/neutral species interaction wherein the projectile ion is dissociated as a result of interaction with the neutral collision partner. This is brought about by conversion of part of the translational energy of the ion to internal energy in the ion during collision. The term collisionally activated dissociation (or decomposition), abbreviated CAD, is also used.

Conversion Dynode A conversion dynode is used to increase the secondary emission characteristics for heavy ions and thus reduce the mass discrimination of the detector. A high potential, of opposite polarity to the ions detected, is used to attract these ions to the dynode. Secondary electrons are produced when the ions hit the dynode and are subsequently recorded via the detector used.

Cyclotron A device that uses an oscillating electric field parallel to a magnetic field to accelerate charged particles.

Dalton Non-SI unit of mass (equal to the atomic mass constant), defined as one-twelfth of the mass of a carbon-12 atom in its ground state and used to express masses of atomic particles, $u \approx 1.660\ 5402(10) \times 10^{-27}$ kg.

Daly Detector A rounded metal surface held at high potential that emits secondary electrons when ions impinge on the surface. Often used in conjunction with photomultipliers.

Detection Limit The detection limit of an instrument should be differentiated from sensitivity. The detection limit reflects the smallest flow of sample or the lowest partial pressure that gives a signal that can be distinguished from the background noise. One must specify the experimental conditions used and give the value of signal-to-noise ratio corresponding to the detection limit.

Detection of Ions The observation of electrical signal due to particular ionic species by a detector under conditions that minimize ambiguities from interferences. Ions may be detected by photographic or suitable electrical means.

Detector Gain The ratio of the detector output current to the input ion current.

Double-focusing Mass Spectrometer An instrument that uses both direction and velocity focusing so that an ion beam, initially diverging and containing ions of different energies, is separated into beams according to the mass/charge ratio.

Dynamic Range A measure of the detection range of a detector. The ratio of the largest to smallest detectable signal.

Dynode One of a series of electrodes in a photomultiplier tube. Such an arrangement is able to amplify the current emitted by the photocathode.

Electron Energy The potential difference through which electrons are accelerated before they are used to bring about electron ionization.

Electron Ionization This term is used to describe ionization of any species by electrons accelerated between 50 and 150 eV. Usually 70 eV electrons are used to produce positive ions. The term "electron impact" is not recommended.

Electron Multiplier A device to multiply current in an electron beam (or in a photon or particle beam after conversion to electrons) by incidence of accelerated electrons upon the surface of an electrode. This collision yields a number of incident electrons. These electrons are then accelerated to another electrode (or another part of the same electrode), which in turn emits secondary electrons, continuing the process.

Electron Volt A non-SI unit of energy (eV) defined as the energy acquired by a particle containing one unit of charge through a potential difference of one volt. An electron-volt is equal to 1.602 177 33(49) × 10^{-19} J

Electrospray Ionization A process whereby ionized species in the gas phase are caused from an analyte containing solution via highly charged fine droplets, by means of spraying the solution from a narrow-bore needle tip at atmospheric pressure in the presence of a high electric field. When a pneumatic assistance is added to the formation of a stable spray, the term *pneumatically assisted electrospray ionization* is also used. (The term "ion spray" is not recommended.)

Exact Mass The calculated mass of an ion or molecule containing a single isotope of each atom.

Faraday Cup A hollow collector, open at one end and closed at the other, used to measure the ion current associated with an ion beam. On arrival at the earthed metal plate, the ions are neutralized either by accepting or donating electrons. A small electric current is produced which can be amplified. The device, while simple and robust, is not sensitive.

Focal Plane Collector These are used for spatially disperse spectra, that is, all ions simultaneously impinge on the detector plane.

Fourier Transform Ion Cyclotron (FT-ICR) Mass Spectrometer A mass spectrometer based on the principle of ion cyclotron resonance. An ion placed in a magnetic field will move in a circular orbit at a frequency characteristic of its m/z value. Ions are excited to a coherent orbit using a pulse of radio frequency energy and their image charge is detected on receiver plates as a time domain signal. Fourier transformation of the time domain signal results in the frequency domain FT-ICR signal which, on the basis of the inverse proportionality between frequency and m/z, can be converted to a mass spectrum.

Hybrid Mass Spectrometer A mass spectrometer that combines m/z analyzers of different types to perform tandem mass spectrometry.

Hydrogen–Deuterium Exchange A technique used for exploring the conformational structure of analyte by means of the surface accessibility with a deuterated reactant. The deuterium-labeling reaction is carried out either in the gas-phase inside a mass spectrometer or in the solution-phase prior to introduce the sample into a mass spectrometer.

Ionization Cross Section This is a measure of the probability that a given ionization process will occur when an atom or molecule interacts with a photon.

Ionization Efficiency The ratio of the number of ions formed to the number of electrons or photons used.

Ionization Energy This is the minimum energy of excitation of an atom, molecule, or molecular species M required to remove an electron in order to produce a positive ion.

Ion–Pair Formation This involves an ionization process in which a positive fragment ion and a negative fragment ion are among the products.

Ion Spray Pneumatically assisted electrospray ionization. The use of this term is not recommended.

Magnetic Field A property of space that produces a force on a charged particle equal to $qv \times B$ where q is the particle charge and v its velocity.

Magnetic Sector A device that produces a magnetic field perpendicular to a charged particle beam that deflects the beam to an extent that is proportional to the particle momentum per unit charge. For a monoenergetic beam, the deflection is proportional to m/z.

Mass Defect The difference between the monoisotopic and nominal mass of a molecule or atom.

Mass Number (m) The sum of the protons and neutrons in an atom, molecule, or ion.

Mass Spectrograph An instrument that separates a beam of ions according to their mass-to-charge ratio in which the ions are directed onto a focal plane detector such as a photographic plate.

Mass Spectrometer An instrument that measures the mass-to-charge ratio and relative abundances of ions.

Mass Spectrometry The branch of science that deals with all aspects of mass spectrometers and the results obtained with these instruments.

Mass Spectrometry-Mass Spectrometry (MS/MS) The acquisition, study, and spectra of the electrically charged products or precursors of a preselected ion or ions.

Mass Spectroscopy Not recommended. Use *mass spectrometry.*

Mass Spectrum The signal detected from a collection of ions plotted as a function of the mass-to-charge ratio (*m/z*).

Mattauch-Herzog Geometry An arrangement for a *double-focusing mass spectrometer* in which a deflection of $\pi/(4\sqrt{(2)})$ radians in a radial *electric field* is followed by a *magnetic deflection* of $\pi/2$ radians.

Micro-Channel Plate (MCP) An incident X-ray photon enters a channel and frees (via "photoelectric emission") an electron from the channel wall. An electron-accelerating potential difference (approx. –1500 Volts) is applied across the length of the channel. The initial electron strikes the adjacent wall, freeing several electrons (via "secondary emission"). These electrons will be accelerated along the channel until they in turn strike the channel surface, giving rise to more electrons. Eventually, this cascade process yields a cloud of several thousand electrons, which emerge from the rear of the plate.

Molecular Mass The mass of one mole of a molecular substance (6.022 1415(10) × 1023 molecules).

Monoisotopic Mass The mass of an ion or molecule calculated using the mass of the most abundant isotope of each element. *m/z*—The symbol *m/z* is used to denote the dimensionless quantity formed by dividing the mass number of an ion by its charge number. The symbol should consist of italicized lower case letters with no spaces.

MS/MS in Space MS/MS data that are recorded in sequential mass analysers, that is, multi-analyzer machines, for example, triple quadrupole instruments and q-tofs. Specific scan functions are designed such that in one section of the instrument, predetermined ions are selected, then dissociated, and the resultant ions are then transmitted to another analyzer for data acquisition.

MS/MS in Time MS/MS data that are recorded in one mass analyser in discreet time steps, for example, in an ion trap. Here, the specific scan functions are designated, and the resultant MS/MS data are subsequently recorded in the same device along a discreet timeline.

MS/MS Spectrum A generic term that should be discouraged. It does not explain which type of MS/MS data is being discussed. A scan-specific term should always be used, for example, product ion spectrum or second-generation product ion spectrum.

MS^n This refers to multistage MS/MS experiments where n is the number of product ion stages (progeny ions). For ion traps, sequential MS/MS experiments can be undertaken where $n > 2$ whereas for a simple triple quadrupole system $n = 2$.

Multiple Reaction Monitoring (MRM) Incorrect term, see Selected Reaction Monitoring (SRM).

Nanoelectrospray This term is used to describe the electrospray interface which achieves a reduction in the sample flow rates (usually less than 100 nL/min). The reduced flow rate is due to a narrower inner diameter of the needle tip (usually 1–3 µm) and is established automatically, dependent on the potential on the needle tip.

Neutral Loss The loss of an electrically uncharged species during a rearrangement process.

Nitrogen Rule The rule that an organic molecule containing the elements C, H, O, S, P, or halogen will have an odd nominal mass if it contains an odd number of nitrogen atoms.

Nominal Mass The mass of an ion or molecule calculated using the mass of the most abundant isotope of each element rounded to the nearest integer value.

Odd-Electron Rule Upon dissociation, odd-electron ions may form either odd or even-electron ions, whereas even-electron ions tend to form even-electron ions.

Orthogonal Extraction The pulsed acceleration of ions perpendicular to their direction of travel into a time-of-flight mass spectrometer. Ions may be extracted from a directional ion source, drift tube, or *m/z* separation stage.

Parent Ion Obsolete term, see precursor ion.

Parent Ion Scan Obsolete term, see precursor ion scan.

Parent Ion Spectrum Obsolete term, see precursor ion spectrum.

Paul Ion Trap Ion trapping device that permits the ejection of ions with an *m/z* lower than a prescribed value and retention of those with higher mass. It depends on the application of radio frequency voltages between a ring electrode and two end-cap electrodes to confine the ion motion to a cyclic path described by an appropriate form of the Mathieu equation. The choice of these voltages determines the *m/z* below which ions are ejected.

Peak A local maximum of signal in a mass spectrum. Although peaks are often associated with particular ions, the terms peak and ion should not be used interchangeably.

Peak Intensity The height or area of a peak in a mass spectrum.

Photographic Plate The recording of ion currents (usually associated with ion beams that have been spatially separated by *m/z* values) by allowing them to strike a photographic plate, which is subsequently developed.

Photomultiplier (Scintillator) Electrons produced following collision of an ion with a conversion dynode strike a phosphorous screen to produce a burst of photons. Cascade amplification of these photons, in a manner similar to electron amplification within an *electron multiplier*, results in an increased signal that is converted into a measurable current. Photomultiplier are sealed units that prevent contamination and hence increase the lifetime of the unit.

Point Collector The ion beam is focused onto a point and the individual ions sequentially arrive at the detector.

Post-Acceleration Detector (PAD) A high voltage bias is applied pre detection and post analyzer to accelerate the ions and produce an improved signal at the detector. This device is particularly useful for large molecules (>100 kDa) and polymers.

Postacceleration Detector An ion detector in which the charged particles are accelerated to a high velocity and impinge on a conversion dynode, emitting secondary electrons. The electrons are accelerated onto a phosphor screen, which emits photons that are in turn detected using a photomultiplier or other photon detector.

Precursor Ion An electrically charged species that may dissociate to form fragments, of which one or more are ions and one or more are neutral species. A precursor ion may be a molecular ion, protonated, or deprotonated molecule, or a fragment ion. A precursor ion may also be any ion undergoing a change in charge state.

Precursor Ion Scan The specific scan function or process that will record the precursor ion or ions of any predetermined product ions.

Precursor Ion Spectrum The specific mass spectrum recorded from any spectrometer in which the appropriate scan function can be set to record the precursor ion or ions of pre-determined product ions.

Product Ion An electrically charged product of the reaction of a particular precursor ion. In general, such ions have a direct relationship to a particular precursor ion and indeed may relate to a unique state of the precursor ion. The reaction need not necessarily involve fragmentation. It could, *e.g.* involve a change in the number of charges carried. Thus, all fragment ions are product ions but not all product ions are necessarily fragment ions.

Product Ion Scan The specific scan function or process that will record the product ion or ions of predetermined precursor ions.

Product Ion Spectrum The specific mass spectrum recorded from any spectrometer in which the appropriate scan function can be set to record the product ion or ions of predetermined precursor ions.

Prolate Trochoidal Mass Spectrometer A mass spectrometer in which the ions of different *m/z* are separated by means of crossed electric and magnetic fields in such a way that the selected ions follow a prolate trochoidal path.

Quistor An abbreviation of quadrupole ion storage trap. This term is equivalent to *Paul ion trap*.

Reflectron A type of time-of-flight mass spectrometer that uses a static electric field to reverse the direction of travel of the ions entering it. A reflectron improves mass resolution by assuring that ions of the same *m/z* but different kinetic energy arrive at the detector at the same time.

Resolution (Mass) The smallest mass difference Äm between two equal magnitude peaks such that the valley between them is a specified fraction of the peak height.

Resolving Power (Mass) In a mass spectrum, the observed mass divided by the difference between two masses that can be separated: m/Äm. The method by which Äm was obtained and the mass at which the measurement was made should be reported.

Sector Mass Spectrometer A mass spectrometer consisting of one or more magnetic sectors *m/z* selection in a beam of ions. Such instruments may also have one or more electric sectors for energy selection.

Selected Reaction Monitoring (SRM) The specific product ions of preselected precursor ions recorded via multistage MS; that is, the precursor ion(s) must undergo a specific reaction or transformation to be detected. Selected reaction monitoring can be preformed "in time" or "in space", that is, in an ion trap or a multi-analyzer instrument, respectively.

Sensitivity Sensitivity is ratio of change in ion current to the sample concentration. It is the slope of the graph for the sample concentration versus the measured signal.

Stable Ion An ion with internal energy below the threshold for dissociation or isomerization (not sufficiently excited) which therefore does not dissociate, rearrange, nor react spontaneously when flying inside a mass spectrometer reaching the detector in its intact form (configuration and connectivity).

Time-of-flight Mass Spectrometer An instrument that separates ions by *m/z* in a field-free region after acceleration to a constant kinetic energy.

Total Ion Current The sum of all the separate ion currents carried by the different ions contributing to the spectrum.

Transmission The ratio of the number of ions leaving a region of a mass spectrometer to the number entering that region.

Transmission Quadrupole Mass Spectrometer A mass spectrometer that consists of four parallel rods whose centers form the corners of a square and whose opposing poles are connected. The voltage applied to the rods is a superposition of a static potential and a sinusoidal radio frequency potential. The motion of an ion in the x and y dimensions is described by the Mathieu equation whose solutions show that ions in a particular *m/z* range can be transmitted along the *z* axis.

Unstable Ion An ion that does not possess sufficient energy to dissociate within the ion source, under stated experimental conditions, but is too energetic to reach the detector as an intact species and hence dissociates before

IUPAC MS Terms and Definitions, First Public Draft, June 2005 by Kermit K. Murray, Robert K. Boyd, Marcos N. Eberlin, G. John Langley, Liang Li, Yasuhide Naito, and Jean Claude Tabe

CHAPTER 4

IMPROVING DRUG DESIGN: CONSIDERATIONS FOR THE STRUCTURAL MODIFICATION PROCESS

ALA F. NASSAR
Chemistry Department, Brandeis University, Waltham, MA

During the drug discovery process, most candidates undergo modifications to adjust absorption, distribution, metabolism, and excretion (ADME) properties so as to maximize both safety and effectiveness for human patients. ADME studies represent a critical component in support of drug discovery and development, and help guide medicinal chemists and metabolism scientists to maximize the candidate's characteristics as they relate to the rule of three, which is the triangular balance between activity, exposure, and toxicity. The study of structural modifications produces valuable information which, throughout the process of drug discovery, keeps on giving direction and focus in the effort to improve and balance these relationships. This is very important, because the therapeutic efficacy of a drug is dependent on its exposure, which in turn is dictated in part by metabolic stability of the molecule. In addition, drug metabolism may lead to the formation of metabolites that can either be pharmacologically active or elicit adverse effects. On this basis, metabolite identification and profiling have become a routine exercise during lead optimization and subsequent development processes. Metabolite characterization has become one of the main drivers of the drug discovery process, helping to optimize ADME properties and increase the success rate for drugs. This chapter will discuss the importance of structural modifications in efforts to improve metabolic stability, enhance pharmacokinetics (PK) properties, and reduce toxicity in drug candidates. The discussion is supported by case studies. Several examples are included to show how metabolic stability, PK, and reactive intermediates have influenced and guided drug design and candidate selection.

Biotransformation and Metabolite Elucidation of Xenobiotics, Edited by Ala F. Nassar

4.1 INTRODUCTION

Prediction and assessment of the ADME/TOX characteristics of compounds as early as possible in the drug discovery process has become a major focus of pharmaceutical research (Kubinyi et al., 1977; Lesko et al., 2000; Taylor et al., 2002; Kantharaj et al., 2003; Kostiainen et al., 2003; Nassar et al., 2003a,b; Plumb et al., 2003; Watt, 2003; Nassar et al., 2004a,b). Pharmaceutical companies are coming under increasing pressure to more precisely prove long-term safety and to provide more data on the safety of their products. Failure to successfully address these concerns causes companies to withdraw drugs from the market, not only those that present human-health consequences but also those with negative economic and public-relations impact on the pharmaceutical industry. Because of human-health or safety concerns, a long list of drugs have been withdrawn from the market. Failures in the drug discovery process account for a substantial part of the cost of drug development because many compounds are discarded after substantial investment owing to unforeseen toxicological effects (Bakke et al., 1995; Pirmohamed et al., 1998; Richard et al., 1998; Meyboom et al., 2000; Greene et al., 2002; Lasser et al., 2002; Tiller et al., 2002; Uetrecht et al., 2002).

The discovery process in the pharmaceutical industry has undergone a dramatic change in the last decade due to advances in molecular biology, high-throughput pharmacological screens, and combinatorial synthesis. The high cost, long development time, and high failure rate in bringing drugs to market have been important factors behind this change. The major reasons for this low success rate include poor pharmacological activity, low bioavailability, or high toxicity. Thus, the design of drugs with optimal potency and PK properties, and lower or no toxicity, is challenging, given the opposing requirements for absorption and metabolism. During this process, we need to avoid high metabolic liability that usually leads to poor bioavailability and high clearance, as well as the formation of active or toxic metabolites, which also have an impact on the pharmacological and toxicological outcomes. One approach to this dilemma would be to thoroughly explore the structure requirements for *in vitro* pharmacological activity through combinatorial synthesis and identify the key structural features that are essential for activity for each chemical series, and then attempt to improve the metabolism, PK, and toxicity properties through structure modifications to the regions of the molecule that have little or no impact on the desired activity. Thus, one can maintain an optimal level of potency while introducing structural features that will improve the metabolism and toxicity characteristics.

4.1.2 Value of Metabolite Characterization in Structure Modification

Optimization of the metabolic stability, PK properties, and toxicity of candidates are among the important considerations during the drug discovery process. The rate and sites of metabolism of new chemical entities by drug metabolizing enzymes are amenable to modulation by appropriate structural changes. Until recently, this optimization has been achieved largely by empirical methods and trial and error. Now, metabolite identification is enabling early identification of potential metabolic liabilities or issues, providing a metabolism perspective to guide synthesis efforts with the aim of either blocking or enhancing metabolism so as to optimize the pharmacokinetic and safety profiles of newly synthesized drug candidates, and assisting the prediction of the metabolic pathways of potential drug candidates. This is crucial to the drug discovery process, as it can be used to investigate the Phase I metabolites likely to be formed *in vivo*, differences between species in drug metabolism, the major circulating metabolites of an administered drug, Phase I and Phase II pathways of metabolism, pharmacologically active or pharmacologicallym toxic metabolites, and to help determine the effects of metabolizing enzyme inhibition and/or induction. The ability to produce, interpret, and act on this information early in the discovery phase has become increasingly important as a basis for judging whether a drug candidate merits further development. This in turn guides the pharmaceutical researcher's strategies to achieve the best balance of (1) metabolic stability, (2) optimal PK properties, and (3) low toxicity. The discussion on each of these three areas will be illustrated with examples from the literature explaining how each of them have influenced drug design strategies, and helped to facilitate improvements.

4.2 (A) ENHANCEMENT OF METABOLIC STABILITY

4.2.1 *In Vitro* Studies

Several *in vitro* methods are available to evaluate metabolic stability. Among the most popular and widely utilized systems are liver microsomes. Microsomes retain activity of key enzymes involved in drug metabolism which reside in the smooth endoplasmic reticulum, such as cytochrome P450s (CYPs), flavin monooxygenases, and glucuronosyltransferases. Isolated hepatocytes offer a valuable whole cell *in vitro* model, and retain a broader spectrum of enzymatic activities, including not only reticular systems, but cytosolic and mitochondrial enzymes as well. Because of a rapid loss of hepatocyte-specific functions, it has been possible to generate useful data only with short-term hepatocyte incubations or cultures. However, significant strides have been made toward enhancing the viability of both cryopreserved and cultured hepatocytes. Liver slices, like hepatocytes, retain a wide array of enzyme activities, and are used to assess metabolic stability. Furthermore, both hepatocytes and liver slices are capable of assessing enzyme induction *in vitro*. The choice of which system to employ in a drug discovery screening program is dependent on numerous factors such as information about a particular chemical series and availability of tissue. For example, if a pharmacological probe has been shown to induce CYP activity, it may be preferable to use liver slices rather than microsomes for metabolic screening in support of this program. Cross-comparison of metabolic turnover rates in various tissues from various species can be quite beneficial. In addition to providing information on potential rates and routes of metabolism, interspecies comparisons may help in choosing species to be used in efficacy and toxicology models. The use of *in vivo* methods to assess metabolism has advantages and disadvantages. On the plus side, data are being generated within a living species, allowing a full early PK profile for the compound to be assessed. However, the methods are generally of low throughput, are time-consuming, and can suffer from marked species differences. In this respect, *in vitro* studies can be used in conjunction with *in vivo* experiments in order to select the animal model with metabolism most similar to human.

Intrinsic Metabolic Clearance Using the concept of intrinsic metabolic clearance (CL_{int}), it has been demonstrated that *in vitro* metabolism rates for a selected set of model substrates correlated well with hepatic extraction ratios determined from isolated perfused rat livers (Rane et al., 1977). Thus, when measured under appropriate conditions, the intrinsic hepatic metabolic clearance (CL_{int}) of human and preclinical species can be predicted by measuring enzyme kinetics parameters: K_m (Michaelis–Menton constant) and V_{max} (maximum velocity) ($CL_{int} = V_{max}/K_m$) (Segel et al., 1993). However, it is time-consuming to measure K_m and V_{max}. In drug discovery, one can use a single substrate concentration ($\ll K_m$) to determine the parent drug disappearance half-life ($T_{1/2}$). The intrinsic hepatic metabolic clearance (CL_{int}) can be calculated based on scaling factors from microsomes as well as hepatocytes (Obach et al., 1999a). Then, the hepatic metabolic clearance (CL_h) and hepatic extraction ratio (E_h) can be calculated (Obach et al., 1999b). Integration of metabolic stability studies with the other high-throughput pharmacology screens and other ADME screens such as inhibition screening is essential. Solving a metabolic stability problem may not necessarily lead to a compound with an overall improvement in activity or even PK properties. For example, compounds with improved metabolic stability may be developed only to find out that absorption becomes a problem. Reduction in metabolic clearance may be accompanied by an increase in renal or biliary clearance of the parent drug. It is also possible that improvement of *in vitro* intrinsic clearance (CL_{int}) may result from self inhibition of one or more drug-metabolizing enzymes. This apparent improvement is only realized during the *in vitro* screening process whereas *in vivo*, the compound may exhibit saturable metabolism, nonlinear PK, or drug–drug interactions. Thus, it is advisable that *in vitro* metabolic stability data be integrated with inhibition screening.

Benefits of Enhancing Metabolic Stability Several benefits to be gained from enhancing metabolic stability have been reported (Ariens et al., 1982):

- Increased bioavailability and longer half-life, which in turn should allow lower and less frequent dosing, thus promoting better patient compliance.

- Better congruence between dose and plasma concentration, thus reducing or even eliminating the need for expensive therapeutic monitoring.
- Reduction in metabolic turnover rates from different species which, in turn, may permit better extrapolation of animal data to humans.
- Lower patient-to-patient and intra-patient variability in drug levels, since this is largely based on differences in drug metabolic capacity.
- Diminishing the number and significance of active metabolites, thus lessening the need for further studies on drug metabolites in both animals and man.

Strategies to Enhance Metabolic Stability To enhance microsomal stability, *in vitro* metabolism studies can be performed to confirm formation of metabolites as well as to provide quantitative analysis of major metabolites. After identifying moieties which contribute to activity (the pharmacophore) and other moieties necessary for activity, several modifications can be used to enhance metabolic stability. In general, metabolism can be reduced by incorporation of stable functions (blocking groups) at metabolically vulnerable sites. Substrate structure–activity relationships of metabolizing enzymes have to be accommodated within the structure–activity relationships of the actual pharmacological target. However, when it is appropriate to enhance metabolic stability in the molecular design, the following strategies have been used (Testa et al., 1990; Humphrey and Smith, 1992):

- Deactivating aromatic rings toward oxidation by substituting them with strongly electron-withdrawing groups (e.g., CF_3, SO_2NH_2, SO_3–).
- Introducing an *N*-t-butyl group to prevent N-dealkylation.
- Replacing a labile ester linkage with an amide group.
- Constraining the molecule in a conformation which is unfavorable to the metabolic pathway, more generally, protecting the labile moiety by steric shielding.
- The phenolic function has consistently been shown to be rapidly glucuronidated. Thus, avoidance of this moiety in a sterically unhindered position is advised in any compound intended for oral use.
- Avoidance of other conjugation reactions as primary clearance pathways would also be advised in the design stage in any drug destined for oral usage.
- Sometimes, the best strategy is to anticipate a likely route of metabolism and prepare the expected metabolite if it has adequate intrinsic activity. For example, often *N*-oxides are just as active as the parent amine, but will not undergo further N-oxidation.

Case studies from the literature for the above strategies to enhance metabolic stability in the molecular design are summarized in Table 4.1 (Genin et al.,

TABLE 4.1 Enhancement of Metabolic Stability through Structural Modification

Category	Approach/ Strategy	Enzyme, Pathway	Lead Compound
I. Lipophilicity and metabolic stability	Reduce logP and logD	NA	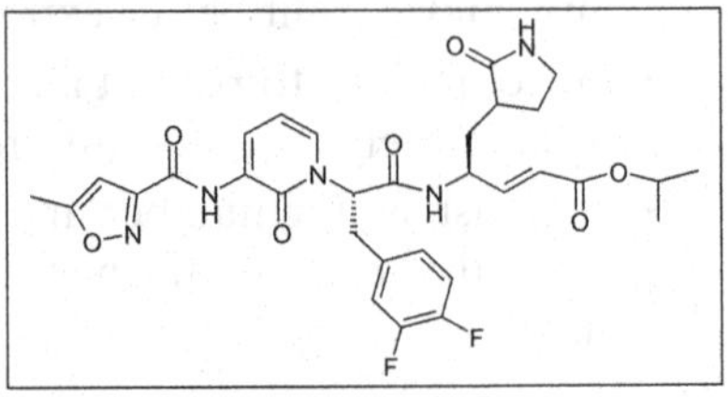 EC_{50} = 0.078 mM, clogP = 2.07, C7h = 0.012 μM
		NA	
	Introduce isosteric atoms or polar functional group	NA	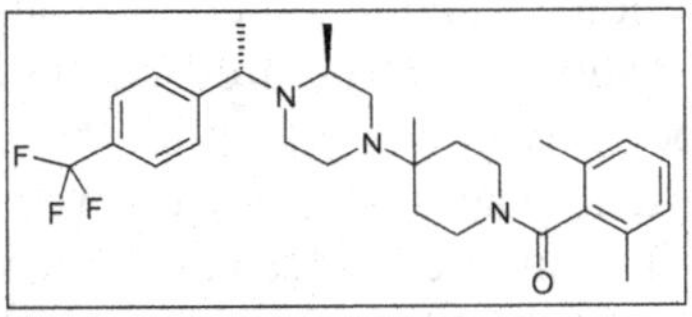 Ki = 1 nM, AUC_{0-6h} = 922 ng/mL h
		NA	
		NA	
		NA	

Optimized Compound	Experimental Model	Therapeutic Class	Reference
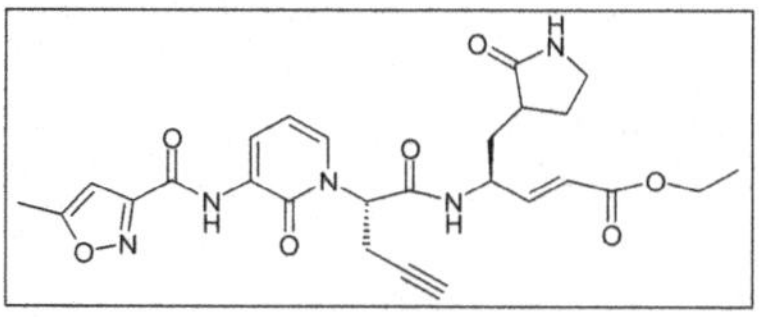 $EC_{50} = 0.058\,\mu M$, clogP = 0.18, C7h = 0.057 μM	Orally dosed monkey	3C protease inhibitor	Dragovich et al. (2003)
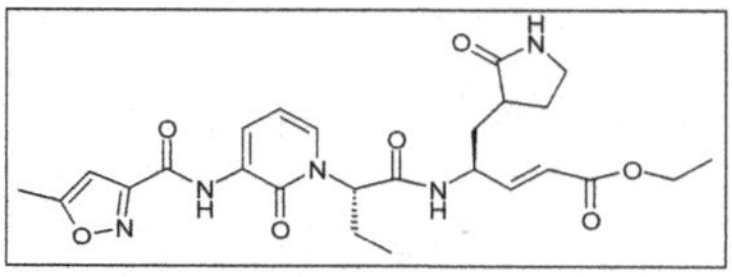 $EC_{50} = 0.047\,\mu M$, clogP = 0.66, C7h = 0.896 μM	Orally dosed monkey	3C protease inhibitor	Dragovich et al. (2003)
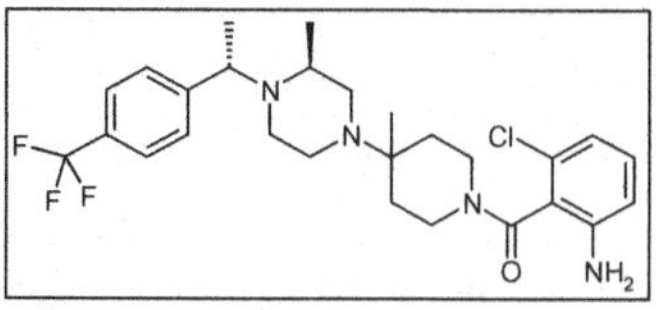 Ki = 5 nM, $AUC_{0\text{-}6h}$ = 1872 ng/mL h	Orally dosed rat	CCR5 antagonists	Tagat et al. (2001)
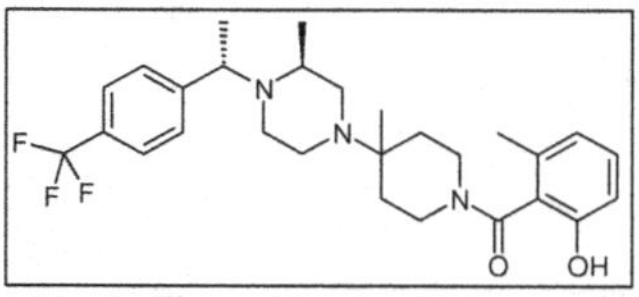 Ki = 5 nM, $AUC_{0\text{-}6h}$ = 2543 ng/mL h	Orally dosed rat	CCR5 antagonists	Tagat et al. (2001)
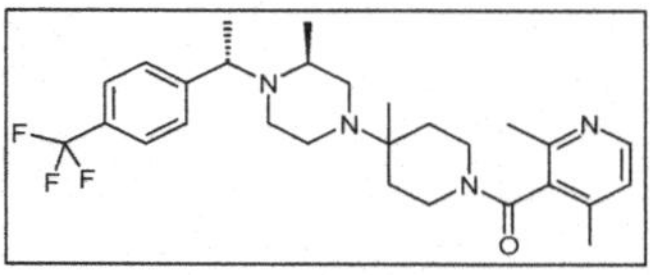 Ki = 2.3 nM, $AUC_{0\text{-}6h}$ = 3905 ng/mL h	Orally dosed rat	CCR5 antagonists	Tagat et al. (2001)
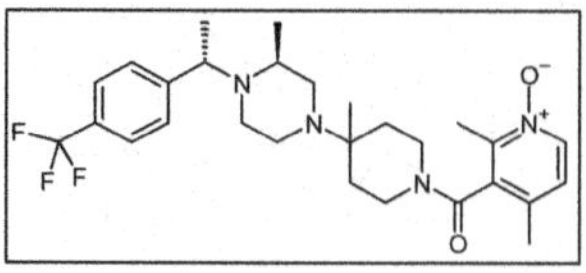 Ki = 7 nM, $AUC_{0\text{-}6h}$ = 9243 ng/mL h	Orally dosed rat	CCR5 antagonists	Tagat et al. (2001)

(*Contined*)

TABLE 4.1 (*Contined*)

Category	Approach/ Strategy	Enzyme, Pathway	Lead Compound
II. Metabolically Labile Groups	Remove or block the vulnerable site of metabolism	Benzylic oxidation	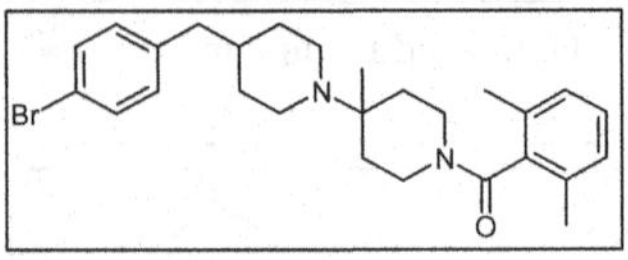 Ki = 66 nM, $AUC_{0\text{–}6h}$ = 40 ng/mL h
		Allylic oxidation 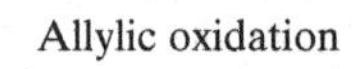	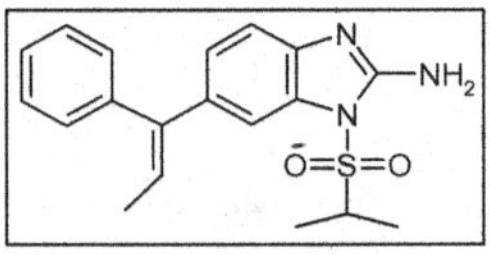 IC_{50} = 0.06 µg/mL, C_{max} = 14–140 ng/mL
		Phenyl oxidation	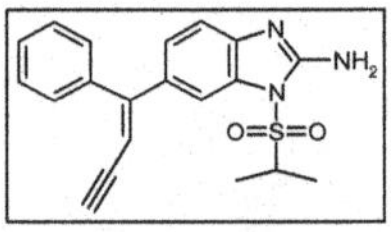 IC_{50} = 0.02 µg/mL, %F = 9
		N-oxidation	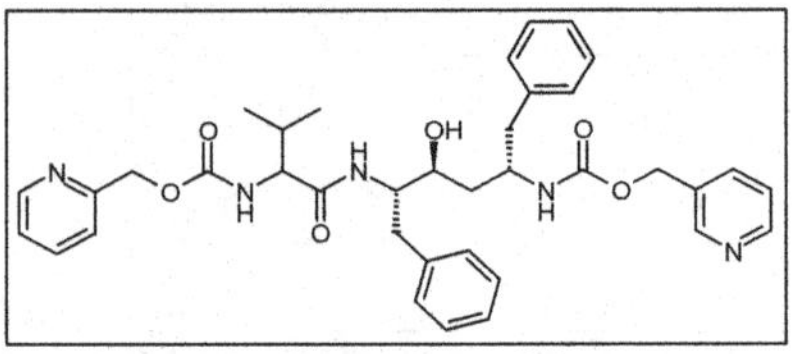 AUC = 1.98 µg.h/mL, %F = 26
		N-demethylation	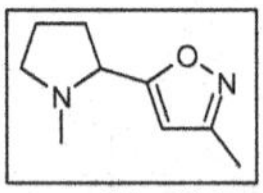 $t_{1/2}$ = 3 h, %F = 1.2
		N-dealkylation	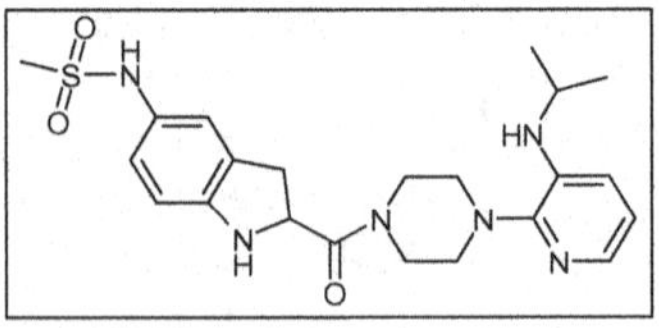 $t_{1/2}$ = 10.8 min
		Ester hydrolysis	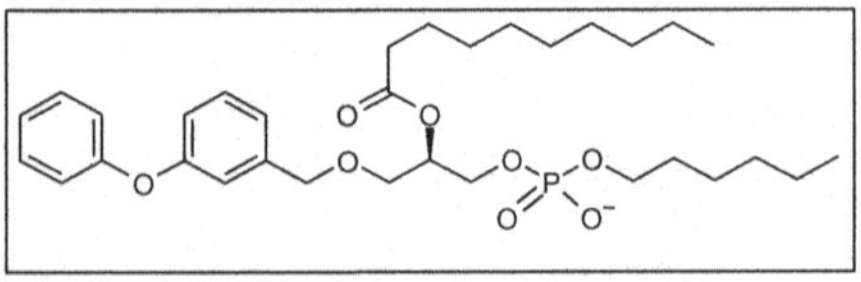 $t_{1/2}$ = 33 min, C_{max} = 465 ng/mL, %F = 4

Optimized Compound	Experimental Model	Therapeutic Class	Reference
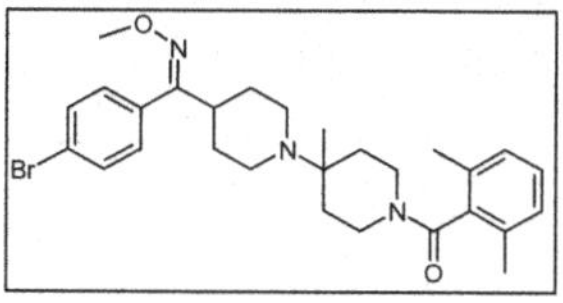 Ki = 2 nM, AUC_{0-6h} = 1400 ng/mL h	Orally dosed rat	CCR5 antagonists	
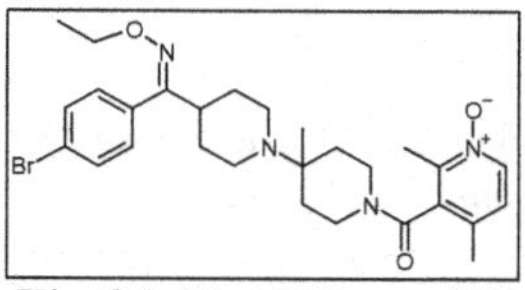 Ki = 2.1 nM, AUC_{0-6h} = 6500 ng/mL h	Orally dosed rat	CCR5 antagonists	Palain et al. (2002)
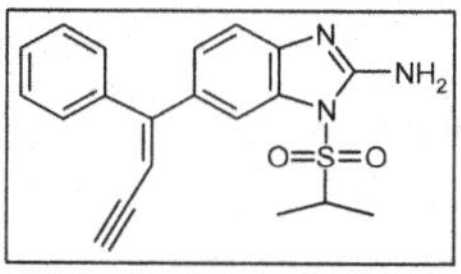 IC_{50} = 0.02 μg/mL, C_{max} = 70–300 ng/mL	Orally dosed monkey	Vinylacetylene antivirals	Victor et al. (1997)
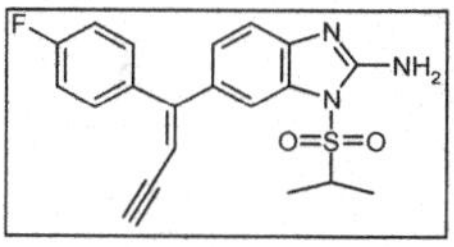 IC_{50} = 0.04 μg/mL, %F = 23	Orally dosed monkey	Vinylacetylene antivirals	Victor et al. (1997)
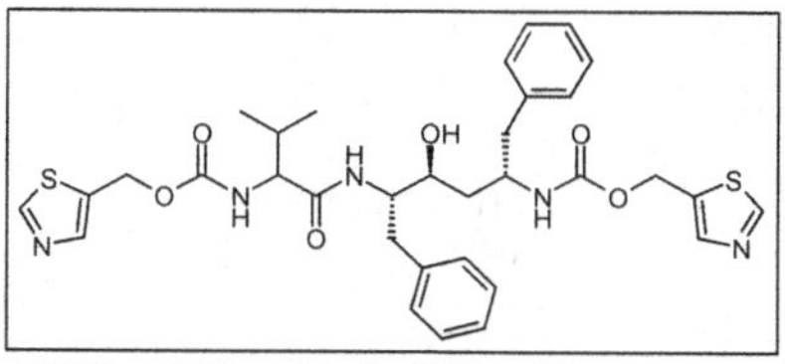 AUC = 4.24 μg.h/mL, %F = 47	Orally dosed rat	HIV protease inhibitors	Kempf et al. (1998)
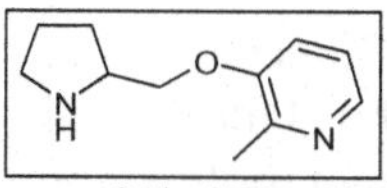 $t_{1/2}$ = 24 h, %F = 61.5	Dog liver slices	NAChR ligands	Lin et al. (1997)
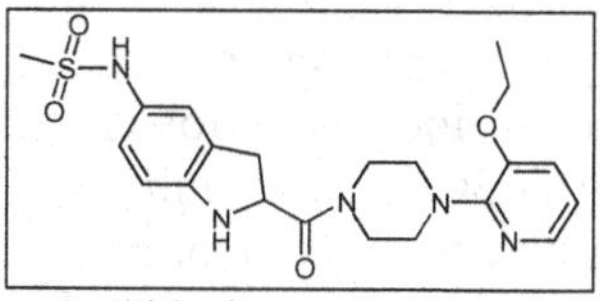 $t_{1/2}$ = 46.8 min	Rat liver microsomes	BHAP reverse transcriptase inhibitors	Genin et al. (1996)
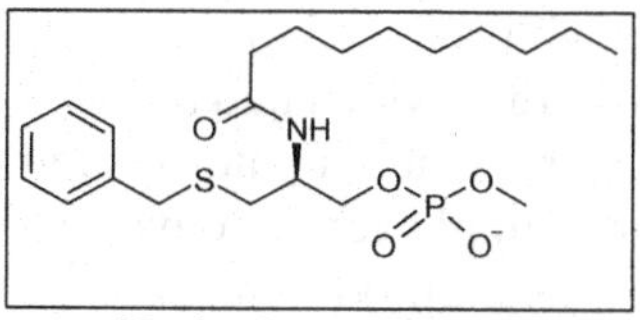 $t_{1/2}$ = 39 min, C_{max} = 3261 ng/mL, %F = 90	Rat blood and plasma, liver microsomes, and homogenate	Phospholipase A inhibitors	Blanchard et al. (1998)

(*Contined*)

TABLE 4.1 (*Contined*)

Category	Approach/ Strategy	Enzyme, Pathway	Lead Compound
		Amide hydrolysis	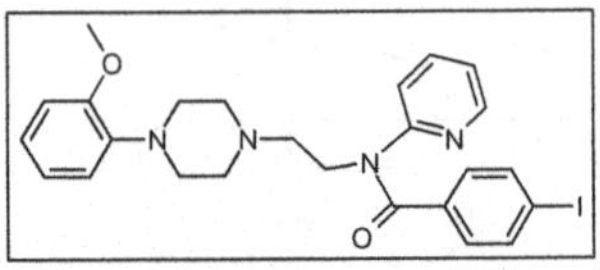 Ki = 0.2 nM, 40% and >60% degradation in human liver cytosole and microsomes, respectively
		Glucuronidation (effect of linker)	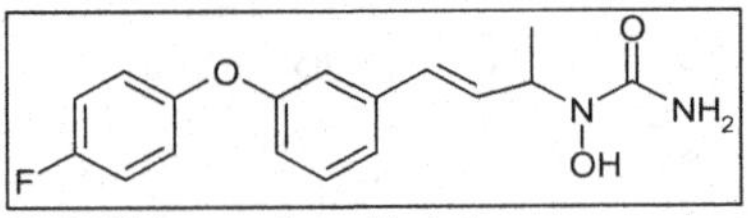 UDPGA rate (nmol/min/mg protein) = 0.19, $t_{1/2}$ = 4.7 h
		Glucuronidation (effect template)	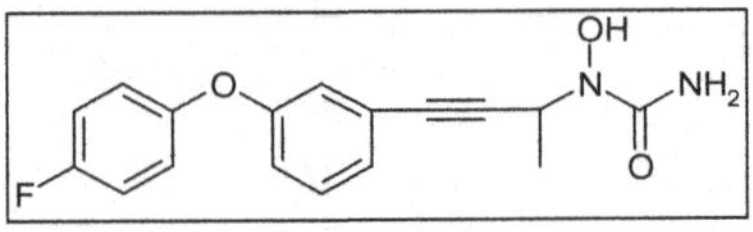 UDPGA rate (nmol/min/mg protein) = 0.05, $t_{1/2}$ = 5.5 h
		Glucuronidation (effect of stereochemistry	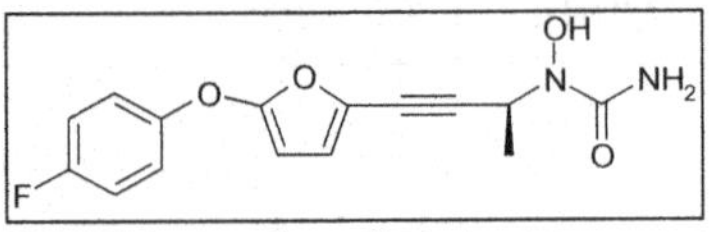 UDPGA rate (nmol/min/mg protein) = 0.02, $t_{1/2}$ = 7.7 h

1996; Bouska et al., 1997; Lin et al., 1997; Victor et al., 1997; Blanchard et al., 1998; Kempf et al., 1998; Zhuang et al., 1998; Tagat et al., 2001; Palani et al., 2002; Dragovich et al., 2003). One strategy for improving the metabolic stability of a compound is to reduce the overall lipophilicity (logP, logD) of the structure. This is due to the fact that the metabolizing enzymes generally have a lipophilic binding site and hence more readily accept lipophilic molecules. As exemplified in a Pfizer study of the human rhinovirus 3C protease inhibitor (Dragovich et al., 2003), the lead compound, which exhibited poor oral bioavailability *in vivo* in the monkey, was subjected to PK optimization with the benzyl group being identified as a position for reducing lipophilicity without compromising activity. Introduction of substituents with reduced lipophilicity led to the propargyl analog 2 and the ethyl analog 3, both of which had reduced clogPs when compared to the parent. These compounds demonstrated improved oral exposure in the monkey, highlighting a simple yet effective way to improve the metabolic profile of a compound. This method is not always

Optimized Compound	Experimental Model	Therapeutic Class	Reference
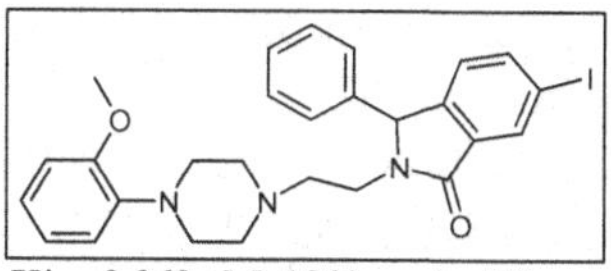 Ki = 0.069 nM, 10% and <5% degradation in human liver cytosole and microsomes, respectively	Human liver cytosole and microsomes	5-HT_{1A} receptor ligand	Zhuang et al. (1998)
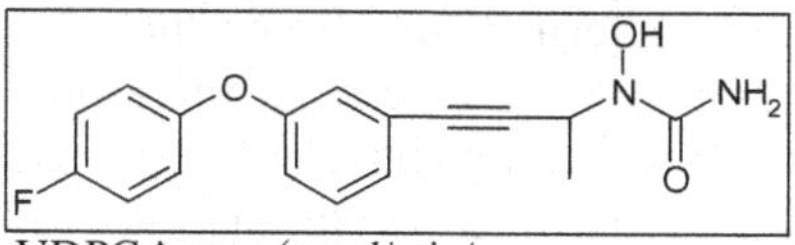 UDPGA rate (nmol/min/mg protein) = 0.05, $t_{1/2}$ = 5.5 h	Monkey liver microsomes and plasma	5-LO inhibitors	Bouska et al. (1997)
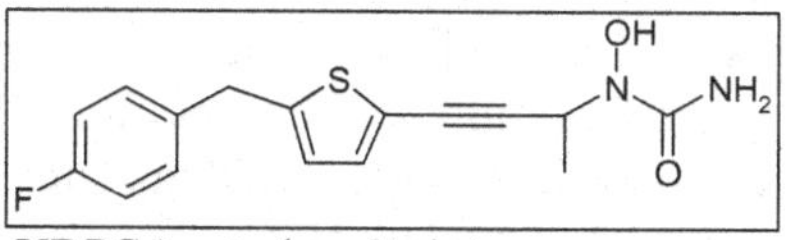 UDPGA rate (nmol/min/mg protein) = 0.012, $t_{1/2}$ = 14.5 h	Monkey liver microsomes and plasma	5-LO inhibitors	Bouska et al. (1997)
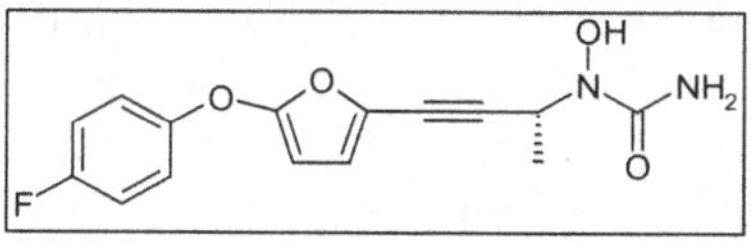 UDPGA rate (nmol/min/mg protein) = 0.01, $t_{1/2}$ = 8.7 h	Monkey liver microsomes and plasma	5-LO inhibitors	Bouska et al. (1997)

successful, however, as lipophilic groups are usually involved in binding to the biological target.

An alternative approach toward lowering the lipophilicity of a lead is to introduce isosteric atoms or functional groups into the molecule that impart increased polarity. This could be in the form of a heteroatom (e.g., a nitrogen introduced into a benzene ring to form the more polar pyridine) or instigating functional group transformations toward more polar functionalities (e.g., conversion of a ketone to the corresponding carboxamide). This strategy was utilized in a Schering-Plough program to enhance the pharmacokinetics of a CCR5 antagonist lead in their HIV-1 inhibitor program (Tagat et al., 2001). The benzamide of the lead compound was identified as a position for reducing lipophilicity, and changes were made to alter the substitution pattern of the ring. The more polar anthranilamide and salicylamide gave significant improvements in the rat oral blood levels, with the area under the curve (AUC) increasing substantially. However, the best results were achieved by introducing a heteroatom into the phenyl ring, with the more polar nicotinamide showing satisfactory oral blood levels in the rat. During this study, the authors noticed a single metabolite of nicotinamide rapidly forming in the plasma samples—they concluded this was due to N-oxidation. Thus, the pyridine *N*-oxide was prepared and found to be an excellent compound both in terms of potency and oral bioavailability in rat, dog, and monkey.

A much more elegant approach toward improving metabolic stability is to remove or block the vulnerable site of metabolism. For example, identification of sites potentially labile toward oxidation (e.g., benzylic or allylic positions) can be addressed by blocking the site of metabolism (introduction of a halogen onto the carbon atom involved, replacing the benzylic CH_2 with an isostere such as O). Using this approach, workers at Schering-Plough conducted a metabolism-driven optimization of a second-generation lead from the aforementioned CCR5 antagonist program (Palani et al., 2002). The lead compound exhibited good potency against the receptor, but showed very poor oral bioavailability in the rat. The benzylic position was identified as a metabolically labile site, and a variety of isosteres were introduced including O, CHOH, and C=O among others. Eventually, methoxime was identified as a potent, metabolically stable derivative and formed the next-generation lead compound.

The work of Victor et al. (1997) provides another example of blocking sites of potential oxidative metabolism to improve metabolic stability. Replacing a methyl group vulnerable to allylic oxidation with acetylene resulted in up to fivefold improvement in blood levels for monkeys. A further improvement of about twofold in oral bioavailabilty in the monkey was achieved when an acetylenyl moiety replaced the methyl and the *p*-fluoro substitution was made on the phenyl ring. It may be that the electron-withdrawing character of fluorine deactivates the aromatic ring toward metabolic oxidation. Or perhaps the somewhat lipophilic character of fluorine coupled with its ability to hydrogen bond provides a unique mixture of physical properties which assist in the absorption process.

Kempf and coworkers employed the principle of bioisosterism in their campaign to synthesize the potent HIV protease inhibitor ritonovir (Kempf et al., 1998). Pyridyl N-oxidation was a limiting route of metabolism in one of their early lead compounds. The authors embarked on a systematic search for other heterocyclic groups which would decrease the rate of metabolism while at the same time maintaining adequate solubility to ensure absorption as well as inhibitory activity. Ultimately, pyridyl groups were replaced with metabolism-resistant thiazole groups, resulting in an analog (ritonovir) that exhibited a nearly twofold improvement in bioavailability while maintaining potent inhibition of HIV protease. Lin et al. (1997) observed up to about a 50-fold improvement in potency at the nicotinic acetylcholine receptor when the *N*-desmethyl analog was used instead of N-methyl analog, which was vulnerable to N-demethylation. Genin and colleagues (Blanchard et al., 1998) realized a significant improvement of about 4-fold when they replaced a metabolically vulnerable 3-isopropylamino substituent on the lead compound with an ethoxy substituent. While one might reason that this substituent would also be rapidly metabolized, in this case it proved to be stable, at least when compared to the lead compound.

Functional groups can be open to other types of metabolism other than oxidation; it is well documented that amidases and esterases are present in the liver and can hydrolyze amides and esters, respectively. Additionally, Phase II metabolism where polar functionality (e.g., glucuronides, sulfates) is imparted onto a molecule in order to make it more water-soluble and hence easier to clear, can be targeted as a method for improving overall metabolic stability. Replacing a labile ester linkage with an amide group is known to impart added stability to esterase activity. In Blanchard's study (Blanchard et al., 1998), this can be illustrated by observing the 22-fold improvement in bioavailability in amide when compared to its corresponding ester.

During a metabolism-driven optimization of a series of new arylpiperazine benzamido derivatives as potential ligands for 5-HT1A receptors, Zhuang et al. (1998) attributed the low brain uptake in the human subject to rapid metabolism (amide hydrolysis), whereas in other species (rats and monkeys), the metabolic pathway (cleavage of the amide bond by liver microsomes) appears to be much slower. The authors reported that the cyclized amide derivatives provided better than twofold improvement in amide stability compared to the open-chain lead compound.

An extensive body of work on N-hydroxyurea inhibitors of 5-lipogenase related to zileuton provides an excellent opportunity to examine structure–activity relationships with respect to conjugation with glucuronic acid (Bouska et al., 1997). It was discovered early on in this investigation that glucuronidation of the N-hydroxyl moiety was an activity-limiting step, so it became of pivotal importance to this program to impart metabolic stability toward this conjugation. The N-hydroxyurea compounds were divided into three areas for structural modifications: the template, the linking group, and the pharmacophore. The N-hydroxyurea was identified in zileuton as an optimal pharmacophore for potency and selectivity; therefore, it was not modified in the search for metabolically stable compounds. The template and link components were the major focus of medicinal chemistry efforts to optimize *in vivo* duration in the cynomolgus monkey. This group systematically modified first the linkage group joining the template with the N-hydroxyurea pharmacophore and then the benzthiophene template. Each compound was screened for stability to glucuronidation of the hydroxyurea. Of the linkage moieties, compounds with acetylene linkage groups were generally found to have lower uridine diphosphate glucuronic acid (UDPGA) rates than any of the links tested. This lower UDPGA rate resulted in the longest *in vivo* duration in monkey. Apparently, the rigid conformation required by the acetylenic group disrupts binding in the active site of uridine diphosphate glucuronyltransferase (UDPGT) and diminishes conjugation.

Many variations of the template group were also tested in the UDPGA assay. Compounds containing the benzthiophene template found in zileuton demonstrated more rapid rates of glucuronidation. Conversely, biaryl templates such as phenoxyphenyl, phenoxyfuran, and phenoxythiophene demonstrated reduced rates of glucuronidation. These different metabolic stability results may be attributed to changes in lipophilicity and/or remote electronic effects.

Finally, the importance of the stereochemistry of the carbon center immediately adjacent to the hydroxyurea was investigated. N-hydroxyurea compounds containing a methyl branch on the linking group had already shown longer *in vivo* durations than their methylen analogs, which encouraged further work with this methyl branch series. The addition of the methyl group introduced a chiral center and resulted in a compound with separable *R*- and *S*-enantiomers. A difference of approximately twofold in the glucuronidation rate, depending on the configuration of the adjacent carbon center, was reported.

Review of Major Points Strategies and tools have evolved to enhance metabolic stability to enable researchers in the drug companies to design drugs that are safer and more robust for humans. Structural information on metabolites is a great help in enhancing as well as streamlining the process of developing new drug candidates, which in turn has great value in several important aspects of drug discovery and development. By improving our ability to identify both helpful and harmful metabolites, suggestions for structural modifications will optimize the likelihood that other compounds in the series are more successful. *In silico* and *in vitro* techniques are available to screen compounds for key ADME characteristics, which, when applied within a rational strategy, can make a major contribution to the design and selection of successful drug candidates. Structural modifications to solve a metabolic stability problem may not necessarily lead to a compound with an overall improvement in PK properties. Solving metabolic stability problems at one site could result in an increase in the rate of metabolism at another site, a phenomenon known as metabolic switching. Further, reduction in hepatic clearance may lead to increased renal or biliary clearance of a parent drug or inhibition of one or more drug-metabolizing enzymes. Therefore, it is imperative that *in vitro* metabolic stability data be integrated with other ADME screening.

4.3 (B) DRUG DESIGN STRATEGIES

The discussion in the first section should begin to make clear the importance for the drug designer, medicinal chemist, and metabolism scientist to produce the optimal combination of diverse properties such as activity, toxicity, and exposure. For any drug candidate to advance in the development process, these characteristics must be determined as early as possible, and then successfully manipulated to maximize safety and efficacy. In addition to optimization of metabolic stability, for the drug metabolism scientist, the task becomes to optimize PK properties such as plasma half-life, drug/metabolic clearance, metabolic stability, and the ratio of metabolic to renal clearance. The effort to achieve these positives must be balanced by the need to minimize or eliminate potential difficulties or dangers such as gut/hepatic-first-pass metabolism, inhibition/induction of drug metabolizing enzymes by metabolites, biologically active metabolites, metabolism by polymorphically expressed drug metabolizing enzymes, and formation of reactive metabolites. The optimal result will be a safer drug that undergoes predictable metabolic inactivation or even undergoes no metabolism. Among the approaches available to the drug design team, as they seek to meet the above goals, are active metabolites, prodrugs, and hard and soft drugs; these will be discussed in this section, with some examples from recent case studies.

4.3.1 Active Metabolites

Some drugs produce metabolites which are pharmacologically active; in some cases these metabolites surpass their parent drugs as candidates for development. Active metabolites are often indicated by an elevated level of pharmacological effect, which is revealed through PK tests on the parent drug candidate. These metabolites often are subject to Phase II metabolism and thus have better safety profiles. Table 4.2 contains the examples of 14-hydroxy-clarithromycin and morphine 6-glucuronide; they are more potent and are less toxic than the parent drugs clarithromycin and morphine, respectively (Dabernat et al., 1991; Bergeron et al., 1992; Burkhardt et al., 2002; Kopecky et al., 2004). Phase I metabolism of clarithromycin produces 14-hydroxy-clarithromycin. Clarithromycin is an important antibacterial drug in the treatment of community-acquired respiratory tract infections. In a double-blind, randomized, 2-period cross-over study, the pharmacokinetic properties of clarithromycin were determined after single and multiple doses in 12 healthy male volunteers. On day 1, the half-life was 3.7 h for clarithromycin and 6.2 h for its metabolite 14-hydroxy-clarithromycin. On day 7, the half-life was longer than on day 1 (5.1 h for clarithromycin and 10.1 h for 14-hydroxy-clarithromycin). Also, the *in vitro* activity of clarithromycin and its main metabolite 14-hydroxy clarithromycin against *Haemophilus influenzae* was evaluated. The 14-hydroxy metabolite was more active than the parent compound against *H. influenzae*. Microdilution broth MICs and MBCs of both clarithromycin and 14-hydroxy-clarithromycin were determined. For clarithromycin, the MIC50 was 4 mg/L and the MIC90 was 8 mg/L, while the hydroxy metabolite was 2–4-fold more active with an MIC50 and MIC90 of 2 mg/L. As a second example of working with an active metabolite, morphine 6-glucuronide is formed by Phase II metabolism of morphine. Morphine-6-glucuronide is more potent than morphine itself without side effects. Morphine-6-glucuronide is a major metabolite of morphine with potent analgesic actions. The PK of morphine and its active metabolite morphine-6-glucuronide were studied in humans, giving AUC_{0-12h} results of

TABLE 4.2 Metabolism-Driven Optimization of PK Properties by Structure Modification

Category	Pre-optimized Compound	Optimized Compound	Optimized Compound	Reference
Active metabolites	Morphine $AUC_{0-12} = 614$ ng.h/mL	Morphine-6-glucuronide, $AUC_{0-12} = 1672$ ng.h/mL		
	Clarithromycin $t_{1/2} = 3.7$ h day 1, $t_{1/2} = 5.1$ h day 7	14-hydroxy-clarithromycin $t_{1/2} = 6.2$ h day 1, $t_{1/2} = 10.1$ h day 7		
Prodrug	%F = 17 (rats), %F = 4 (monkeys)	%F = 30 (rats), %F = 23 (monkeys)		Stearns et al. (2002)

	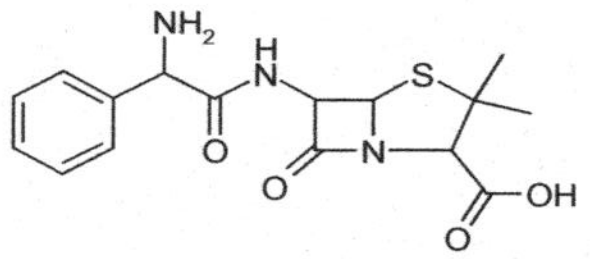 Ampicillin, mean peak serum level = 3.7 ug/mL, %F = 47–49	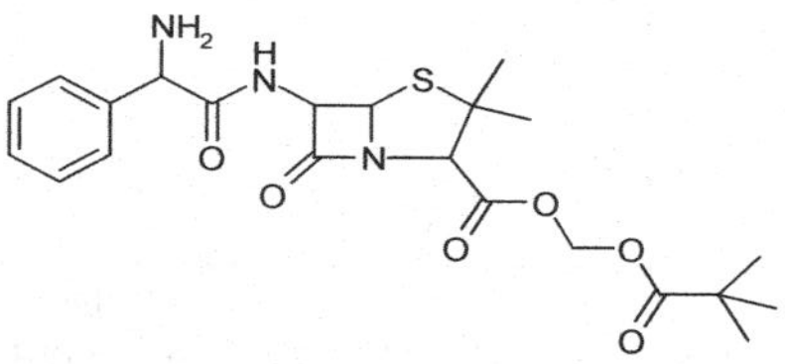 Pivampicillinmean, peak serum level = 7.1 ug/mL, %F = 82–89	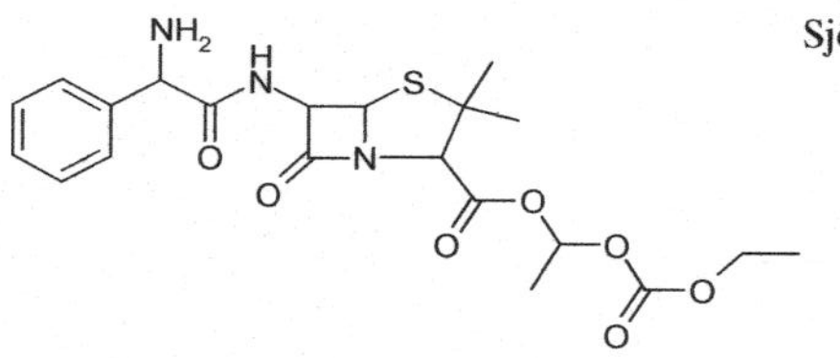 Bacampicillinmean, peak serum level = 8.3 ug/mL, %F = 87–95	Sjövall et al. (1978)
First-pass effect	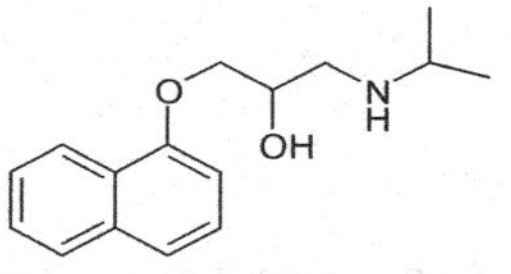 Propranolol AUC_{0-6} = 132 ng/mL.h	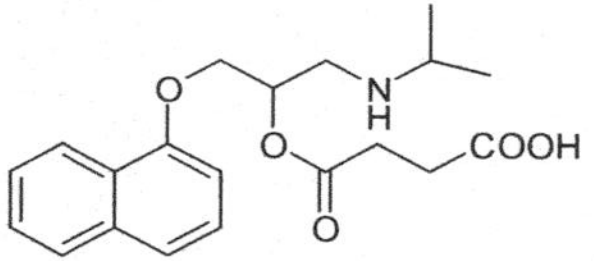 Hemisuccinate ester of propranolol AUC_{0-6} = 1075 ng/mL.h		
	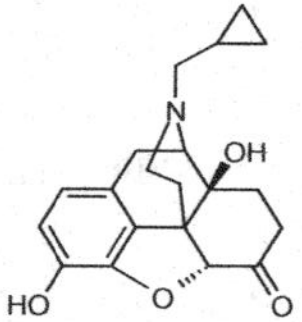 Naltrexon %F = 1.1	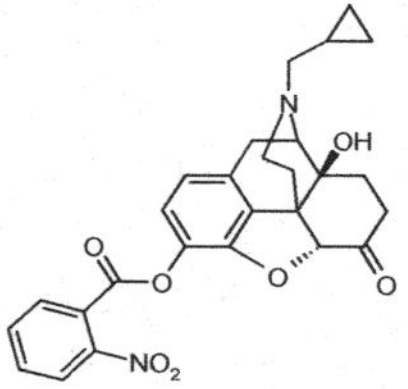 Anthranilate %F = 49.2	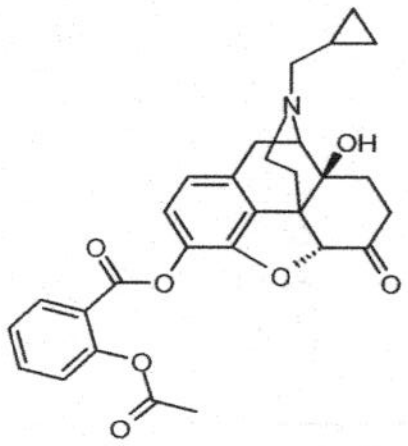 Acetylsalicylate %F = 31.0	Hussain et al. (1987)
	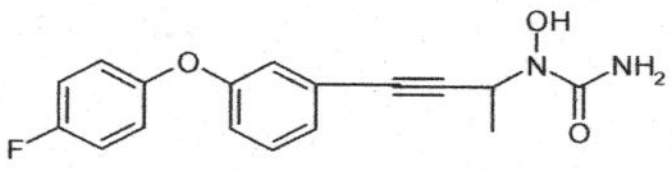 5-LO inhibitors $t_{1/2}$ = 5.5 h	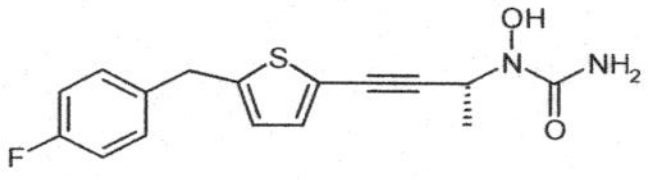 $t_{1/2}$ = 16 h		Bouska et al. (1997)

(*Contined*)

TABLE 4.2 (*Contined*)

Category	Pre-optimized Compound	Optimized Compound	Optimized Compound	Reference
Half-life	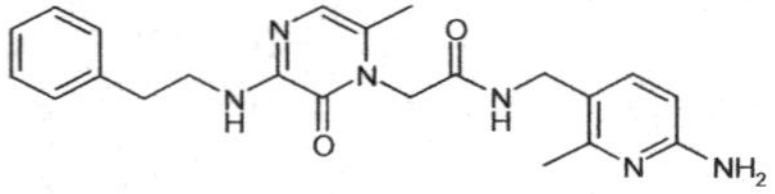 Thrombin $t_{1/2} = 3.8\,h$	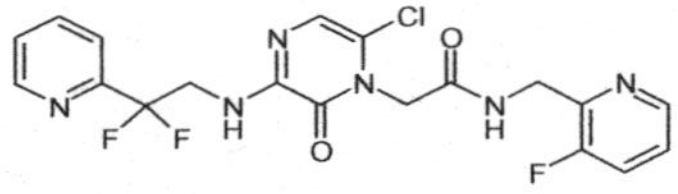 $t_{1/2} = 6.6\,h$	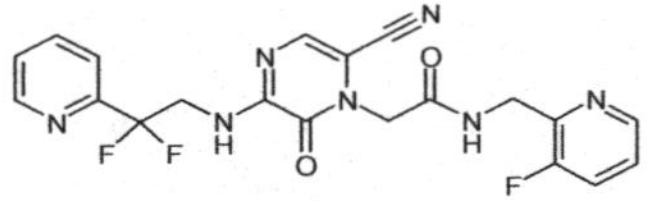 $t_{1/2} = 9.7\,h$	Burgey et al. (2003)
	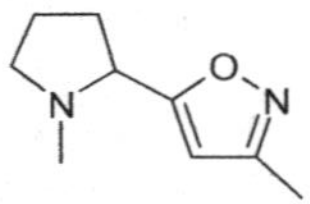 5 pyrollidine ABT-418 $t_{1/2} = 0.2\,h$	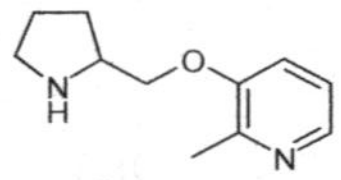 ABT-089 $t_{1/2} = 1.6\,h$		Lin et al. (1997)
PO and IV	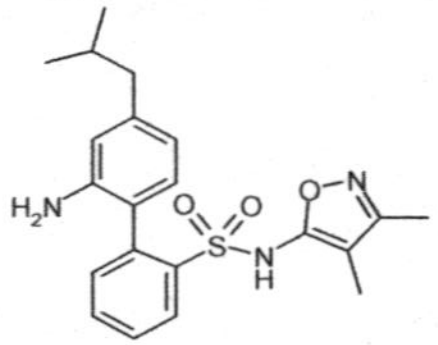 BMS 187308, $AUC_{0\text{-inf.}}$ (uM.h) = 343 (iv) and 166 (po) @100 umol/kg	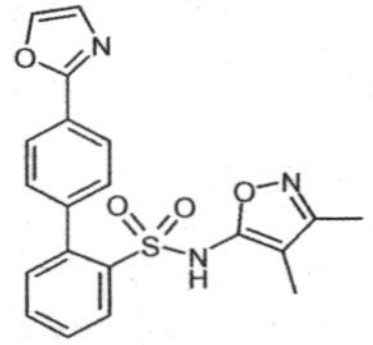 BMS 193884, $AUC_{0\text{-inf.}}$ (uM.h) = 840 (iv) @100 umol/kg		Humphreys et al. (2003)

614.3 ng.h/mL and 1672.3 ng.h/mL for morphine and morphine-6-glucuronide, respectively. Thus, it becomes clear that morphine-6-glucuronide is a highly potent metabolite, and there is strong evidence that it is important to the overall effect of morphine.

4.3.2 Prodrugs

The prodrug approach has been widely used in drug design. Prodrugs were discovered when it was demonstrated that the antibacterial agent protosil was active *in vivo* only when it was metabolized to the actual drug sulfanilamide. Prodrugs most commonly use either oxidative or reductive activation; for example, protosil is activated by reduction of its azo linkage to the amine sulfa drug. Another example of a prodrug is the antipyretic agent phenacetin, which provides its activity after conversion to acetaminophen. Carbamazepine is an anticonvulsant drug that is the metabolic precursor of the active agent carbamazepine 10,11-oxide. Minoxidil is a potent vasodilator that also induces hypertrichosis of facial and body hair (Buhl et al., 1990). Research data show that sulfation is a critical step for the hair-growth effects of minoxidil and that it is the sulfated metabolite that directly stimulates hair follicles.

As an example of the effort to improve PK properties through structure modification, the oral bioavailability of a 3-pyridyl thiazole benzenesulfonamide adrenergic receptor agonist (Table 4.2) was investigated in rats, dogs, and monkeys (Stearns et al., 2002). It was shown that the limited bioavailability of this compound was due to poor oral absorption in rats and monkeys (%F = 17 and 4 for rats and monkeys, respectively). A slight modification to the structure of this compound resulted in a significant improvement to its oral bioavailability. The linkage to the pyridine moiety was changed from the 3- to the 2-position so that the pyridyl-nitrogen atom was positioned to the hydrogen bond with the ethanolamine hydroxyl group; this minimized intermolecular interactions that may limit the oral absorption of this compound class. Bioavailability in rats and monkeys improved to %F = 30 and 23 for rats and monkeys, respectively.

The PK evaluation of various structural modifications of ampicillin is another example of how the prodrug strategy can improve PK properties (see Table 4.2). Pivampicillin, talampicillin, and bacampicillin are prodrugs of ampicillin, each of which is produced from the esterification of the polar carboxylate group to form lipophilic, enzymatically labile esters (Sjövall et al., 1978; Ehrnebo et al., 1979; Ensink et al., 1996). These prodrugs result in nearly total absorption, whereas that of ampicillin is <50%. A randomized cross-over study on 11 healthy volunteers to determine the pharamcokinetics of bacampicillin, ampicillin, and pivampicillin showed that the relative bioavailability of bacampicillin (%F = 87–95) and pivampicillin (%F = 82–89) was comparable, whereas ampicillin was less than two-thirds that of the others (%F = 47–49). Additionally, the mean of the individual peak concentrations in serum was 8.3 μg/mL for bacampicillin, 7.1 μg/mL for pivampicillin, and 3.7 μg/mL for ampicillin.

4.3.3 Hard and Soft Drugs

Hard drugs have been defined as drugs that are biologically active and non-metabolizable *in vivo*. Soft drugs are defined as those that produce predictable and controllable *in vivo* metabolism to form nontoxic products after they have achieved their therapeutic role. For example, quaternary ammonium compounds, such as benzalkonium chloride, are hard antibacterial agents. They are toxic to humans and animals, and their chemical stability makes them unsuitable for general environmental sanitation. Soft analogs of such hard antibacterial agents have been developed, and are less toxic. Although the soft analogs have been shown to possess antibacterial activity in *in vitro* studies, a problem for researchers is that it is likely that their *in vivo* activity will be hampered by their chemical instability. When considering toxicology and pharmacology, a safer drug is one that undergoes predictable metabolic inactivation or even undergoes no metabolism (Schuster et al., 1999; Fromigue et al., 2000; Bos et al., 2002).

4.3.4 Hard Drugs

The hard drug approach is very attractive to drug design teams because the problem of toxicity due to reactive or active metabolites is eliminated, and the pharmacokinetics also are simplified because the drugs are excreted primarily through either the bile or kidney. When excretion of a drug occurs mainly through the kidney, differences in its elimination between humans and animals will depend mainly on the renal function, which is a strong predictor of PK profiles. Examples of hard drugs include biophosphonates and certain ACE inhibitors. Because they are not metabolized in animals or humans, and are only eliminated through renal excretion, biophosphonates are very safe with no significant systemic toxicity. Similarly, ACE inhibitors undergo very limited metabolism and are exclusively excreted by the kidney. It is worth noting that biophosphonates and carboxyalkyldipeptide ACE inhibitors were not intentionally designed as hard drugs; rather, their hard qualities are the result of structural improvement.

4.3.5 Soft Drugs

Soft drugs are those that produce predictable *in vivo* metabolism which forms nontoxic products after achieving their therapeutic role. The soft drug approach also aims to produce safer drugs with an increased therapeutic index by integrating metabolism considerations into the drug design process (Bodor and Kaminski, 1980; Bodor, 1984; Bodor and Buchwald 2000, 2004; Huang et al., 2003). Because most oxidative reactions of drugs are produced by hepatic CYP enzyme systems that are often affected by age, sex, disease, and environmental factors, biotransformation becomes more complex and PK is variable. P450 oxidative reactions have the potential to form reactive intermediates and active metabolites which can produce toxicity. The distribution and clearance of such metabolites may be different from those of the parent drug. The design of soft drugs aims to avoid oxidative metabolism as much as possible and to use hydrolytic enzymes to achieve predictability and control of drug metabolism. There are a number of drugs currently on the market, such as esmolol, remifentanil, or loteprednol etabonate, which are products of the soft drug design approach. Other promising drug candidates are currently under investigation in a variety of fields, including possible soft antimicrobials, anticholinergics, corticosteroids, beta-blockers, analgetics, ACE inhibitors, antiarrhythmics, and others. An example of this concept in drug design is the isosteric soft analog of the hard antifungal drug cetylpyridinium chloride. While cetylpyridinium is quite effective against both Gram-positive and Gram-negative bacteria, it is also quite toxic (LD_{50} = 108 mg/kg). The soft analog has a metabolically soft spot (the ester group) built into the structure to replace the ethylene group for detoxification as well as 3-methyl imidazol derivative to replace the pyridine moiety. The soft nature of the new antimicrobial is low toxicity with LD_{50} = 4110 mg/kg.

4.4 PK PARAMETERS

To increase the chances of success for a drug's development, it is essential that a drug candidate has good bioavailability and a desirable half-life, while avoiding problems such as poor absorption and extensive first-pass metabolism. Therefore, an accurate measurement of the pharmacokinetic parameters and a good understanding of the factors that affect the pharmacokinetics will guide drug design. Below we will discuss strategies for structural modification of drug candidates to improve PK properties such as first-pass effect, half-life, and IV versus oral absorption.

4.4.1 First-Pass Effect/Prodrug

Table 4.2 shows three examples of how the first-pass effect can be minimized through structure modification. The first example, propranolol, shows the potential usefulness of the prodrug approach when a highly metabolized drug is protected from first-pass elimination in order to improve bioavailability. Oral dosage of propranolol produces a low bioavailability and a wide variation from patient to patient when compared to intravenous administration; this difference is attributed to first-pass elimination of the drug. Hemisuccinate ester of propranolol was selected as a potential prodrug with the hypothesis that propranolol hemisuccinate ester administration would avoid glucuronide formation during absorption and subsequently be released in the blood by hydrolysis. Following 80 mg po of propranolol hydrochloride and an equivalent dose of its hemisuccinate ester derivative, the AUC_{0-6h} was 1075 ng/mL.h and 132 ng/mL.h for hemisuccinate ester and propranolol, respectively (Garceau et al., 1978; Shameem et al., 1993; Imai et al., 2003).

A similar approach was successfully employed to enhance the bioavailabilty of naltrexon [17-(cyclopropylmethyl)-4,5 alpha-epoxy-3,14-

dihydroxymorphinan-6-one]. Naltrexon is used in the treatment of opioid addiction and absorbs well from the gastrointestinal tract. However, its systemic bioavailability is very poor, as it undergoes extensive first-pass metabolism when administered orally. Employing the prodrug strategy, a number of prodrug esters on the 3-hydroxyl group were prepared: the anthranilate, acetylsalicylate, benzoate, and pivalate. The oral bioavailability of these prodrugs was determined in dogs. The anthranilate and acetylsalicylate compounds exhibited the greatest enhancement of naltrexon bioavailability (45 and 28 times greater than naltrexon, respectively). No correlation was found between the rates of plasma hydrolysis and bioavailability. Naltrexone-3-acetylsalicylate hydrolyzed in human and dog plasma, with a fast deacetylation step to naltrexon salicylate followed by a slower hydrolysis step to naltrexon (Hussain et al., 1987; Reuning et al., 1989).

An extensive body of work on N-hydroxyurea inhibitors of 5-lipoxygenase related to zileuton provides an excellent opportunity to examine structure-activity relationships (SARs) with respect to conjugation with glucuronic acid (Bouska et al., 1997). In zileuton, the N-hydroxyurea moiety was identified as an optimal pharmacophore for potency and selectivity; therefore, it was not modified in the search for metabolically stable compounds. Medicinal chemistry efforts to optimize *in vivo* duration in the cynomolgus monkey were focused on the template and link components. In general, compounds with acetylene linker groups were found to have lower uridine 5-diphosphoglucuronic acid (UDPGA) rates than any of the links tested. Biaryl templates such as phenoxyphenyl, phenoxyfuran, and phenylthiophene also demonstrated reduced rates of glucuronidation and longer *in vivo* duration in monkeys. Finally, the importance of the stereochemistry of the carbon center immediately adjacent to N-hydroxyurea was investigated. It had already been shown that N-hydroxyurea compounds containing a methyl branch on the linking group had longer *in vivo* durations than their methylene analogs, which led to additional research on this methyl branch series. By adding the methyl group, a chiral center was created, and resulted in a compound with separable *R*- and *S*-enantiomers. The glucuronidation rate showed a ~twofold difference, depending on the configuration of the adjacent carbon center. The result was the identification of a clinical candidate, ABT-761, which demonstrated the longest *in vivo* duration ($t_{1/2} = 16$ h) compared to that of zileuton ($t_{1/2} = 0.4$ h).

4.4.2 Half-Life

Most drugs are administered as a fixed dose given at regular intervals to achieve therapeutic objectives. The plasma half-life is a measure of the duration of drug action. Thus, the half-life of drugs in plasma is one of the most important factors that determine the selection of a dosage regimen. Shorter half-life means more frequent dosing, which tends to reduce the likelihood that the patient will maintain the regimen. The two factors affecting half-life are volume of distribution and elimination clearance. An increase in the first or a reduction of the second will improve the drug's half-life. Modification of the chemical structure is one viable approach to slowing a drug's clearance.

A metabolism-based approach to the optimization of the 3-(2-phenethylamino)-6-methylpyrazinone acetamide template for thrombin is an example of these efforts, which resulted in the identification of several potent thrombin inhibitors with high levels of oral bioavailability and long plasma half-lives. Thrombin is vital to hemostasis because it mediates conversion of fibrinogen to fibrin and activates platelets. By developing the aminopyrazinone acetamide thrombin-inhibitor template, an efficacious, orally bioavailable pre-optimized compound with plasma half-life of 3.8 h was identified. This compound was found to metabolize through oxidation at the benzylic positions and Phase II conjugation of the amino group. These structural areas of metabolism represented major means of clearance of the drug. Their elimination enabled generation of compounds with improved pharmacokinetics. Both the chloro ($t_{1/2}$, 6.6 h) and cyano ($t_{1/2}$, 9.7 h) modifications imparted significantly improved half-lives versus the pre-optimized compound (Burgey et al., 2003).

As a second example, structural modification of ABT-418 was required to achieve further improvements in the margin of safety and to identify compounds with favorable oral bioavailability. ABT-418, an analogue of (S)-nicotine in which the pyridine ring is replaced by the 3-methyl-5-isoxazole moiety, has been shown to possess cognitive-enhancing and anxiolytic-like activities in animal models with an improved safety profile compared to that of nicotine (Lin et al., 1997). One shortcoming of ABT-418 was its very poor bioavailability (%F = 1.2), with a short plasma half-life ($t_{1/2}$ = 0.21 h). Research on structural modification led to the identification of ABT-089, 2-methyl-3-(2(S)-pyrrolidinylmethoxy)pyridine, with a vastly improved oral bioavailability (%F = 61.5) with $t_{1/2}$ = 1.6 h.

4.4.3 Oral Absorption and Intravenous Dose

The oral bioavailability of a drug is defined as the fraction of an oral dose of the drug that reaches the systemic circulation. An important consideration for orally administered drugs is that the entire blood supply of the upper gastrointestinal tract passes through the liver before reaching the systematic circulation. This means that the drug may be metabolized by the gut wall and/or the liver during the first passage of drug absorption. A drug with high metabolic clearance is always subject to an extensive first-pass effect, thus reducing its bioavailability. By contrast, administering a drug intravenously ensures that all of it enters the blood. Rapid injection will promptly achieve elevated concentrations of the drug in the blood; infusion at a controlled rate produces a constant concentration which can be maintained for any desired length of time. The lipophilicity of a drug is important to its membrane permeability as well as metabolic activity. Higher lipophilicity for a drug generally results in higher permeability and greater clearance, and thereby higher first-pass metabolism.

Bristol-Myers Squibb Pharmaceutical Research Group (Humphrey and Smith, 1992; Humphreys et al., 2003) presents a good example of working with structure–metabolism relationships to produce favorable pharmacokinetic properties. They identified an initial lead (2′-amino-*N*-(3,4-dimethyl-5-isoxazolyl)-40-(2-methylpropyl)[1,10-biphenyl]-2-sulfonamide) for endothelin (ET) receptor antagonists. However, testing in preclinical animal species and human *in vitro* systems revealed the compound was extensively metabolized due to oxidative biotransformation. The site of metabolism of the candidate was determined and structure–activity and structure–metabolism studies were aimed at optimizing this structural class by finding more metabolically stable analogues that maintained potency. This resulted in the identification of an analogue (*N*-(3,4-dimethyl-5-isoxazolyl)-40-(2-oxazolyl)[1,10-biphenyl]-2-sulfonamide) with improved *in vitro* properties; further studies revealed that the new compound had improved pharmacokinetic properties (Table 4.2).

4.4.4 Review of Major Points

The clinical success of a drug candidate depends upon desirable ADME/TOX properties. An accurate measurement of the pharmacokinetic parameters and a good understanding of the factors that affect the pharmacokinetics will help to guide drug design. By seeking a balance based upon the rule of three (exposure–activity–toxicity), and taking pharmaceutical research beyond the initial candidate or parent drug, drug metabolism scientists have opened up a whole new front in the effort to develop safer and more efficacious drugs. New approaches, evolving technologies, and the wisdom gained from examination of previous difficulties and failures have aided the drug development team to focus their efforts. Improvement in the drug discovery process will be facilitated most by learning how and when to use each of these tools and approaches to design drugs to achieve the maximum benefit, and by increasing cooperation and dialogue among the disciplines involved in the entire process. The following suggestions should be considered when developing a new drug:

- Considering the risks and costs of new drug development, it is critical to eliminate high-risk compounds as early as possible. Fast and reliable *in silico* screens are needed to filter out problematic molecules at the earliest stages of discovery.
- The metabolism of new drugs should be studied *in vitro* before the initiation of any clinical studies. Early information on *in vitro* metabolic processes in humans, such as the identification of the enzymes responsible for drug metabolism and sources of potential enzyme polymorphism, can be useful in the design of clinical studies, particularly those that examine drug–drug interactions. High metabolic liability usually leads to poor bioavailability and high clearance, and formation of active or toxic metabolites will have an impact on the pharmacological and toxicological outcomes.

- Appropriate structural modifications to the drug candidate help to optimize the metabolic liability, drug–drug interactions with co-administered drugs due to inhibition and/or induction of drug metabolism pathways, and the rate and sites of metabolism of new chemical entities by drug metabolizing enzymes.
- Good *in vitro* activity does not correlate to good *in vivo* activity unless a drug has a good bioavailability and a desirable duration of action. These qualities are essential to successful development.
- In the case of a candidate that shows high potency *in vitro* but does not metabolize well, first, it is important to determine if the drug is available to the target. Should the exposure be acceptable, but pharmacological effect is absent, the next step is to determine if a high level of protein binding is limiting the free concentration. Should protein binding be the limiting variable, efforts should be directed to improve the free concentration by eliminating the source(s) of protein binding.
- To avoid exclusion of good compounds, the selection of animal species and the experimental design of studies are important in providing a reliable prediction of drug absorption and elimination in humans. A good compound could be excluded on the basis of results from an inappropriate animal species or poor experimental design.
- Drug candidates should have little or none of the following: gut/hepatic-first-pass metabolism, inhibition/induction of drug-metabolizing enzymes, biologically active metabolites, metabolism by polymorphically expressed drug-metabolizing enzymes, and formation of reactive metabolites. Also, it is important to have the most desirable plasma half-life and ratio of metabolic to renal clearance.
- One of the most important keys to successful drug design and development is finding the right combination of multiple properties such as activity, toxicity, and exposure. To design good drug candidates, the rule of three should be applied regardless of the therapeutic index.

4.5 (C) REDUCING TOXICITY

Toxicity problems, especially those which may only occur under unusual or idiosyncratic conditions during the late stages of drug development, are one of the most devastating surprises for pharmaceutical companies. Variations in human drug metabolizing enzymes can produce subtle evidence of potential toxicity, or none at all, during preclinical safety studies. Such problems are also unlikely to show up in all but the largest clinical trials, but if the side effects are serious, can result in product withdrawal. There are indications that some substructures found in drugs can form reactive metabolites that are involved in toxicities in humans. Reactive metabolites are unstable, and are intermediates to more stable metabolites. Table 4.3 shows several examples

TABLE 4.3 Examples of Chemical Structures Activating to Produce Toxic Metabolites

Chemical Class	Bio-transformation	Toxic Metabolite	Compound		Biological Effects	Reference
			Name	Clinical Use		
Quinone	Oxidation	Quinone-type	Tacrine	Alzheimer's disease	Hepatic toxicity	Madden et al. (1993)
			Troglitazone	Treat Type II diabetes	Hepatic toxicity	Kassahun et al. (2001); Tettey et al. (2001)
			Minocycline	Antibiotics	Hepatic toxicity Lupus-like syndrome	Shapiro et al. (1997)
			Acetaminophen	Analgesic agent	Hepatic toxicity	Isaacson et al. (1999)
			Aminosalicylic acid	Inflammatory bowel disease	Lupus-like syndrome Pancreatic toxicity Hepatic toxicity Renal toxicity	Liu et al. (1995)
			Amodiaquine	Treat malaria	Hepatic toxicity Agranulocytosis	Ruscoe et al. (1995)
			Phenytoin	Anticonvulsant	Drug-induced hypersensitivity Teratogenicity	Munns et al. (1997)
			Carbamazepine	Anticonvulsant	Teratogenicity	Ju et al. (1999)
			Vesnarinone	Phosphodiesterase inhibitor	Agranulocytosis	Uetrecht et al. (1994)

			Prinomide	Anti-inflammatory	Agranulocytosis	Parrish et al. (1997)
			Estrogens	NSAID	Breast cancer Uterine cancer	Mohsin et al. (2004)
			Tamoxifen	NSAID	Endometrial cancer	Fan et al. (2001)
			Fluperlapine	Antipsychotic agent	Agranulocytosis	Mann et al. (1987)
Aryl nitro	Reduction	Nitroso	Tolcapone	Parkinson's disease	Liver toxicity	Borges et al. (2000)
			Chloramphenicol	Antibiotic	Aplastic anemia Bone marrow toxicity	Yunis et al. (1980)
			Dantrolene	Muscle relaxant	Liver toxicity	Utili et al. (1977)
			Nimesulide	COX-2 inhibitors	Liver toxicity	Merlani et al. (2001)
Nitrogen-containing aromatic	Oxidation	Nitrenium ion Free radical	Clozapine	Antipsychotic agent	Agranulocytosis Liver toxicity Myocarditis	Idanpaan-Heikkila et al. (1977); Haack et al. (2003)
			Aminopyrine	Painkiller	Agranulocytosis CNS toxicity	Uetrecht et al. (1995)
			Dipyrone	Painkiller	Agranulocytosis	Sabbaga et al. (1993)

(*Contined*)

TABLE 4.3 (*Contined*)

Chemical Class	Bio-transformation	Toxic Metabolite	Compound		Biological Effects	Reference
			Name	Clinical Use		
Aryl amines	Oxidation to hydroxylamine	Nitroso	Sulfamethoxazole	Antibacterial agent	Hepatotoxicity Agranulocytosis Lupus-like syndrome Skin rashes	Cribb et al. (1992)
			Dapsone	Antiparasitic	Agranulocytosis Flu-like syndrome Hemolytic anemia Methemoglobinemia	Coleman et al. (1994)
			Procainamide	Cardiac antiarrhythmic	Lupus-erythematosis Agranulocytosis Fever	Woosley et al. (1978)
			Nomifensine	Antidepressant	Hemolytic anemia Allergic reactions	Salama et al. (1983)
			Sulfasalazine	Ulcerative colitis	Abnormal liver function Decreased blood counts Allergic reactions	Senturk et al. (1997)
			Aminoglutethimide	Breast cancer	Skin rashes Fever Agranulocytosis Thrombocytopenia Liver toxicity	Stuart-Harris et al. (1984)

Michael acceptors	Hydrolysis Oxidation	Aldehyde Co-A conjugate	Felbamate	Anticonvulsant	Aplastic anemia Liver toxicity	Thompson et al. (1996); Dieckhaus et al. (2002)
			Terbinafine	Antifungal agent	Bone marrow toxicity Liver toxicity Skin rashes	Gupta et al. (1998)
			Valproic acid	Anticonvulsant	Liver toxicity	Grillo et al. (2001)
			Mianserin	Antidepressant	Agranulocytosis	Inman et al. (1988)
			Leflunomide	Inflammatory arthritis	Liver toxicity Agranulocytosis	Jardine (2002)
Carboxylic acids	Glucuronidation	Acyl glucuronides	Diclofenac	NSAID	Liver toxicity Agranulocytosis	Ware et al. (1998)
			Zomepirac	NSAID	Liver toxicity	Bailey et al. (1996)
			Ibufenac	NSAID	Liver toxicity	Prescott et al. (1986)
			Bromfenac	NSAID	Liver toxicity	Moses et al. (1999); Skjodt et al. (1999)
			Benoxaprofen	NSAID	Liver toxicity	Halsey et al. (1982)
			Indomethacine	NSAID	Bone marrow toxicity	Godessart et al. (1999)

of drugs that undergo metabolic activation and cause adverse reaction in humans, which have been withdrawn from the market or restricted in use with toxicity warnings. Clearly, a drug candidate which does, or may, metabolize to such substructures would increase the risk of failure or withdrawal.

4.5.1 Importance of Reactive Intermediates

One key to success in clinical development is to minimize reactive intermediates in drug metabolism. There is a growing consensus that idiosyncratic drug reactions (IDRs) have enormous consequences for patients and the pharmaceutical industry; it is estimated that they account for ~5% of all hospital admissions and occur in 10–20% of hospital inpatients (Li et al., 2002; Matzinger et al., 2002; Uetrecht et al., 2002; Nassar et al., 2004a). Of the new prescription drugs approved in the United States during the period between 1975 and 2000, 10.2% acquired a new "black box" warning or had to be withdrawn from the market because of IDRs (Bakke et al., 1995; Lasser et al., 2002). Idiosyncratic drug toxicity is generally believed to be a phenomenon that cannot be readily evaluated experimentally because it is a rare event (usually <1 in 5000 cases), making clinical trials impractical since an extremely large patient population would be required. The current data suggest that IDRs are human-specific events that may not be detectable with experimental animals currently used. IDRs may have a complex pathology and can lead to a large number of symptoms, ranging from nonspecific rashes to specific organ damage, such as agranulocytosis or cholestasis. Immunological reactions can affect several organs including the liver, kidney, or lung, or they may have systemic effects. Hepatotoxicity is the most common type of idiosyncratic reaction. The time to onset of liver reactions varies from several days to almost 1 year, suggesting that the rare cases of liver injury may be caused by a metabolic idiosyncrasy (Li et al., 2002). Immune system responses may account for between 3 and 25% of all IDRs. For example, the antimalarial drug amodiaquine produces life-threatening agranulocytosis and hepatotoxicity in approximately 1 in 2000 patients during prophylactic administration. The unexpected occurrence of IDRs during late clinical trials has led to severe restrictions on use or failure to launch, and in the case for launched drugs, even withdrawal from the market. During clinical development, toxicity of drug candidates accounts for a significant portion of attrition (~20%). Some examples of drugs withdrawn from the market due to IDRs are troglitazone, practolol, benoxaprofen, ticrynafen, zomepirac, and nomifensine. While all pharmaceutical companies consider IDRs to be a significant issue and are making efforts to predict them, the elusive nature and mechanisms of IDRs hinders the development of a clear and universal approach (Richard et al., 1998; Greene et al., 2002; Uetrecht et al., 2002; Nassar et al., 2004a).

In this section, reactive metabolites and their effects on the safety profile of new drug candidates will be discussed, along with recently developed tools for identifying and determining their potential toxicity. The use of these tools is evaluated at several stages of the process, along with their potential for improving drug safety. Also, some of the chemical substructures which form reactive metabolites and are involved in toxicities in humans are discussed.

4.5.2 Idiosyncratic Drug Toxicity and Molecular Mechanisms

Adverse Drug Reactions From a clinical perspective, adverse drug reactions (ADRs) can be classified as type A (predictable) and type B (unpredictable or idiosyncratic) reactions. Type A reactions, which account for ~80% of all ADRs, are dose-dependent and are predictable based on the pharmacology of the drug. Type B, or IDRs, cannot be predicted on the basis of the pharmacology of the drug and lack simple dose dependency. Type B reactions, which include anaphylaxis, blood dyscrasias, hepatotoxicity, and skin reactions, may result from a combination of genetic predisposition and environmental factors, and are reviewed elsewhere (Uetrecht et al., 2002).

Types of Reactive Metabolites There is significant evidence to indicate that some substructures found in drugs can form reactive metabolites that are involved in toxicities in humans. These substructures include arylacetic and arylpropionic acids, anilines, anilides, hydrazines, hydantoins, quinones, quinone methides, nitroaromatics, heteroaromatics, halogenated hydrocarbons, some halogenated aromatics, chemical groups that can be oxidized to acroleins, and medium-chain fatty acids. The thiazolidinedione ring, a relatively new substructure, can form reactive metabolites that may cause hepatotoxicity (Nassar et al., 2004a). Reactive metabolites are unstable, and are intermediates to more stable metabolites. For the purpose of this discussion, it is sufficient to categorize the three major types of metabolites as electrophiles, polarized double bonds and free radicals. Previously published reports have provided significant evidence that chemically reactive metabolites may be responsible for serious forms of toxicity, including cellular necrosis, mutagenesis, carcinogenesis, teratogenesis, hypersensitivity, and blood disorders (Figure 4.1).

Hypotheses for Idiosyncratic Reactions Several hypotheses have been proposed to explain potential mechanisms for IDRs (Uetrecht, 2008; Li and Uetrecht, 2010). These include the hapten, the P-I and the danger hypotheses, which are discussed briefly below. Alternatively, the theory of tolerance toward an immune reaction may explain why the majority of individuals are not susceptible to IDRs.

The hapten hypothesis proposes that modification of an endogenous protein by a reactive metabolite or directly by a reactive parent drug generates a "foreign" protein that, in some cases, leads to an immune-mediated adverse reaction. Immune-mediated mechanisms have been proposed for several drugs including halothane, phenytoins, sulfonamides, and tienilic acid. Failure to downregulate harmful immune responses due to a "foreign" protein may

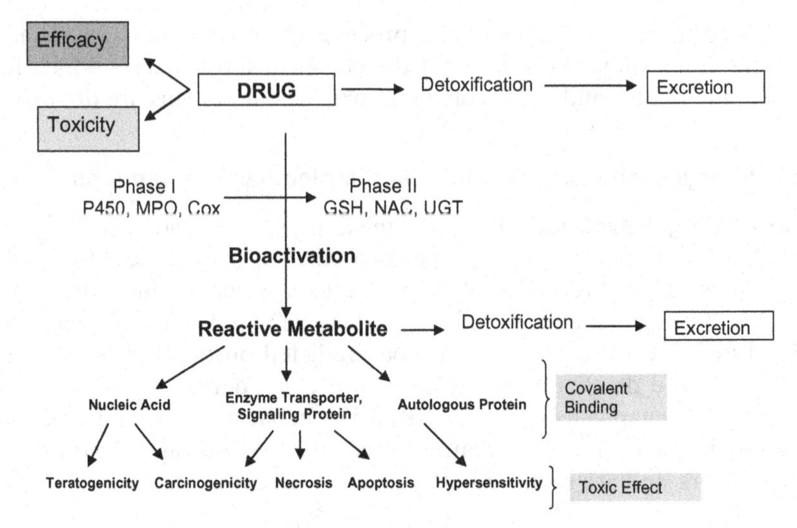

Figure 4.1 Reactive intermediate theory and idiosyncratic reactions/toxic effects.

cause IDRs in susceptible individuals as described in the danger hypothesis (Uetrecht, 2008; Li and Uetrecht, 2010).

The P-I hypothesis provides a new explanation for the occurrence of IDRs in the absence of a reactive intermediate. It is proposed that T cells do not necessarily bind to protein, but may reversibly bind to major histocompatibility complex/T-cell receptors (MHC/TCRs). There is a lack of evidence linking such direct interactions with immune system responses (Uetrecht, 2008; Li and Uetrecht, 2010).

The danger hypothesis proposes that, rather than the "foreignness" of a reactive metabolite triggering the immune system, cell damage produced by reactive metabolites results in a "danger signal." If the reactive metabolites are not toxic enough to cause cell injury, or if the target cells are resistant to the stress induced by reactive metabolites, protein adducts will have a lower probability of escaping the target cells and inducing a specific immune response. This could be one of the reasons why not all drugs that form reactive metabolites cause IDRs, and not all people are susceptible to these reactions (Nassar et al., 2004a).

Currently, the risk assessment of a given toxic effect in humans is usually based on a safety margin of the compound, which is the ratio of the no observable adverse effect level (NOAEL) in the most sensitive species and the expected therapeutic dose in man. The NOAEL is typically derived from a rodent and non-rodent species. Clinical and pathological measurements are the gold standard for identification of organ toxicity in animals. The nonclinical studies are conducted in healthy animals, and the risk assessment is extrapolated to healthy and health-compromised patients. This traditional approach to risk assessment has performed reasonably well in predicting most ADRs;

however, it has poor predictability for IDRs (Li and Uetrecht, 2010). Therefore, strategies need to be developed or adapted from current drug-safety assessment methods to consider reactive intermediates in the overall assessment of the potential risk profile during drug discovery and development.

Considerations To illustrate how to reduce or eliminate reactive metabolites, consider drug-induced adverse reactions, particularly IDRs (referred to as type B reactions). Such reactions are a major issue because, given current techniques and approaches, they often go undetected until late in the process, or even after the drug has been released onto the market, and the consequences are disastrous for all concerned. IDRs mediated through a reactive metabolite may be associated with several mechanisms (Nassar et al., 2004a; Li and Uetrecht, 2010). Currently, there is no general approach or "one-size-fits-all" screen that addresses idiosyncratic reactions, because of the difficulties in understanding the mechanisms of these reactions and in accurately predicting clinical results. Some of the strategies under consideration to improve our understanding of these mechanisms and to develop safer drugs include:

- Avoidance or flagging of chemical functional groups that are known to cause toxicity during drug design (e.g., aromatic and hydroxy amines, phenols, epoxides, acyl halides, acyl glucuronides, thiopenes, furans, fatty acid-like compounds, hydroxylated metabolites, and quinines).
- Development of suitable *in vitro* and *in vivo* systems to elucidate the role of short-lived, potentially toxic metabolites in the pathogenesis of idiosyncratic toxicity.
- Identification of chemical functional groups those are associated with low or no toxicity, and development of more metabolically stable drugs to potentially avoid metabolic interactions.

IDRs mediated through a reactive metabolite may be associated with several mechanisms, and it is unlikely that one test or screen will accurately predict all such mechanisms. Although there is no evidence that safer drugs will be identified by addressing the following questions in the drug discovery/development stages, it is probable that drugs without these potential liabilities may have a better safety profile:

- Does the candidate have the potential to form reactive metabolites based on chemical "structure alerts"?
- Does the compound form reactive metabolites in liver, blood, or skin tissue or cells?
- Is the binding of the drug >50 pmol/mg of microsomal protein?
- Does the candidate form reactive intermediates that are able to "travel" and react covalently with other tissues?

- What proteins are affected and what are the effects of the modified proteins?
- Which genes are affected? Could these affected genes generate a potential IDR?
- What cellular functions are affected by the reactive intermediate?
- Bioaccumulation in liver?
- Glutathione depletion?
- Drug–drug interactions?
- Ames and/or micronucleus positive responses?

It remains to be determined whether the answers to these questions will allow the pharmaceutical industry to reduce or eliminate IDRs. Since the major organs involved in IDRs are the liver, skin, and bone marrow, preclinical approaches should consider these target organs to study the effects of reactive intermediates. Thus, a sound strategy should evaluate three major elements: reactive intermediate characterization, covalent binding, and biological impact and/or function.

4.6 KEY TOOLS AND STRATEGIES TO IMPROVE DRUG SAFETY

The exact mechanism(s) for IDRs are unknown, and because many drugs form reactive metabolites, screening processes may eliminate potentially good candidates (false negatives). Current safety screening methods with animal models do not accurately predict IDRs, according to the Multinational Pharmaceutical Company Survey and the outcome of the International Life Sciences Institute Workshop. Many tools are in the exploratory stage, encompassing computational approaches and experimental assays to predict the formation of reactive metabolites *in vivo* and their role in IDRs (Nassar et al. 2004a,b). For example, during *in vitro* testing, if significant protein binding is found, an attempt is made to discover the chemical basis for the binding so that it can be designed out of the structure. Another potential technique is to trap reactive intermediates by incubating them in human liver microsomes. To evaluate the biological effects of reactive intermediates initially, methods for cytotoxicity testing including tests for cell viability, membrane integrity, protein synthesis, DNA synthesis, glutathione (GSH) level, apoptosis, free radical production, lipid peroxidation, enzyme inhibition, and other enzymatic activities due to oxidative stress are being developed for early drug discovery. Results suggest that these types of screens can be used to differentiate known toxicants from relatively safe drugs. Below we will summarize the advantages and disadvantages of the current tools and relate them to the problems facing those in drug discovery and development, and attempt to synthesize these tools and strategies into an alternative strategy.

4.6.1 *In-Silico* Screens as Filtering Process

A promising new trend in early drug discovery is the collaboration between medicinal and computational chemists and drug metabolism scientists seeking to identify and eliminate potential toxic chemical groups. Structure–activity relationships are well-known as a means to identify several toxic end points. One way to improve screening for such potentially toxic compounds would be to increase the use of *in silico* testing methods at earlier stages in drug discovery (Nassar et al., 2004a,b). Several programs such as TOPKAT, CASE/MULTI-CASE, DEREK, HazardExpert, and OncoLogic are commercially available for the prediction of mutagenicity and carcinogenicity based on chemical structure. These available commercial systems for mutagenicity and/or carcinogenicity prediction differ in their specifics, yet most fall into two major categories. One is automated approaches that rely on the use of statistics for extracting correlations between structure and activity. The others are knowledge-based expert systems that rely on a set of programmed rules distilled from available knowledge and human expert judgment. The advantages of *in silico* techniques include time and money savings, reduced use of laboratory animals, and the ability to rapidly screen large numbers of structures even before synthesis occurs. This high capacity improves prioritization in early discovery for toxicology testing and highlighting toxophores for easy identification. Because it may be difficult to avoid some of these functionalities in the design of new compounds, in some cases, their presence should be considered as a "structural alert" for the drug metabolism scientist. In turn, studies can be conducted at an earlier stage to determine whether the compound in question undergoes metabolism at the site to generate a potentially toxic intermediate. However, *in silico* methods still cannot completely replace conventional toxicity testing. These systems are not designed/validated to screen for potential IDRs but could be useful as a first step in eliminating compounds with other potential toxic effects.

4.6.2 *In Vitro* Assays

Most *in vitro* assays are amenable to high-throughput screening (HTS), and their use is preferred in early drug discovery. To try to identify reactive metabolites *in vitro*, human tissue, cells or purified/recombinant enzymes have been used under similar conditions to those in typical drug metabolism studies. Since liver, blood, and skin are the most common targets for the majority of IDRs, these systems have been considered. The human liver and blood systems are the most metabolically active in forming reactive intermediates, while the skin has a lower metabolic activity. Although these *in vitro* assays can identify reactive metabolites, it has been difficult to extrapolate the results to *in vivo* settings. Many drugs have reactive intermediates and *in vitro* assays may result in a significant number of false negatives for IDRs. Metabolic models of cytotoxicity and methods to assess cell death have been reviewed elsewhere

(Uetrecht, 2008). These methods of testing cell viability can be useful to evaluate potential toxic mechanisms. For hepatocellular necrosis, hepatocyte culture and crude cytotoxicity or cell death end points (i.e., enzyme leakage and dye exclusion) can be considered toxicologically relevant. Hepatocyte incubation/cultures have received the most attention (Nassar et al. 2004a), while fresh hepatocytes are of limited use. Cryopreserved human hepatocytes are commercially available, and technological improvements in preservation will facilitate studies to further understand the mechanisms of IDRs.

4.6.3 Reactive Intermediates: Phase I and Phase II

Phase I and/or Phase II enzymes can be involved in drug bioactivation. Several Phase I enzymes are recognized as important in forming reactive intermediates such as CYPs, myeloperoxidases (MPOs), cyclooxygenases (COXs), aldehyde oxidase, and flavin-containing monooxygenases (FMOs) (Nassar et al., 2004c). In theory, any enzyme is capable of forming reactive intermediates, and the involvement of CYPs has received most attention. Metabolism by FMOs has a lower potential for the formation of reactive metabolites (except for certain S or N groups) because substrates do not bind directly to the enzyme and electrons are not transferred, avoiding formation of free radicals. In general, Phase II metabolism, involving enzymes such as glucuronyl transferases, sulfatase, and GSH transferases, are less frequently recognized as being involved in reactive intermediate formation, despite being responsible for the formation of many reactive intermediates. The study of acyl glucuronides has attracted much interest, as their reactivity is strongly related to covalent binding and potential toxicity.

Trapping Reactive Intermediates Reactive electrophile screenings generated from *in vitro* experiments in several tissues (i.e., liver microsomes) have been designed to react with GSH and allow for subsequent analysis of the GSH adducts by liquid chromatography–mass spectrometry (LC-MS) (Nassar et al., 2004b). Several types of reactive metabolites (epoxide, arene oxide, quinone, quinone imine, quinone methide, iminoquinone methide, nitroso, nitrenium ion, nitro reduction, nitro radical, iminium ion, free radical, S-oxidation, Michael acceptor, *S*-oxide, aliphatic aldehyde, or hydrolysis/acetylation) can be trapped in the presence of either GSH or an equimolar mixture of *N*-acetyl-cysteine and *N*-acetyl-lysine. This method is relatively simple and rapid, and has been implemented for HTS to identify reactive metabolites generated by bioactivation with Phase I and Phase II enzymes. A preliminary evaluation of this method was conducted using 20 commercially available compounds with known toxicological profiles in liver microsomes at substrate concentrations of 10 μM. The results indicated that the method is unlikely to produce false negative responses because relatively safe compounds did not generate GSH conjugates, while 8 of the 10 compounds that are known to generate reactive metabolites resulted in positive responses. The two compounds that are known to generate reactive metabolites, but did not produce positive responses in the current assay, were valproic acid and phenytoin. The reason for this was that the formation of the valproic acid reactive metabolite (2,4-dieneVPA) requires P450-oxidation and β-oxidation catalyzed by a mitochondrial coenzyme A (CoA)-dependent process rather than microsomal enzymes (Nassar, 2009). The reactive metabolite of phenytoin may be a free radical instead of an epoxide that cannot form a stable GSH conjugate.

This method was further evaluated using 43 compounds as positive controls and 16 compounds as negative controls. The results indicated that 40 of 43 compounds tested gave positive results as expected from the literature. Those producing false negatives were felbamate, trimethoprim, and sulfamethoxazole. The potential reason for this was a relatively low *in vitro* concentration used in the assay (100 μM). On the other hand, all 16 of the compounds used as negative controls were found to be negative as expected, indicating that this method may be useful in determining the potential for a compound to form reactive intermediates. Although feasible for HTS, the strategy has limited applications to trap the most stable GSH adducts. Those unstable GSH adducts can spontaneously regenerate GSH and the reactive metabolite. A highly reactive metabolite can react with an active site residue of the enzyme that forms it and thus may not be trapped by GSH. If the reactive metabolite is a free radical, it will most likely abstract a hydrogen atom from GSH rather than react with it. Alternative trapping reagents may be used for more reactive intermediates (Nassar et al., 2004b).

Since thousands of compounds must be screened in a typical pharmaceutical industry setting, it is almost impossible, and very expensive, to have all of them radiolabeled. Using radiolabeled reagents such as GSH and potassium cyanide (KCN) is easier and is more cost-effective to obtain an accurate measurement of adduct formation. The reactive intermediates may be formed in other subcellular fractions such as liver S9, mitochondria, and cytosol, and thus, screening for reactive intermediates in these fractions will minimize false negatives. As an example, valproic acid forms reactive metabolites in subcellular mitochondria, as explained above. A comparison was done using *in vitro* cellular and subcellular models for identifying drug candidates with the potential to produce reactive metabolites. For purposes of evaluating these models, six known compounds currently in the clinic with dose limitations due to hepatotoxicity were selected. Labetalol, acetaminophen, niacin, iproniazid, 8-hydroxyquinoline, and isoniazid were incubated separately with human liver mitochondria, S9, microsomes, cytosol, and hepatocytes. Liquidchromatography–quadrupole time-of-flight–mass spectrometry (LC-QTOF-MS) methods were used to elucidate metabolic profiles of selected compounds. The Phase II metabolites were in good agreement with previously published results. Without supplementation of reagents in subcellular fractions, Phase II metabolites were not detected. Hepatocytes also formed GSH conjugates, but the relative abundance was lower, indicating that the best model to identify potential reactive metabolites may involve the subcelluar fractions. Using the results for Acetaminophen, for example, when reactive intermediates are produced, they can be trapped by GSH, while in the case of labetalol, no reactive intermediates were produced, and subsequently, no GSH adducts were detected. More work needs to be done to improve our knowledge of, and hence techniques used for, screening with the use of radiolabeled reagents, which would give an accurate measurement of the amount of reactive intermediates. Also, we should investigate reactive intermediate formation in subcellular fractions, which will aid in reducing false results.

Trapping Free Radicals Techniques for trapping free radicals are well established. Free radicals resulting from reduction and oxidation can be trapped with spin-trapping reagents (Timmins et al., 1999; Nassar, 2009). For example, a method for trapping free radicals was used on hydrazine analogs and their derivatives that are responsible for hemolytic and hepatotoxic events, presumably via the alkyl or aryl free radicals that oxidize essential cysteinyl residues in proteins or covalently react with biomacromolecules. Another example is the bioactivation of phenytoin to a free radical species, proposed to be mediated by COX-1, that can be trapped with the spin trapping agent α-phenyl-*N*-t-butylnitrone (PNB) *in vitro* in embryo cell cultures. The techniques to trap free radicals need further evaluation to develop high-throughput methodology and establish their role in IDRs.

Trapping Iminium Ions Trapping iminium ions with cyanide is also a well-known technique. The technique can be implemented as a screen for the detection of reactive iminium intermediates by trapping them with radiolabeled cyanide. With (S)-nicotine as the reference compound, the extent of radiolabeled cyanide incorporated into the test compounds can be quantified. If compounds have higher cyanide incorporation than (S)-nicotine, it is considered indicative of the formation of iminium intermediates. However, the iminium ions can also form GSH adducts depending on their relative stability. For example, U-89843 incubated in liver microsomes formed *N*-acetylcysteinyl and GSH adducts in NADPH-supplemented rat liver microsomes supporting a bioactivation pathway potentially involving an iminium intermediate (Pritsos et al., 1985; Gorrod et al., 1991; Zhao et al., 1996a,b; Sanders et al., 2000). The reactive metabolites of both DMP 406 and mianserin reacted with a range of nucleophiles, but in many cases the reaction was reversible. The best nucleophile for trapping these reactive metabolites was cyanide (Gorrod et al., 1991). Taken together, the above results suggest the need for a combination of trapping methods to evaluate the formation of reactive intermediates.

4.7 PEROXIDASES

Peroxidases are present in polymorphonuclear granulocyte (PMN) and play a significant role in the bioactivation of many drugs in humans. Several *in vitro* systems have been used to try to mimic bioactivation processes *in vivo* (i.e., myeloperoxidase (MPO)/H_2O_2/Cl_2, HOCl, chloroperoxidase, horseradish peroxidase (HRP)/H_2O_2). A preliminary evaluation of the nucleophile-activated system was conducted to assess its ability to predict drug-induced agranulocytosis. In this study, DMP-406 (a clozapine analog), which caused agranulocytosis in dogs, was used as the testing agent, with clozapine and mianserin as positive controls, and olanzapine as a negative control. Clozapine is thought to cause agranulocytosis via a reactive nitrenium ion metabolite produced by neutrophils. It has been demonstrated that the products of clozapine oxidation by HRP/H_2O_2 induced apoptosis in neutrophils at therapeutic concentrations (Miyamoto et al., 2003). The major reactive intermediate of DMP-406 is an imine ion in activated neutrophils and is similar to the reactive intermediate responsible for mianserin-induced agranulocytosis. DMP-406 did not increase apoptosis at concentrations <50 μM, while mianserin increased apoptosis at a concentration 10 μM above its therapeutic concentration, indicating that this assay lacks predictability (resulting in false negatives) for both drugs. Olanzapine increased apoptosis at the same concentration as clozapine (1 μM), but this concentration is above the therapeutic concentration of olanzapine, consistent with *in vivo* results. There was no increase in apoptosis with any drug in the absence of HRP/H_2O_2, which forms the reactive intermediates of these drugs. These preliminary results indicate that this assay may be unable to reliably predict the ability of different types of drugs to cause agranulocytosis and that different drugs may induce agranulocytosis by different mechanisms. Drugs that form nitrenium ion intermediates might be better predicted by this system.

4.8 ACYL GLUCURONIDATION AND *S*-ACYL-COA THIOESTERS

Phase II biotransformation is generally considered to be a bioinactivation pathway, but some drugs become bioactivated during this phase. Since acyl glucuronide metabolites of some carboxylic acids might cause fatal IDRs in some patients, much attention has been focused on the formation of reactive intermediates through glucuronidation and the methods to screen these conjugates to select for safer drugs. Some acyl glucuronides are reactive intermediates, which bind covalently to protein by mechanisms that may or may not result in the cleavage of the glucuronic acid moiety. Carboxylic acids can be metabolized by alternative pathways: (1) by glucuronide conjugation to form *O*-acyl glucuronides; and (2) by CoA conjugation to form *S*-acyl-CoA thioesters. Both pathways result in the formation of potentially reactive, electrophilic intermediates, due to the reactivity of the carbonyl carbon. The electrophilic carbon center can react with nucleophilic targets on macromolecules, forming covalent adducts with biomacromolecules, for example, proteins. Three major factors have been suggested to determine the reactivity of acyl glucuronides *in vivo*. The first is the relative stability of a given acyl glucuronide in aqueous buffer at pH 7.4/37°C. Some compounds with short, first-order half-lives (<0.5 h) result in a relatively high concentration of drug–protein adducts in human tissues and are associated with a high risk for causing IDRs in patients. In contrast, acyl glucuronides with longer half-lives (>5h) have rarely been reported to cause IDRs. Secondly, the degree of substitution at the α-carboxyl carbon of the aglycone (an increase in degree of substitution leads to a decrease in acyl glucuronide reactivity). Thirdly, the overall drug exposure (including the up-concentration of the acyl glucuronides) in particular tissue compartments. A recent example of apparent success using this screening strategy is the assessment of telmisartan 1-*O*-acyl glucuronide, the principal metabolite of telmisartan in humans. Telmisartan 1-*O*-acyl glucuronide exhibited a long, first-order degradation half-life of 26h compared with the short half-life of diclofenac (0.5h), suggesting telmisartan has a low potential for covalent binding. The current literature indicates a low incidence/absence of ADRs for telmisartan, supporting the potential use of this screening approach.

S-Acyl-CoA thioesters have received significant attention as reactive intermediates involved in IDRs. They can be formed with carboxylic acid drugs, and the resultant thioesters can be much more reactive than the acyl glucuronides due to the nature of the thioester bond. Similarly to acyl glucuronides, the reactivity of synthetic *S*-acyl-CoA thioesters at pH 7.4/37°C may be an important indicator of their potential to covalently bind to biomolecules. The covalent binding of seven structurally different carboxylic acid drugs for GSH was compared at pH 7.4/37°C. The results indicated that hydrolysis rates might be a good predictor for GSH conjugation, and similarly to acyl glucuronides, substitution at the α-carbon to the thioester bond affects reactivity (ISSX 2003). An HTS assay could be easily implemented to assess reactivity or stability of *S*-acyl-CoA thioesters, but additional studies are necessary to further evaluate the role of this metabolic pathway in IDRs.

4.9 COVALENT BINDING

Several methods are available to detect or quantify covalent binding of drugs and their metabolites to macromolecules, including radiochemical and immunological methods, and protein analysis (LC-MS, matrix-assisted laser desorption ionization time-of-flight [MALDI-TOF], and quadrupole time-of-flight [Q-TOF]). A large number of studies indicate that there is a good correlation between drug-induced toxicities and protein covalent binding (Nassar et al., 2004c). The alkylation of certain proteins has been suggested to lead to formation of neoantigens that may trigger immunomediated hepatotoxicity. Immunochemical techniques have been used to identify the protein adducts of hepatotoxic compounds. In an effort to understand the significance of *in vitro* covalent binding, studies have been carried out in animals, where GSH adduct and/or protein adducts were detected in biological fluids (i.e., in samples of bile). Subsequent experiments with radiolabeled compound can reveal more detail on the propensity of the reactive intermediate to covalently bind cellular proteins *in vitro* and *in vivo*. From a quantitative viewpoint, it has been suggested that as a rule-of-thumb, compounds with potential toxic profiles have a binding affinity of 50-pmol/mg microsomal protein; however, in order to quantify the covalent binding, radiolabeled drug is required. An example of the use of this approach in drug discovery has been presented recently using tritiated analogs to rank compounds with lower covalent binding in microsomal protein. Another example of this approach has been discussed using four radiolabeled compounds including imipramine and diclophenac, which exhibited >50 pmol/mg, suggesting potential risk for documented IDRs (Evans, 2004). However, additional studies may be necessary to further evaluate the clinical relevance of the 50-pmol/mg value since only liver microsome proteins have been considered. There are many other enzymes which can also form reactive metabolites in other cellular fractions (cytosol or mitochondria), blood, lung, and so on.

The question remains as to what is an acceptable level of covalent binding to proteins/macromolecules in the liver (or other organs) for a drug candidate to be taken into development. Another important consideration is that covalent binding might be an important detoxification pathway for some reactive intermediates, and it should therefore not always be perceived as a negative attribute for a drug in development; without a sound understanding of the potential implications/mechanisms of covalent binding, a safe and effective drug could be eliminated from further development.

4.9.1 Reactive Oxygen Species (ROS)

It has been proposed that some IDRs may be the result of oxidative stress (increases in the intracellular levels of ROS) (Takeyama et al. 1993; Miyamoto et al., 2003). Generally, ROS are generated as by-products from cellular metabolism, primarily in the mitochondria. When the cellular production of ROS exceeds the antioxidant capacity of the cell, cellular macromolecules such as lipids, proteins, and DNA can be damaged. To prevent oxidative stress under normal physiological conditions, these free radicals are neutralized by an elaborate antioxidant defense system consisting of enzymes (e.g., catalase, superoxide dismutase, and GSH peroxidase), and numerous non-enzymatic antioxidants (e.g., vitamins A, E, and C, GSH, ubiquinone, and flavonoids). GSH-associated metabolism is a major mechanism for cellular protection against agents or reactive intermediates. It has been reported that several mechanisms could contribute to cell death associated with oxidative stress, which can be summarized in three steps. The first stage is the recognition of stress by sensitive protein(s), (e.g., the depletion of GSH), mitochondrial damage, inactivation of critical cellular functions, activation of transcription factors, defense gene expression, protein expression, protein function, and release of pro-inflammatory cytokinase. Stage 2 is the subsequent activation of cellular defenses through Phase II metabolism enzymes (e.g., uridine diphosphate-glucuronyltransferase [UDP-GT]), GSH-related enzymes (e.g., glutathione S-transferase [GST]), heat shock proteins (e.g., Hsp72), antioxidants and cell cycle inhibitors. At stage 3, the cells of tolerant individuals are able to dynamically protect themselves from continued stress, but the cells of susceptible individuals may not be able to do so, leading to premature apoptotic death. Strategies for antioxidative defense are the transition metals can be inactivated by chelating proteins (e.g., ferritin), and ROS can be reduced enzymatically (e.g., by the GSH peroxidase) or non-enzymatically by antioxidants (e.g., by vitamin E, vitamin C, and GSH).

4.10 MECHANISTIC STUDIES

Screening for cytotoxicity, covalent binding, and oxidative stress may help to characterize reactive intermediates. However, a greater understanding of the role of the reactive metabolite will allow for better decisions to be made in the final assessment of quality candidates. The apparent lack of predictability of IDRs by animal models may be due to (1) differences in drug metabolism (and, therefore, reactive metabolites being formed) between species; (2) species differences in immune-mediated IDRs; and (3) predisposing factors in man that do not usually exist in animal models (e.g., alcohol, co-administration of drugs, or disease). Some of the following mechanistic studies have furthered our understanding of the safety profile of a drug candidate with reactive intermediates.

4.10.1 Covalent Binding to Albumin/Plasma Proteins

In general, covalently modified proteins can be repaired or degraded; if not, they may impair important cellular functions, which could be directly pathogenic. Several acyl glucuronides have been demonstrated to covalently bind albumin/plasma proteins *in vitro* and *in vivo* (human volunteers). Studies of covalent binding to albumin can be performed with or without radiolabeled compounds. To facilitate these studies, radiolabeled material is preferred and can be useful to identify plasma proteins. Synthetic standards of intermediate metabolites (i.e., acyl glucuronides) are required to assess covalent binding to these proteins. Although non-radiolabeled material can be used to identify albumin covalent binding, quantitation via this approach is difficult. The purpose of these *in vitro* or *in vivo* studies is to characterize the disposition of reactive intermediates in plasma; these studies cannot answer the question of whether plasma proteins can function as protectants of reactive intermediate toxicity or are pathogenic after modification. For some acyl glucuronides, current, indirect evidence suggests a link between covalent plasma binding of tissues and potential for IDRs.

4.10.2 Time-Dependent Inhibition

CYP time-dependent inhibition studies can potentially identify reactive intermediates when they are covalently bound to metabolizing enzymes, as this interaction decreases the activity of the latter. CYP time-dependent inhibition assays are widely used and implemented in a high-throughput manner during early drug discovery as part of the candidate selection criteria. However, this data may be used only as preliminary information since some reactive intermediates do not exhibit time dependency. For example, acetaminophen (APAP) is not a CYP inhibitor, but its reactive intermediate *N*-acetyl-4-benzoquinone imine (NAPQI) is formed by CYP2E1. The stability of the reactive intermediate may be an important factor that determines the effect on time dependency. A highly reactive intermediate may react with the metabolizing enzyme, rendering it unable to distribute to other cell compartments or tissues, while a reactive intermediate with moderate reactivity may diffuse into additional cell compartments. For example, APAP reactive metabolite appears to distribute from the endothelium reticulum into the mitochondria where it reacts with critical proteins. On the other hand, tienilic acid induces hepatotoxicity, and patients form anti-LKM2 (liver kidney microsomes type 2) antibodies in the serum which recognizes CYP2C9. Tienilic acid covalently binds to CYP2C9, the major CYP metabolizing enzyme, and inhibits its activity in a time-dependent manner. The time inhibition assay may record reactive intermediates with moderate reactivity as false negatives, and may be limited to the activity of the enzyme that is being measured.

4.10.3 Antioxidants/Trapping Reagents

In vivo and *in vitro* studies have been conducted with supplementation/depletion of antioxidants to elucidate toxicity mechanisms. The results of these studies indicate that antioxidants can prevent covalent binding of several types of reactive intermediates to biomolecules, and they can also prevent oxidative stress generated by non-covalent interactions. For example, addition of antioxidants such as GSH or ascorbic acid resulted in significant prevention of CYP1A2-dependent cytotoxicity and protein-reactive metabolite formation of tacrine. In another *in vitro* example during early discovery, the amount of reactive intermediate covalently bound to microsomal protein was reduced on addition of GSH to the microsomal incubation, as determined using radiolabeled compounds. Another interesting approach is the use of the spin-trap agent PNB *in vivo*. This trapping agent was used with thalidomide, which initiates embryonic DNA oxidation and teratogenicity in rabbits, both of which were abolished by pretreatment with PNB.

Several potential biomarkers of oxidative stress can be monitored to evaluate the potential toxic effects of drugs. Drugs that undergo redox cycling or form free radicals can generate toxic effects through oxidative stress without forming covalent adducts with biomolecules. This suggests that certain groups of compounds can be screened out in the drug design. The overall data in this area of research suggest that more study is required to select the appropriate biomarker(s) of oxidative stress that can aid in the assessment of the safety profile of a new potential drug candidate. The potential advantage of an oxidative stress biomarker is that it can be monitored during *in vitro* and/or *in vivo* studies.

Biological Markers (Biomarkers) Biomarkers can be measured and quantified, providing useful information for a wide range of clinical and preclinical uses. Some potential examples of biological parameters which can be measured include concentration of specific enzyme(s) and/or specific hormones; specific gene phenotype distribution in a population; presence of biological substances that are useful as indicators for health and physiology-related assessments such as disease risk, psychiatric disorders, environmental exposure and its effects; disease diagnosis; metabolic processes; substance abuse; pregnancy; cell line development; and epidemiologic studies. These and other parameters can be used to identify a toxic effect in an individual organism and can be used to extrapolate between species. Biomarkers can serve to confirm diagnoses, monitor treatment effects or disease progression, and predict clinical results.

Biomarkers are clearly indicated as having important roles in drug development for a number of situations, such as their ability to provide a rational basis for selection of lead compounds, or as a help in determining the ability to work toward qualification and use of a biomarker as a surrogate end point. Changes in a biomarker can provide useful indicators for pathophysiology, which in turn are important in identifying a suitable therapeutic target. For example, the association of elevated serum cholesterol levels with an increased incidence of coronary heart disease provides an underlying rationale for developing drugs that lower cholesterol by inhibiting 3-hydroxy-3-methylglutaryl CoA reductase. Thus, total cholesterol is a good example of a clinical biomarker that has been qualified for use as a surrogate end point.

Biomarkers are also important to the preclinical assessment of the potential benefits and harmful effects of a new drug candidate. Screening tests in animals using biomarkers, such as reduction of blood pressure, provide important demonstration that a compound is likely to produce the intended therapeutic activity in patients. By measuring blood levels during adverse events, such as seizures, in animal toxicology studies may help guide the design of dose escalation studies in humans and serve as a surrogate for preventing or reducing the likelihood of similar adverse events in humans. Biomarkers for potential toxicity play an equally important role. For example, a drug found to prolong the QT interval in animals may warn of potential cardiovascular risk in subsequent clinical studies. Also, biomarkers could be GSH conjugates and/or glutathionylated or oxidized proteins detectable both *in vitro* and *in vivo*. Biomarkers of oxidative stress involving gene/protein expression at least require a whole cell system. It is highly unlikely that a single compound/gene/protein/function will be an effective biomarker; however, combinations of these may prove more successful. Several potential biomarkers of oxidative stress can be monitored to evaluate the potential toxic effects of drugs. Drugs that undergo redox cycling or form free radicals can generate toxic effects through oxidative stress without forming covalent adducts with biomolecules. This suggests that certain groups of compounds can be screened out in the drug design phase. The potential advantage of an oxidative stress biomarker is that it can be monitored during *in vitro* and/or *in vivo* studies. Pharmacokinetic-pharmacodynamic studies using biomarkers may be particularly useful; for example, one such study showed good correlation between the hypotensive effects of an antiarrhythmic drug in dogs and humans. One shortcoming is that most biomarkers, used singly, are unlikely to capture all the effects of a drug, and thereby fulfill the most stringent criterion for a surrogate end point, although by using several of them in combination, it is more likely, and desirable, to produce evidence which is consistent enough to point in a particular direction.

4.11 PRECLINICAL DEVELOPMENT

The main goal in preclinical development is the extrapolation of efficacy and safety data from animal and *in vitro* models to humans. This approach should be considered when a metabolic pathway is suspected to play a major role in this extrapolation. One example of the usefulness of this technique is illustrated by tamoxifen, which results in the development of hepatic tumors in rats, although these tumors were not detected in mice in the rodent carcinogenicity assay (Bouska et al., 1997). The proposed metabolic pathway responsible for tumorigenesis involves sequential bioactivation of tamoxifen via α-hydroxylation and *O*-sulfonation, and the resultant reactive metabolite reacts with DNA. *In vitro* bridging studies with hepatic subcellular fractions, which formed part of the human risk assessment, demonstrated that rat microsomes were threefold more active than human microsomes in forming α-hydroxytamoxifen, and fivefold more active than human hepatic cytosol with regard to *O*-sulfonation (bioactivation). In contrast, the rate of *O*-glucuronidation (bioinactivation) in human hepatic microsomes was at least 100-fold greater than for rat hepatic microsomes. The dose of tamoxifen required to induce tumors in rats was 40 mg/kg, which is ~150-fold greater than the therapeutic dose in man (0.3 mg/kg). This data suggest a safety margin of 150,000 for risk tumors, which is consistent with the clinical experience of the drug to date, and is consistent with the mouse data. This example, along with others, illustrates the need for a better understanding of the connection between exposure of reactive intermediates and their reactivity in the assessment of a safety profile. However, a reliable animal model to predict IDRs is needed to be able to justify the use of metabolic profiles in the safety profile assessment. Currently, species differences in metabolic pathways may be considered in the attempt to bridge safety data between animals and humans. Additional knowledge of biomarkers may also help improve the development of this approach.

4.12 CLINICAL DEVELOPMENT: STRATEGY

Drugs dosed at ≤10 mg/day appear to have a low incidence of IDRs, while low potent drugs (β-lactam antibiotics, sulfonamides, phenobarbital, phenytoin, carbamazepine, tricyclic antidepressants, and nonsteroidal anti-inflammatory drugs [NSAIDs] administered at unusually high doses) administered at doses of >10 mg/day may exhibit a higher incidence of IDRs. It has been suggested that more potent drugs should be developed to reduce the risk of IDRs. For example, olanzapine forms a reactive nitrenium ion similar to that formed by clozapine, which is considered to be responsible for clozapine-induced agranulocytosis. Olanzapine, however, is not associated with a significant incidence of agranulocytosis. A critical difference between the two drugs is that of the dose delivered: clozapine is administered at a dose of several hundred mg/day while olanzapine is administered at a maximum dose of 10 mg/day. Another example is troglitazone, which, when administered at 200–600 mg/day, resulted in serious hepatic injury and was withdrawn from the market by the FDA in 2000. Rosiglitazone, an analog of troglitazone, is administered at 4–8 mg/day and is not hepatoxic. The *in vitro* induction of CYP enzymes by rosiglitazone and troglitazone suggests that other thiazolidinediones may have the potential to cause clinically significant drug interactions if administered at sufficiently high doses.

In an effort to characterize the spectrum of IDRs more accurately, the following issues may be considered during early-to-late clinical phases:

- Study of high-risk patients to identify pharmacokinetic and pharmacodynamic factors that influence susceptibility to drug toxicity.
- Development of computer-based schemes to monitor for adverse reactions and adverse events in primary and secondary care.
- Encouragement to report ADRs to regulatory agencies.
- Identification of risk factors for different types of drug toxicity using pharmacoepidemiological approaches.
- Identification of multigenic predisposing factors to permit the prediction of individual susceptibility.

4.13 REVIEW OF MAJOR POINTS AND FUTURE POSSIBILITIES

During the past decade, there has been an enormous increase in our understanding of how cells and organisms respond to the generation of metabolites, which are chemically reactive. An important consideration is that any screening process will produce false positive and false negative results, and our goal must be to reduce or eliminate these, which in turn will improve drug safety.

There are two major interrelated points of emphasis in the effort to understand and manage IDRs. First, research should focus on the process or mechanism by which IDRs occur. Unfortunately, due to their relative rarity and unpredictability, these have proven difficult to determine with any certainty. Second, further studies are required to improve our knowledge in and hence techniques used for screening, and this will be aided by continued improvement in cooperation and dialogue between pharmaceutical companies and academia. For example, greater advances could be gained if researchers involved in this field shared data to correlate human toxicity with animal toxicity or functional assays. Future screens could focus on biomarkers for oxidative stress. Since there is a general inconsistency in the correlation of toxicity with covalent binding, further validation of *in vitro* covalent binding with regard to *in vivo* toxicity is required. Future implementation plans should focus on:

- Correlating covalent binding in different *in vitro* systems (animal models and human);
- Defining biomarkers for oxidative stress;
- Correlating covalent binding *in vitro* with findings in animals;
- Continued improvement of databases of genomics/proteomics; and
- Extrapolating data to humans.

REFERENCES

Ariens EJ et al. 1982. *Strategy in Drug Research.* Amsterdam: Elsevier Scientific Publishing Company.

Bailey MJ et al. 1996. Chemical and immunochemical comparison of protein adduct formation of four carboxylate drugs in rat liver and plasma. *Chem Res Toxicol* 9:659–666.

Bakke OM et al. 1995. Drug safety discontinuations in the United Kingdom, the United States and Spain from 1974 through 1993: A regulatory perspective. *Clin Pharmacol Ther* 58(1):108–117.

Bergeron MG, Bernier M, L'Ecuyer J. 1992. In vitro activity of clarithromycin and its 14-hydroxy-metabolite against 203 strains of Haemophilus influenzae. *Infection* 20:164–167.

Blanchard SG et al. 1998. Discovery of bioavailable inhibitors of secretory phospholipase A2. *Pharm Biotechnol* 11:445–463.

Bodor N. 1984. Soft drugs: Principles and methods for the design of safe drugs. *Med Res Rev* 4:449–469.

Bodor N, Buchwald P. 2000. Soft drug design: General principles and recent applications. *Med Res Rev* 20:58–101.

Bodor N, Buchwald P. 2004. Designing safer (soft) drugs by avoiding the formation of toxic and oxidative metabolites. *Mol Biotechnol* 26:123–132.

Bodor N, Kaminski JJ. 1980. Soft drugs: 2. Soft alkylating compounds as potential antitumor agents. *J Med Chem* 23:566–569.

Borges N et al. 2000. Tolcapone-related liver dysfunction: Implications for use in Parkinson's disease therapy. *Drug Saf* 26(11):743–747.

Bos H, Henning RH, De Boer E, Tiebosch AT, De Jong PE, De Zeeuw D, Navis G. 2002. Addition of AT1 blocker fails to overcome resistance to ACE inhibition in adriamycin nephrosis. *Kidney Int* 61:473–480.

Bouska JJ et al. 1997. Improving the in vivo duration of 5-lipoxygenase inhibitors: application of an in vitro glucuronosyltransferase assay. *Drug Metab Dispos* 25(9):1032–1038.

Bouska JJ, Bell RL, Goodfellow CL, Stewart AO, Brooks CD, Carter GW. 1997. Improving the in vivo duration of 5-lipoxygenase inhibitors: Application of an in vitro glucuronosyltransferase assay. *Drug Metab Dispos* 25:1032–1038.

Buhl AE, Waldon DJ, Baker CA, Johnson GA. 1990. Minoxidil sulfate is the active metabolite that stimulates hair follicles. *J Invest Dermatol* 95:553–557.

Burgey CS et al. 2003. Metabolism-directed optimization of 3-aminopyrazinone acetamide thrombin inhibitors: Development of an orally bioavailable series containing P1 and P3 pyridines. *J Med Chem* 46:461–473.

Burkhardt O, Borner K, Stass H, Beyer G, Allewelt M, Nord C, Lode H. 2002. Single- and multiple-dose pharmacokinetics of oral moxifloxacin and clarithromycin, and concentrations in serum, saliva and faeces. *Scand J Infect Dis* 34:898–903.

Coleman MD et al. 1994. Reduction of dapsone hydroxylamine to dapsone during methaemoglobin formation in human erythrocytes in vitro. IV: Implications for the development of agranulocytosis. *Biochem Pharmacol* 48(7):1349–1354.

Cribb AE et al. 1992. Sulfamethoxazole is metabolized to the hydroxylamine in humans. *Clin Pharmacol Ther* 51:522–526.

Dabernat H, Delmas C, Seguy M, Fourtillan JB, Girault J, Lareng MB. 1991. The activity of clarithromycin and its 14-hydroxy metabolite against *Haemophilus influenzae*, determined by in-vitro and serum bactericidal tests. *J Antimicrob Chemother* 27:19–30.

Dieckhaus CM et al. 2002. Mechanisms of idiosyncratic drug reactions: The case of felbamate. *Chem Biol Interact* 142(1–2):99–117.

Dragovich P et al. 2003. Structure-based design, synthesis, and biological evaluation of irreversible human rhinovirus 3C protease inhibitors: 8. Pharmacological optimization of orally bioavailable 2-pyridone-containing peptidomimetics. *J Med Chem* 46(21):4572–4585.

Ehrnebo M, Nilsson SO, Boreus LO. 1979. Pharmacokinetics of ampicillin and its prodrugs bacampicillin and pivampicillin in man. *J Pharmacokinet Biopharm* 7:429–451.

Ensink JM, Vulto AG, van Miert AS, Tukker JJ, Winkel MB, Fluitman MA. 1996. Oral bioavailability and in vitro stability of pivampicillin, bacampicillin, talampicillin, and ampicillin in horses. *Am J Vet Res* 57:1021–1024.

Evans DC. 2004. Drug-protein adducts: An industry perspective on minimizing the potential for drug bioactivation in drug discovery and development. *Chem Res Toxicol* 17(1):3–16.

Fan PW et al. 2001. Bioactivation of tamoxifen to metabolite E quinone methide: Reaction with glutathione and DNA. *Drug Metab Dispos* 29:891–896.

Fromigue O, Lagneaux L, Body JJ. 2000. Bisphosphonates induce breast cancer cell death in vitro. *J Bone Miner Res* 15:2211–2221.

Garceau Y, Davis I, Hasegawa J. 1978. Plasma propranolol levels in beagle dogs after administration of propranolol hemisuccinate ester. *J Pharm Sci* 67:1360–1363.

Genin M et al. 1996. Synthesis and bioactivity of novel bis(heteroaryl)piperazine reverse transcriptase inhibitors: Structure-activity relationships and increased metabolic stability of novel substituted pyridine analogs. *J Med Chem* 39(26): 5267–5275.

Godessart N et al. 1999. Role of COX-2 inhibition on the formation and healing of gastric ulcers induced by indomethacin in the rat. *Adv Exp Med Biol* 469: 157–163.

Gorrod JW, Whittesea CMC, Lam SP. 1991. Trapping of reactive intermediates by incorporation of 14C-sodium cyanide during microsomal oxidation. *Adv Exp Med Biol* 283:657–664.

Greene N et al. 2002. Computer systems for the prediction of toxicity: An update. *Adv Drug Deliv Rev* 54(3):417–431.

Gupta AK et al. 1998. Severe neutropenia associated with oral terbinafine therapy. *J Am Acad Dermatol* 38:765–767.

Grillo MP et al. 2001. Effect of alpha-fluorination of valproic acid on valproyl-S-acyl-CoA formation in vivo in rats. *Drug Metab Dispos* 29:1210–1215.

Haack MJ et al. 2003. Toxic rise of clozapine plasma concentrations in relation to inflammation. *Eur Neuropsychopharmacol* 13(5):381–385.

Halsey JP et al. 1982. Benoxaprofen: Side-effect profile in 300 patients. *Br Med J (Clin Res Ed)* 284:1365–1368.

Huang F, Browne CE, Wu WM, Juhasz A, Ji F, Bodor N. 2003. Design, pharmacokinetic, and pharmacodynamic evaluation of a new class of soft anticholinergics. *Pharm Res* 20:1681–1689.

Humphrey MJ, Smith DA. 1992. Role of metabolism and pharmacokinetic studies in the discovery of new drugs—Present and future perspectives. *Xenobiotica* 22(7): 743–755.

Humphreys WG, Obermeier MT, Barrish JC, Chong S, Marino AM, Murugesan N, Wang-Iverson D, Morrison RA. 2003. Application of structure-metabolism relationships in the identification of a selective endothelin A antagonist, BMS-193884, with favourable pharmacokinetic properties. *Xenobiotica* 33:1109–1123.

Hussain MA, Koval CA, Myers MJ, Shami EG, Shefter E. 1987. Improvement of the oral bioavailability of naltrexone in dogs: A prodrug approach. *J Pharm Sci* 76:356–358.

Idanpaan-Heikkila J et al. 1977. Agranulocytosis during treatment with chlozapine. *Eur J Clin Pharmacol* 11:193–198.

Imai T, Yoshigae Y, Hosokawa M, Chiba K, Otagiri M. 2003. Evidence for the involvement of a pulmonary first-pass effect via carboxylesterase in the disposition of a propranolol ester derivative after intravenous administration. *J Pharmacol Exp Ther* 307:1234–1242.

Inman WH et al. 1988. Blood disorders and suicide in patients taking mianserin or amitriptyline. *Lancet* 2(8602):90–92.

Isaacson J et al. 1999. Index of suspicion: Case 3. Acetaminophen overdose. *Pediatr Rev* 20:309–310, 312–313.

Jardine DL. 2002. Hodgkin's disease following methotrexate therapy for rheumatoid arthritis. *N Z Med J* 115(1156):293–294.

Ju C et al. 1999. Detection of 2-hydroxyiminostilbene in the urine of patients taking carbamazepine and its oxidation to a reactive iminoquinone intermediate. *J Pharmacol Exp Ther* 288:51–56.

Kantharaj E et al. 2003. Simultaneous measurement of drug metabolic stability and identification of metabolites using ion-trap mass spectrometry. *Rapid Commun Mass Spectrom* 17(23):2661–2668.

Kassahun K. et al. 2001. Studies on the metabolism of troglitazone to reactive intermediates in vitro and in vivo. Evidence for novel biotransformation pathways involving quinone methide formation and thiazolidinedione ring scission. *Chem Res Toxicol* 14:62–70.

Kempf D et al. 1998. Discovery of ritonavir, a potent inhibitor of HIV protease with high oral bioavailability and clinical efficacy. *J Med Chem* 41(4):602–617.

Kopecky EA, Jacobson S, Joshi P, Koren G. 2004. Systemic exposure to morphine and the risk of acute chest syndrome in sickle cell disease. *Clin Pharmacol Ther* 75:140–146.

Kostiainen R et al. 2003. Liquid chromatography/atmospheric pressure ionization-mass spectrometry in drug metabolism studies. *J Mass Spectrom* 38(4):357–372.

Kubinyi H et al. 1977. Quantitative structure-activity relationships: 7. The bilinear model, a new model for nonlinear dependence of biological activity on hydrophobic character. J Med Chem 20:625–629.

Lasser KE et al. 2002. Timing of new black box warnings and withdrawals for prescription medications. *J Am Med Assoc* 287(17):2215–2220.

Lesko LL et al. 2000. Optimizing the science of drug development: Opportunities for better candidate selection and accelerated evaluation in humans. *Pharmaceutical Res* 14:1335–1343.

Li AP et al. 2002. A review of the common properties of drugs with idiosyncratic hepatotoxicity and the "multiple determinant hypothesis" for the manifestation of idiosyncratic drug toxicity. *Chem Biol Interact* 142(1–2):7–23.

Li J, Uetrecht JP. 2010. The danger hypothesis applied to idiosyncratic drug reactions. *Handb Exp Pharmacol* 196:493–509.

Liu ZC et al. 1995. Oxidation of 5-aminosalicylic acid by hypochlorous acid to a reactive iminoquinone. Possible role in the treatment of inflammatory bowel diseases. *Drug Metab Dispos* 23:246–250.

Lin NH et al. 1997. Structure-activity studies on 2-methyl-3-(2(S)-pyrrolidinylmethoxy) pyridine (ABT-089): an orally bioavailable 3-pyridyl ether nicotinic acetylcholine receptor ligand with cognition-enhancing properties. *J Med Chem* 40(3):385–390.

Madden S et al. 1993. An investigation into the formation of stable, protein-reactive and cytotoxic metabolites from tacrine in vitro. Studies with human and rat liver microsomes. *Biochem Pharmacol* 46:13–20.

Mann K et al. 1987. Differential effects of a new dibenzo-epine neuroleptic compared with haloperidol. Results of an open and crossover study. *Pharmacopsychiat* 20:155–159.

Matzinger P et al. 2002. The danger model: A renewed sense of self. *Science* 296 (5566):301–305.

Merlani G et al. 2001. Fatal hepatoxicity secondary to nimesulide. *Eur J Clin Pharmacol* 57:321–326.

Meyboom RHB et al. 2000. An ABC of drug-related problems. *Drug Saf* 22(6): 415–423.

Miyamoto Y et al. 2003. Oxidative stress caused by inactivation of glutathione peroxidase and adaptive responses. *Biol Chem* 384(4):567–574.

Mohsin SK et al. 2004. Progesterone receptor by immunohistochemistry and clinical outcome in breast cancer: A validation study. *Mod Pathol* 17(12):1545–1554.

Moses PL et al. 1999. Severe hepatotoxicity associated with bromfenac sodium. *Am J Gastroenterol* 94(5):1393–1396.

Munns AJ et al. 1997. Bioactivation of phenytoin by human cytochrome P450: Characterization of the mechanism and targets of covalent adduct formation. *Chem Res Toxicol* 10:1049–1058.

Nassar A-EF. 2009. Minimizing the potential for drug bioactivation of drug candidates to success in clinical development. In *Drug Metabolism Handbook: Concepts and Applications*, ed. Nassar A-EF, Hollenberg P, Scatina J. Hoboken, NJ: John Wiley & Sons.

Nassar A-EF et al. 2003a. Metabolite characterization in drug discovery utilizing robotic liquid-handling, quadrupole time-of-flight mass spectrometry and in-silico prediction. *Curr Drug Metab* 4:259–271.

Nassar A-EF et al. 2003b. Online hydrogen-deuterium exchange and a tandem quadrupole time-of-flight mass spectrometer coupled with liquid chromatography for metabolite identification in drug metabolism. *J Chromatog Science* 41:398–404.

Nassar AF et al. 2004a. Strategies for dealing with reactive intermediates in drug discovery and development. *Curr Opin Drug Discov Devel* 7:126–136.

Nassar A-EF et al. 2004b. Strategies for dealing with metabolite elucidation in drug discovery and development. *Drug Discov Today* 9:317–327.

Nassar AE et al. 2004c. Detecting and minimizing reactive intermediates in R&D. *Curr Drug Discov* July:20–25.

Obach RS et al. 1999a. Prediction of human clearance of twenty-nine drugs from hepatic microsomal intrinsic clearance data: An examination of in vitro half-life approach and nonspecific binding to microsomes. *Drug* 27(11):1350–1359.

Obach RS et al. 1999b. Nonspecific binding to microsomes: Impact on scale-up of in vitro intrinsic clearance to hepatic clearance as assessed through examination of warfarin, imipramine, and propranolol. *Drug Metabolism* 25(12):1359–1369.

Palani A et al. 2002. Synthesis, SAR, and biological evaluation of oximino-piperidino-piperidine amides: 1. Orally bioavailable CCR5 receptor antagonists with potent anti-HIV activity. *J Medicinal Chemistry* 45(14):3143–3160.

Parrish DD et al. 1997. Activation of CGS 12094 (prinomide metabolite) to 1,4-benzoquinone by myeloperoxidase: Implications for human idiosyncratic agranulocytosis. *Fundam Appl Toxicol* 35:197–204.

Pirmohamed M et al. 1998. Adverse drug reactions. *Br Med J* 316(7140):1295–1298.

Plumb RS et al. 2003. Use of liquid chromatography/time-of-flight mass spectrometry and multivariate statistical analysis shows promise for the detection of drug metabolites in biological fluids. *Rapid Commun Mass Spectrom* 17(23):2632–2638.

Prescott LF et al. 1986. Effects of non-narcotic analgesics on the liver. *Drugs* 32(4): 129–147.

Pritsos CA, Constantinides PP, Tritton TR, Heimbrook DC, Sartorelli AC. 1985. Use of high-performance liquid chromatography to detect hydroxyl and superoxide radicals generated from mitomycin C. *Anal Biochem* 150(2):294–299.

Rane A et al. 1977. Prediction of hepatic extraction ratio from in vitro measurement of intrinsic clearance. *J Pharmacol Exp Ther* 200(2):420–424.

Reuning RH, Ashcraft SB, Wiley JN, Morrison BE. 1989. Disposition and pharmacokinetics of naltrexone after intravenous and oral administration in rhesus monkeys. *Drug Metab Dispos* 17:583–589.

Richard AM et al. 1998. Structure-based methods for predicting mutagenicity and carcinogenicity: are we there yet? *Mutat Res* 400(1–2):493–507.

Ruscoe JE. et al. 1995. The effect of chemical substitution on the metabolic activation, metabolic detoxication, and pharmacological activity of amodiaquine in the mouse. *J Pharmacol Exp Ther* 273:393–404.

Sabbaga J et al. 1993. Acute agranulocytosis after prolonged high-dose usage of intravenous dipyrone—A different mechanism of dipyrone toxicity? *Ann Hematol* 66(3):153–155.

Salama A et al. 1983. The role of metabolite-specific antibodies in nomifensine-dependent immune hemolytic anemia. *N Engl J Med* 313:469–474.

Sanders SP, Bassett DJ, Harrison SJ, Pearse D, Zweier JL, Becker PM. 2000. Measurements of free radicals in isolated, ischemic lungs and lung mitochondria. *Lung* 178(2):105–118.

Schuster C, Reinhart WH, Hartmann K, Kuhn M. 1999. Angioedema induced by ACE inhibitors and angiotensin II-receptor antagonists: Analysis of 98 cases. *Schweiz Med Wochenschr* 129:362–369.

Segel IH et al. 1993. Kinetics of unireactant enzymes. In *Enzyme Kinetics*, ed. Segel IH, 18–89. New York: John Wiley & Sons, Inc.

Senturk T et al. 1997. Seizures and hepatotoxicity following sulphasalazine administration. *Rheumatol Int* 17(2):75–77.

Shameem M, Imai T, Otagiri M. 1993. An in-vitro and in-vivo correlative approach to the evaluation of ester prodrugs to improve oral delivery of propranolol. *J Pharm Pharmacol* 45:246–452.

Shapiro LE et al. 1997. Comparative safety of tetracycline, minocycline, and doxycycline. *Arch Dermatol* 133:1224–1230.

Sjövall J, Magni L, Bergan T. 1978. Pharmacokinetics of bacampicillin compared with those of ampicillin, pivampicillin, and amoxycillin. *Antimicrob Agents Chemother* 13:90–96.

Skjodt NM et al. 1999. Clinical pharmacokinetics and pharmacodynamics of bromfenac. *Clin Pharmacokinet* 36(6):399–408.

Stearns RA et al. 2002. The pharmacokinetics of a thiazole benzenesulfonamide beta 3-adrenergic receptor agonist and its analogs in rats, dogs, and monkeys: Improving oral bioavailability. *Drug Metab Dispos* 30:771–777.

Stuart-Harris RC et al. 1984. Aminoglutethimide in the treatment of advanced breast cancer. *Cancer Treat Rep* 11:189–204.

Tagat JR et al. 2001. Piperazine-based CCR5 antagonists as HIV-1 inhibitors: II. Discovery of 1-[(2,4-dimethyl-3-pyridinyl)carbonyl]-4- methyl-4-[3(S)-methyl-4-[1(S)-[4-(trifluoromethyl)phenyl]ethyl]-1-piperazinyl]- piperidine N1-oxide (Sch-350634), an orally bioavailable, potent CCR5 antagonist. *J Medicinal Chemistry* 44(21):3343–3346.

Takeyama N et al. 1993. Oxidative damage to mitochondria is mediated by the Ca(2+)-dependent inner-membrane permeability transition. *Biochem J* 15(294):719–725.

Taylor EW et al. 2002. Accelerating the drug optimization process: identification, structure elucidation, and quantification of in vivo metabolites using stable isotopes with LC/MSn and the chemiluminescent nitrogen detector. *Anal Chem* 74(13):3232–3238.

Testa B et al. 1990. Drug metabolism and pharmacokinetics: Implications for drug design. *Acta Pharm Jugosl* 40(3):315–350.

Tettey JN et al. 2001. Enzyme-induction dependent bioactivation of troglitazone and troglitazone quinone in vivo. *Chem Res Toxicol* 14(8):965–974.

Thompson CD et al. 1996. Synthesis and in vitro reactivity of 3-carbamoyl-2-phenylpropionaldehyde and 2-phenylpropenal: Putative reactive metabolites of felbamate. *Chem Res Toxicol* 9:1225–1229.

Tiller PR et al. 2002. Liquid chromatography/tandem mass spectrometric quantification with metabolite screening as a strategy to enhance the early drug discovery process. *Rapid Commun Mass Spectrom* 16:1225–1231.

Timmins GS, Liu KJ, Bechara EJ, Kotake Y, Swartz HM. 1999. Trapping of free radicals with direct in vivo EPR detection: a comparison of 5,5-dimethyl-1-pyrroline-N-oxide and 5-diethoxyphosphoryl-5-methyl-1-pyrroline-N-oxide as spin traps for HO* and SO4. *Free Radic Biol Med* 27(3–4):329–333.

Uetrecht J. 2008. Idiosyncratic drug reactions: Past, present, and future. *Chem Res Toxicol* 21(1):84–92.

Uetrecht JP et al. 1994. Metabolism of vesnarinone by activated neutrophils: implications for vesnarinone-induced agranulocytosis. *J Pharmacol Exp Ther* 270:865–872.

Uetrecht JP et al. 1995. Oxidation of aminopyrine by hypochlorite to a reactive dication: Possible implications for aminopyrine-induced agranulocytosis. *Chem Res Toxicol* 8(2):226–233.

Uetrecht JP et al. 2002. Screening for the potential of a drug candidate to cause idiosyncratic drug reactions. *Drug Discov Today* 5:832–837.

Utili R et al. 1977. Dantrolene-associated hepatic injury: Incidence and character. *J Gastroenterol* 72:610–616.

Victor F et al. 1997. Synthesis, antiviral activity, and biological properties of vinylacetylene analogs of enviroxime. *J Med Chem* 40(10):1511–1518.

Ware JA et al. 1998. Immunochemical detection and identification of protein adducts of diclofenac in the small intestine of rats: Possible role in allergic reactions. *Chem Res Toxicol* 11:164–171.

Watt AP. 2003. Metabolite identification in drug discovery. *Curr Opin Drug Discov Devel* 6(1):57–65.

Woosley RL et al. 1978. Effect of acetylator phenotype on the rate at which procainamide induces antinuclear antibodies and the lupus syndrome. *N Engl J Med* 298:1157–1159.

Yunis AA et al. 1980. Chloramphenicol toxicity: Pathogenetic mechanisms and the role of the p-NO2 in aplastic anemia. *Clin Toxicol* 17(3):359–373.

Zhao Z et al. 1996a. Bioactivation of 6,7-dimethyl-2,4-di-1-pyrrolidinyl-7H-pyrrolo[2,3-d]pyrimidine (U-89843) to reactive intermediates that bind covalently to macromolecules and produce genotoxicity. *Chem Res Toxicol* 9(8):1230–1239.

Zhao Z et al. 1996b. In vitro and in vivo biotransformation of 6,7-dimethyl-2,4-di-1-pyrrolidinyl-7H-pyrrolo[2,3-D]pyrimidine (U-89843) in the rat. *Drug Metab Dispos* 24(2):187–198.

Zhuang ZP et al. 1998. Isoindol-1-one analogues of 4-(2′-methoxyphenyl)-1-[2′-[N-(2″-pyridyl)-p-iodobenzamido]ethyl]pipera zine (p-MPPI) as 5-HT1A receptor ligands. *J Med Chem* 41(2):157–166.

CHAPTER 5

CASE STUDY: THE UNANTICIPATED LOSS OF N_2 FROM NOVEL DNA ALKYLATING AGENT LAROMUSTINE BY COLLISION-INDUCED DISSOCIATION: NOVEL REARRANGEMENTS

ALA F. NASSAR, JING DU, DAVID ROBERTS, KEVIN LIN,
MIKE BELCOURT, and IVAN KING
Vion Pharmaceuticals, Inc. New Haven, CT

TUKIET T. LAM
Yale University, WM Keck Foundation Biotechnology Resource Laboratory,
New Haven, CT

5.1 INTRODUCTION

Laromustine is an active member of a relatively new class of sulfonylhydrazine prodrugs under development as antineoplastic alkylating agents (Penketh et al., 2004; Baumann et al., 2005). Structure of laromustine and VNP4090CE are shown in Scheme 5.1. VNP4090CE generates chloroethylating (DNA-reactive) species that alkylate DNA at the O6-position of guanine residues and rearrange to form G-C interstrand cross-links (Penketh et al., 2000, 2004). Laromustine injection induces durable complete remissions in elderly patients with *de novo* poor-risk acute myelogenous leukemia (AML) and offers an important therapeutic option for this patient population. The overall response rate of 34% was based on independent review of data across a total of 140

Biotransformation and Metabolite Elucidation of Xenobiotics, Edited by Ala F. Nassar

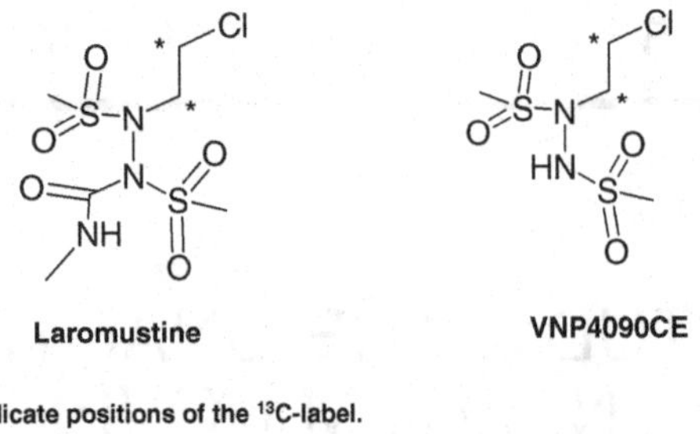

Laromustine

VNP4090CE

Asterisk indicate positions of the ^{13}C-label.

Scheme 5.1 Chemical structure of laromustine and VNP4090CE.

patients 60 years of age or older with *de novo* poor-risk AML who were enrolled in two Phase II studies. Thirty-seven percent of patients achieving complete remission had a duration of response lasting 6 months or longer. Twenty percent of patients achieving complete remission had a duration of response lasting a year or longer.

To assess the mass spectral rearrangement of laromustine, we used mass spectrometry (MS) with an electrospray ionization interface. MS plays a critical role in the analysis of structure information in complex biological matrices (Baillie, 1992; Mutlib et al., 1995; Pochapsky and Pochapsky, 2001; Watt et al., 2003; Nassar and Talaat, 2004; Castro-Perez, 2007). Some of the most effective mass spectrometers are time-of-flight–mass spectrometry (TOF-MS), Fourier transform ion cyclotron resonance–mass spectrometry (FTICR-MS), and Oribtrap. Recent advances in FTICR-MS have made it possible to routinely generate mass spectra with parts per million mass accuracy and ultrahigh resolving power. Mass accuracy and resolving power are critical to determining fragmentation ions for metabolite identification for drug metabolism studies. The online hydrogen-deuterium (H-D) method was developed for metabolite identifications for small molecules where online exchange on column without any further sample preparation is used for metabolite identification and characterization for small molecules. H-D has become common practice for online analysis of metabolite identification of complex matrices. The combination of modern MS with H-D experiments has become ideal for the analysis of structure confirmation (Nassar, 2003; Nassar and Adams, 2003; Nassar et al., 2003; Wạng et al., 2003).

This case study presents the development, application, and analysis of a proposed mechanism for the novel rearrangement of laromustine by collision-induced dissociation. The purpose of this work was to elucidate the fragmentation pathways of laromustine. High mass accuracy and ultrahigh resolution measurements, (H-D) exchange, stable-isotope-labeled analogue (^{13}C-labeled laromustine), and detailed analyses of the LC-MS^n experiments were used to assist with the assignments of these fragments and possible mechanistic rearrangements.

5.2 MATERIALS AND METHODS

5.2.1 Materials

The following were purchased from Sigma-Aldrich (St. Louis, MO): acetic acid, ammonium acetate, and formic acid. Acetonitrile and methanol were purchased from ACROS (Morris Plains, NJ). HPLC grade water was generated from a MilliQ PF plus system from Millipore (Molsheim, France). Deuterium oxide was purchased from Cambridge Isotope Laboratories (Andover, MA). All reagents and solvents were of analytical grade. The analytical column was a Prodigy C18 (250 × 2.0 mm, 5 μm), obtained from Phenomenex Inc. (Torrance, CA).

5.2.2 Test Articles

Laromustine Lot number 04-12-0085b, obtained from Vion Pharmaceuticals, Inc. (New Haven, CT). [^{13}C]-laromustine, [chloroethyl-1,2-^{13}C] Lot PR16480, obtained from Cambridge Isotope Laboratories.

5.2.3 FTICR-MS

The mass spectral data were obtained with a Bruker (Billerica, MA) 9.4 Tesla Apex-Qe Hybrid Qq-FTICR-MS equipped with an Apollo II electrospray ionization source (ESI) (Keck Laboratories at Yale University, New Haven, CT). Laromustine reference standard was infused directly into the FTICR-MS by nanoESI (nESI) with a 30 μm i.d. fused silica tip (New Objective, Inc, Woburn, MA) at a 15 μL/hr flow rate. The nESI tip was grounded and a potential of approximately 1600 V was applied on the glass capillary end-cap. The instrument (running Compass Software with APEX control acquisition component (v.1.2)) was programmed to acquire single free induction decay (FID) (1 M) data and with a mass range (*m/z*) from 85 to 2000.

5.2.4 Liquid Chromatography–Mass Spectrometry (LC-MS) Methods

LC-MS and LC-MS^n experiments were carried out by coupling a Surveyor Liquid Chromatography system to a Finnigan linear ion trap mass spectrometer, LTQ and LTQ-Orbitrap (Thermo Quest, CA). The LTQ ion trap and LTQ-Orbitrap mass spectrometer were equipped with an ESI. For this study, the instruments were operated in both ESI positive and negative ion modes. For MS/MS experiments, the normalized collision energy used was 50 eV and the collision gas was helium. Chromatographic separation was achieved at 40°C using a Prodigy C18 column (250 × 2.0 mm, 5 μm) and a linear gradient using increasing proportions of 0.1% formic acid in methanol and aqueous ammonium acetate. The HPLC gradient program was executed over a period of 60.0 min with water containing 5.0 mM ammonium acetate (mobile phase

A) and methanol/0.1% formic acid (mobile phase B). Both solvents were degassed online. The gradient program was conducted as follows:

Time, min	Solvent A%	Solvent B%	Flow rate, μL/min	Event
0.0	98	2	150	Initial
10.0	95	5	150	Linear gradient
30.0	85	15	250	Linear gradient
40.0	75	25	250	Linear gradient
48.0	2	98	250	Linear gradient
52.0	2	98	250	Hold
53.0	98	2	250	Linear gradient
60.0	98	2	250	Equilibration

5.2.5 H-D Exchange

H-D experiments were used to identify the potential labile protons in the decomposition products. H-D exchange combined with MS is an efficient tool for studying metabolite identification. Because H-D exchange is a low-energy reaction, it can be used to determine the labile protons on the molecule. For H-D experiments, 5 mM ammonium acetate in deuterium oxide mobile Phase "A" and formic acid (0.1%) in methanol were used.

5.3 RESULTS AND DISCUSSION

5.3.1 LC-MS Analysis

When laromustine reference standard (1 μg/mL) was infused in LTQ-Orbitrap MS in a positive mode using mobile Phases A and B, several adducts were tested to improve the sensitivity of MS signal as shown in Fig. 5.1. The ammonium adduct of laromustine produces $[M+NH_4]^+$, observed at *m/z* 325. The potassium adduct of laromustine produces $[M+K]^+$, observed at *m/z* 346. The proton adduct of laromustine produces $[M+H]^+$, observed at *m/z* 308. The optimal MS conditions using the ammonium adduct of laromustine (observed at *m/z* 325) produce a better signal for the sequential MS^n experiments. During metabolism studies of laromustine utilizing LC-MS, LC-MS was developed to identify laromoustine degradation/metabolite product in human liver microsomes (HLMs). The metabolite characterizations for laromoustine were submitted for publication. Laromustine was eluted at 42 min using the gradient program described in the experimental section.

5.3.2 FTICR-MS

Each sample was analyzed in triplicate. Exact mass measurement of the peak of interest (observed at *m/z* at 325) was initially collected. Subsequently, the

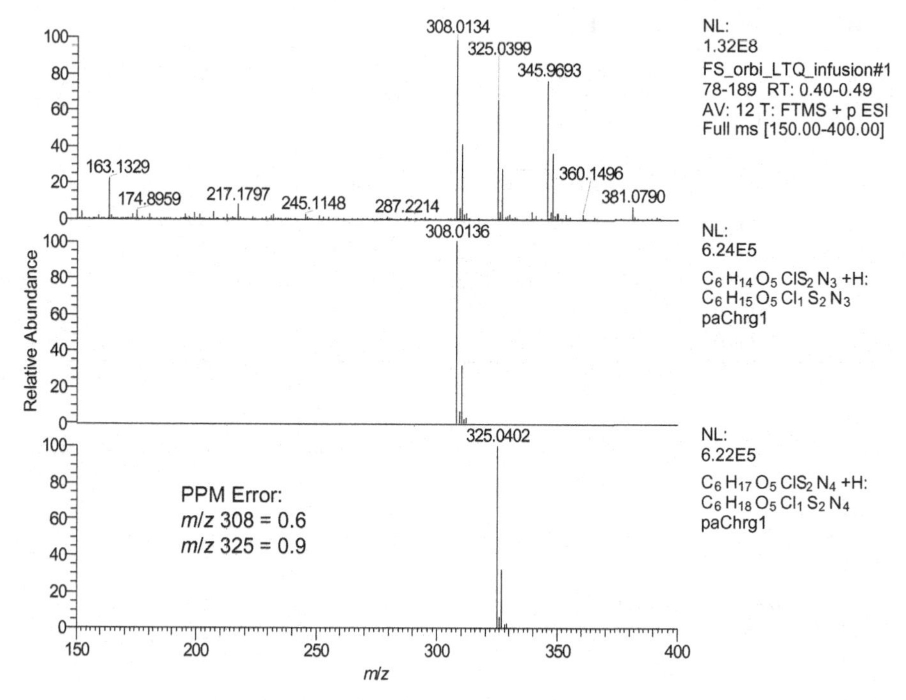

Figure 5.1 Infusion of laromustine reference standard in LTQ-Orbitrap MS.

peak of interest was selected and fragmented by collision induced dissociation (CID) to produce product ions. Bruker Daltonics DataAnalysis software (v. 3.4) was used to analyze the data, and assignments were made based on exact mass measurements and fit of isotopic peaks to that of theoretical isotopic patterns. After generation of *m/z* mass spectrum, data was deconvoluted to determine monoisotopic masses. The data are shown in Table 5.1 and in Fig. 5.2. The mass errors were between −1.64 and 2.12 ppm as shown in Table 5.1. These data allowed us to determine the elemental compositions of the fragmentation ions of laromustine and provided unambiguous fragmentation ion pathways.

The fragmentation processes are typically categorized as direct cleavage or rearrangement. Cleavage reactions are simply the breaking of a bond to produce two fragments. These reactions usually produce an even electron ion. The even electron ion is detected as an odd *m/z* value (assuming no nitrogen) and a neutral odd electron radical. Since the radical is a neutral fragment, it is not observed in the mass spectrum. Rearrangements are more complex reactions that involve both making and breaking bonds. These reactions are thermodynamically favorable because they require less energy. However, they also require a concerted mechanism that is not as kinetically favorable when compared to a simple cleavage reaction (McLafferty and

TABLE 5.1 Results from FTICR-MS

Peak Nominal Mass	Experimental			Average Mass, *m/z*	Formula	Error, ppm
	Trial 1, *m/z*	Trial 2, *m/z*	Trial 3, *m/z*			
325	325.0408	325.0407	325.0405	325.0407	C6H18ClN4O5S2	−1.64
308	308.0141	308.0140	308.0140	308.0141	C6H15ClN3O5S2	−1.57
251	250.9923	250.9924	250.9923	250.9923	C4H12ClN2O4S2	−0.58
143	142.9925	142.9926	142.9926	142.9925	C3H8ClO2S	2.12
107	107.0161	107.0161	107.0160	107.0161	C3H7O2S	0.25
93	93.0214	93.0214	93.0214	93.0214	C2H6ClN2	0.03

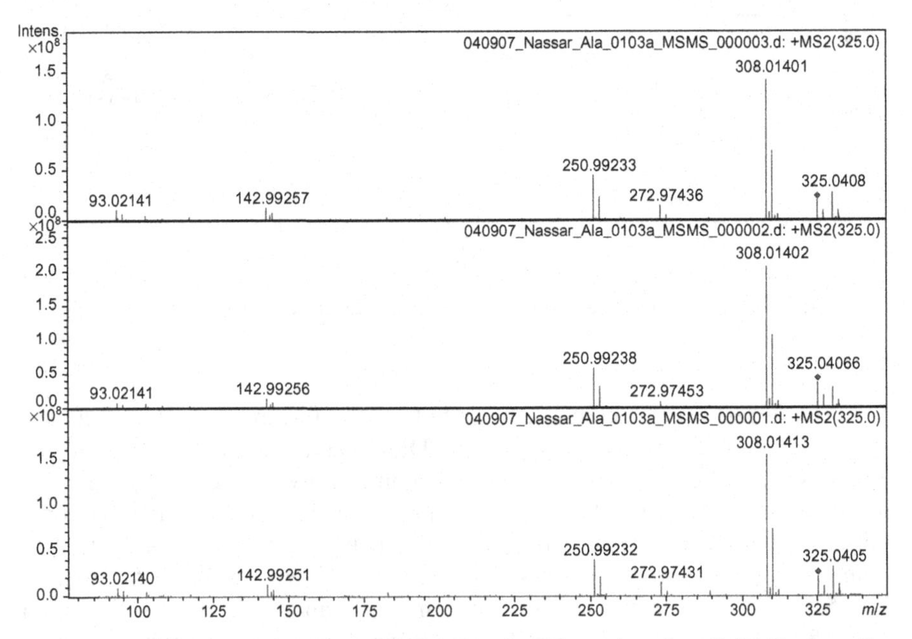

Figure 5.2 FTMS results for MS/MS of *m/z* 325 peak using 9.4T Bruker Qe FTICR-MS.

Turecek, 1993). The novel rearrangement fragment of laromustine was observed and provided important clues about the location and identity of functional groups of laromustine. This was very useful for elucidating ionic structures of laromustine Phase I reactions.

5.3.3 Confirmation of the Fragmentation Ions of Laromustine

Sequential LC-MS^n experiments were carried out on an LTQ ion trap mass spectrometer to gain extensive structural information for laromustine in full-scan MS mode. The fragmentation ions from the initial MS^n experiments on laromustine using LTQ suggested that rearrangement of laromustine took place. Figure 5.3 shows the sequential MS^n experiments up to MS^5. Under the experimental conditions presented here, and on the basis of the MS^5 results, direct cleavage fragment ions were not observed; then, it was speculated that laromustine undergoes rearrangement during the fragmentation of ion *m/z* 171 to *m/z* 143. The MS signature of chlorine isotope (^{35}Cl and ^{37}Cl) in laromustine was observed and verified in the fragmentation ions. The unique MS signature of chlorine isotope (^{35}Cl and ^{37}Cl) confirmed the formation of *m/z* 143. Also the fragmentation ions were further confirmed using H-D exchange and stable-isotope experiments.

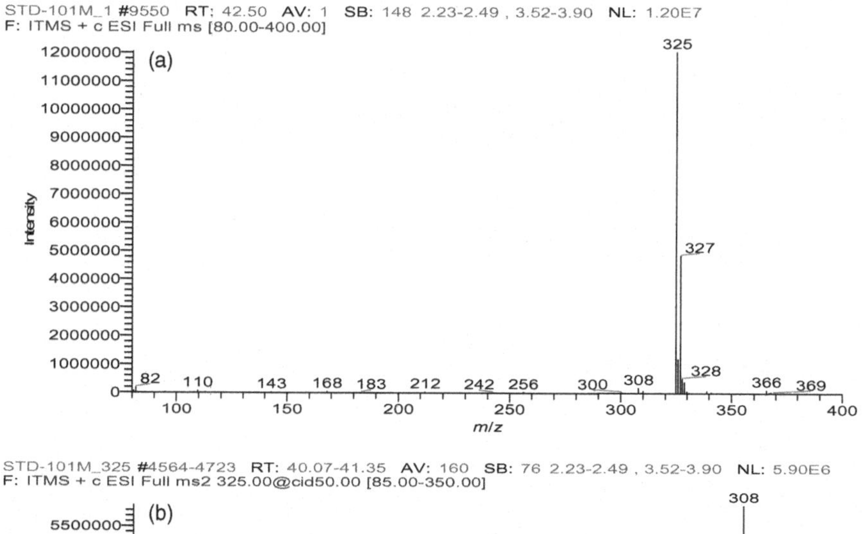

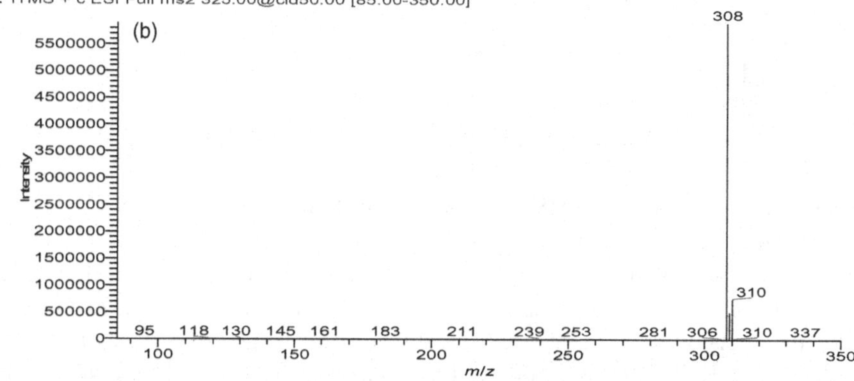

Figure 5.3 LC-MS spectrum of $[M+NH4]^+$, *m/z* 325, laromustine in LTQ (a) full scan, (b) MS^2.

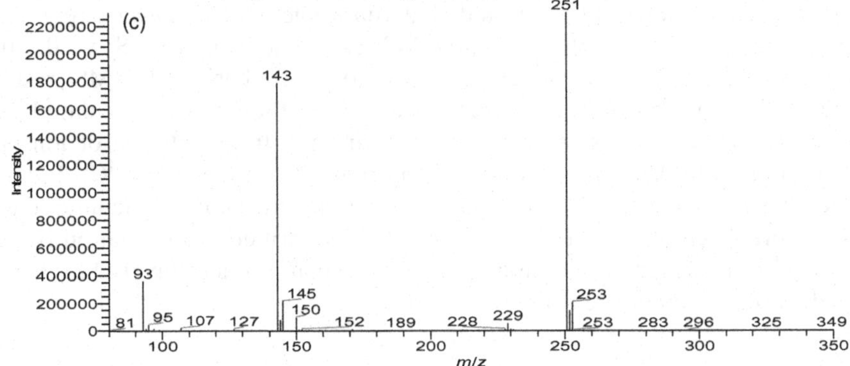

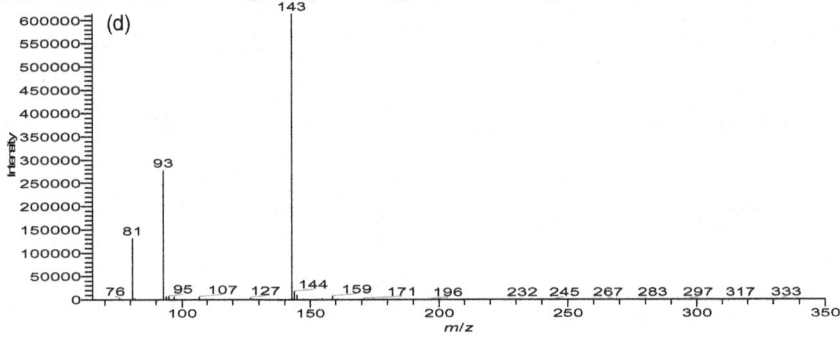

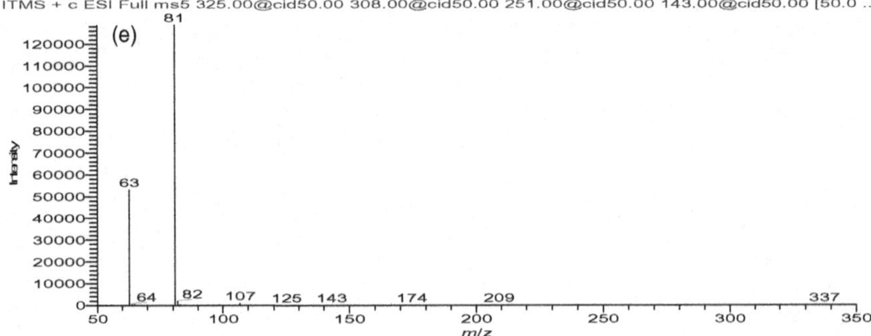

Figure 5.3 (*Continued*) (c) MS^3, (d) MS^4, and (e) MS^5.

The proposed mechanism of formation for the fragmentation ions *m/z* 251, *m/z* 171, *m/z* 143, *m/z* 107, *m/z* 81, and *m/z* 63 is shown in Fig. 5.4. The product-ion spectra of *m/z* 325 showed the fragment ions of *m/z* 308, 273, 251, 143, 107, 93, 81, and 63 (Figs. 5.2 and 5.3). The fragment ion at *m/z* 308 resulted from the loss of ammonia from *m/z* 325. The fragment ion at *m/z* 273 resulted from the losses of HCl from *m/z* 308. The fragment

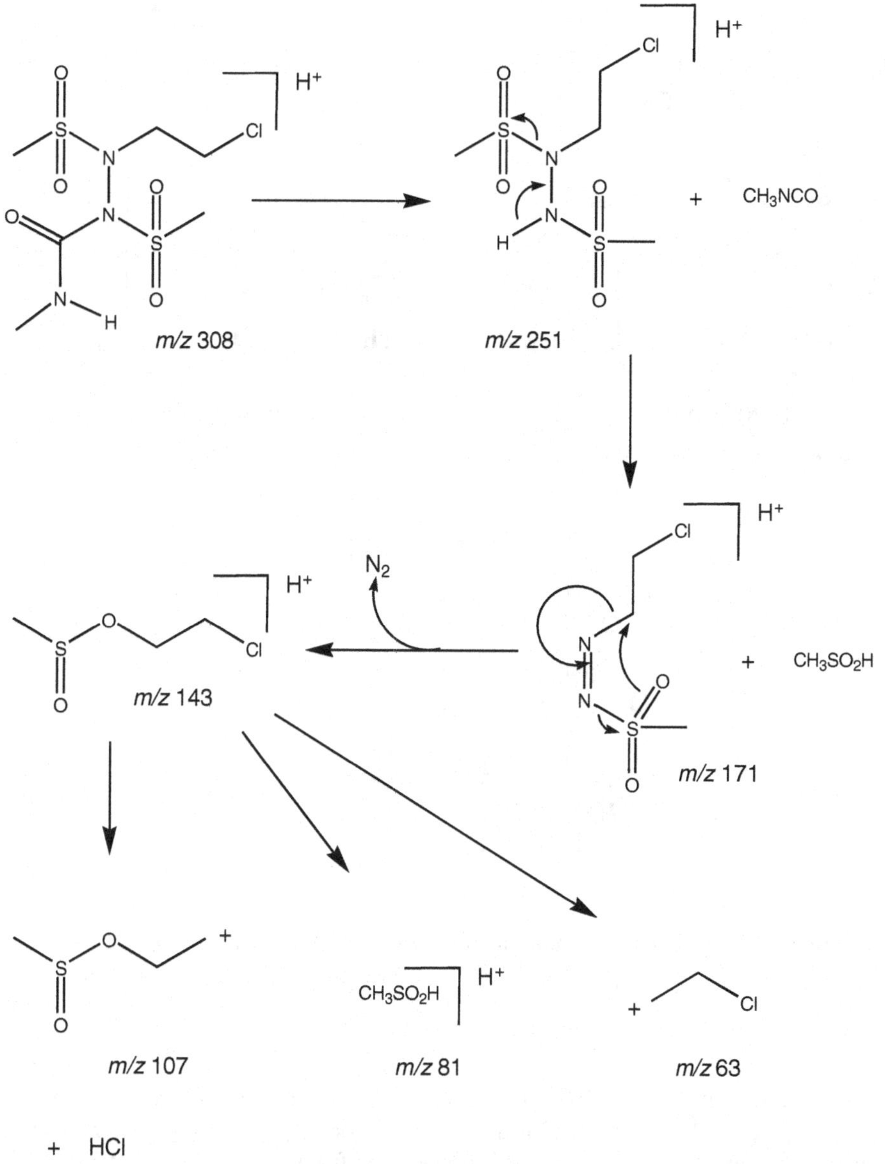

m/z 308

m/z 251

+ CH_3NCO

N_2

m/z 143

+ CH_3SO_2H

m/z 171

m/z 107

m/z 81

m/z 63

+ HCl

Figure 5.4 Proposed mechanism of formation fragmentation ion of *m/z* 143, 107, 81, and 63.

ion at *m/z* 251 resulted from cleavage of the C–N bond. The fragment ion at *m/z* 143 resulted from the loss of CH_3SO_2H and N_2 from *m/z* 251. The fragment ion at *m/z* 107 was due to losses of HCl from *m/z* 143. The fragment ion at *m/z* 81 was due to losses of C_2H_3Cl from *m/z* 143. The product-

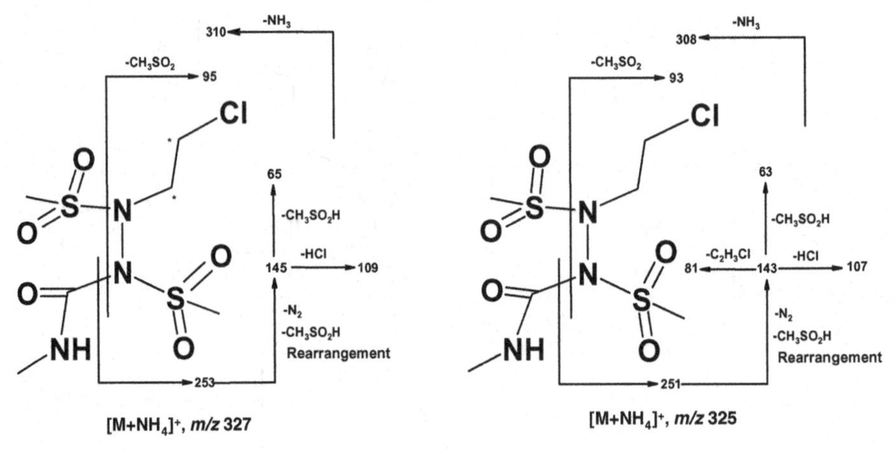

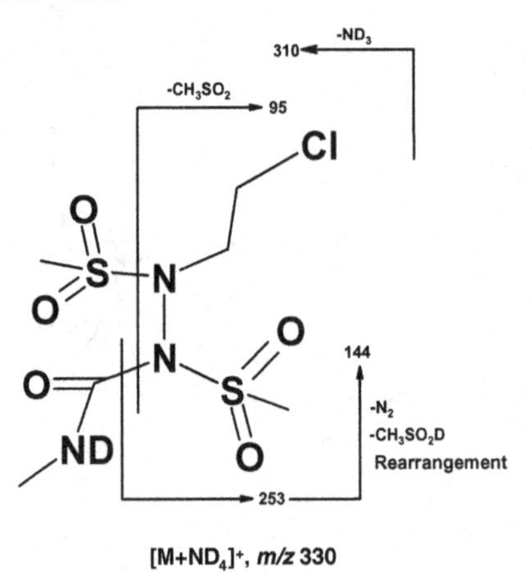

Figure 5.5 Proposed fragmentation ions of laromustine of *m/z* 325, *m/z* 327, and *m/z* 330.

ion spectra of *m/z* 327 (laromustine, $^{13}C_2$) showed the fragment ions of *m/z* 310, 253, 145, 109, 95, and 65. The product-ion spectra of *m/z* 330 using H-D exchange experiments showed the fragment ions of *m/z* 310, 253, 144, and 95. Detailed analyses of the LC-MS^n mass spectra suggest that the fragmentation process involves a rearrangement. This rearrangement fragment provided important clues about the location and identity of functional groups of laromustine. This was very useful for elucidating ionic structures of laromustine Phase I reactions. Proposed fragmentation mechanisms for *m/z* 325 (laromustine), *m/z* 327 (laromustine, $^{13}C_2$), and *m/z* 330 (laromustine, H-D) are shown in Fig. 5.5.

5.4 CONCLUSIONS

Laromustine undergoes a rearrangement which was determined by a combination of different methods such as exact mass measurements, stable isotope, H-D exchange, and MS^n experiments. H-D exchange experiments were useful to determine the number of exchangeable protons in the fragmentation ions. The data of high mass measurement accuracy from FTICR-MS facilitated the determination of the elemental compositions of the fragmentation ions of laromustine and provided unambiguous fragmentation ion pathways. The mechanism of formation for fragmentation ions of *m/z* 251, *m/z* 171, *m/z* 143, *m/z* 107, *m/z* 81, and *m/z* 63 were proposed. Also, fragmentation ions for *m/z* 325 (laromustine), *m/z* 327 (laromustine, $^{13}C_2$), and *m/z* 330 (laromustine, H-D) were proposed. During the fragmentation of ion *m/z* 171 to *m/z* 143, direct cleavage fragment ions were not observed; then, it was speculated that laromustine undergoes rearrangement. This rearrangement fragment provided important clues about the location and identity of functional groups of laromustine. This was very useful for structure elucidation of metabolite and decomposition products of laromustine in *in vitro* and *in vivo* systems.

REFERENCES

Baillie TA. 1992. Advances in the application of mass spectrometry to studies of drug metabolism, pharmacokinetics and toxicology. *Int J Mass Spectrom Ion Proc* 118/119:289–314.

Baumann RP, Seow HA, Shyam K, Penketh PG, Sartorelli AC. 2005. The antineoplastic efficacy of the prodrug Cloretazine is produced by the synergistic interaction of carbamoylating and alkylating products of its activation. *Oncol Res* 15(6):313–325.

Castro-Perez JM. 2007. Current and future trends in the application of HPLC-MS to metabolite-identification studies. *Drug Discov Today* 12:249–256.

McLafferty FW, Turecek F. 1993. *Interpretation of Mass Spectra*, 4th ed. Sausalito, CA: University Science Books.

Mutlib AE, Strupczewski JT, Chesson SM. 1995. Application of hyphenated LC/NMR and LC/MS techniques in rapid identification of in vitro and in vivo metabolites of iloperidone. *Drug Metab Dispos* 23:951–964.

Nassar A-E F. 2003. Online hydrogen-deuterium exchange and a tandem quadrupole time-of-flight mass spectrometer coupled with liquid chromatography for metabolite identification in drug metabolism. *J Chromatogr Sci* 41:398–404.

Nassar A-E F, Adams PE. 2003. Metabolite characterization in drug discovery utilizing robotic liquid-handling, quadrupole time-of-flight mass spectrometry and in silico prediction. *Curr Drug Metab* 4:259–271.

Nassar A-E F, Talaat R. 2004. Strategies for dealing with metabolite elucidation in drug discovery and development. *Drug Discov Today* 9:317–327.

Nassar A-E F, Bjorge SM, Lee DY. 2003. On-line liquid chromatography-accurate radioisotope counting coupled with radioactivity detector and mass spectrometer for metabolite identification in drug discovery and development. *Anal Chem* 75:785–790.

Penketh PG, Shyan K, Sartorelli AC. 2000. Comparision of DNA lesions produced by tumor-inhibitory 1,2-bis(sulfonyl)hydrazines and chloriethylnitrosoureas. *Biochem Pharmacol* 59(3):283–291.

Penketh PG, Shyam K, Baumann RP, Remack JS, Brent TP, Sartorelli AC. 2004. 1,2-Bis(methylsulfonyl)-1-(2-chloroethyl)-2-[(methylamino)carbonyl]hydrazine (VNP40101M): I. Direct inhibition of O6-alkylguanine-DNA alkyltransferase (AGT) by electrophilic species generated by decomposition. *Cancer Chemother Pharmacol* 53(4):279–287.

Pochapsky SS, Pochapsky TC. 2001. Nuclear magnetic resonance as a tool in drug discovery, metabolism and disposition. *Curr Top Med Chem* 5:427–441.

Wang Z, Hop CE, Kim MS, Huskey SE, Baillie TA, Guan Z. 2003. *Rapid Commun Mass Spectrom* 17(1):81–86.

Watt AP, Mortishire-Smith RJ, Gerhard U, Thomas SR. 2003. Metabolite identification in drug discovery. *Curr Opin Drug Discov Devel* 1:57–65.

CHAPTER 6

CASE STUDY: IDENTIFICATION OF *IN VITRO* METABOLITE/DECOMPOSITION PRODUCTS OF THE NOVEL DNA ALKYLATING AGENT LAROMUSTINE

ALA F. NASSAR, JING DU, DAVID ROBERTS, KEVIN LIN,
MIKE BELCOURT, and IVAN KING
Vion Pharmaceuticals, Inc., New Haven, CT

TUKIET T. LAM
Yale University, WM Keck Foundation Biotechnology Resource Laboratory, New Haven, CT

6.1 INTRODUCTION

Laromustine (1,2-bis(methylsulfonyl)-1-(2-chloroethyl)-2-(methylamino)carbonylhydrazine; aka Cloretazine) is an active member of a relatively new class of sulfonylhydrazine prodrugs under development as antineoplastic alkylating agents (Penketh et al., 2004; Baumann et al., 2005). Structures of laromustine and VNP4090CE (90CE) are shown in Scheme 6.1. The 90CE generates chloroethylating (DNA-reactive) species that alkylate DNA at the O6-position of guanine residues and rearrange to form G-C interstrand cross-links (Penketh et al., 2004).

The pharmacological effect of laromustine is briefly summarized above, but the decomposition/metabolite products of laromustine have not yet been reported. Nuclear magnetic resonance spectroscopy (NMR) and mass spectrometry (MS) play a critical role in the analysis of structure information in complex biological matrices (Baillie, 1992; Mutlib et al., 1995; Pochapsky and Pochapsky, 2001; Watt et al., 2003; Nassar and Talaat, 2004; Castro-Perez, 2007). However, these studies pose considerable challenges due to the

Biotransformation and Metabolite Elucidation of Xenobiotics, Edited by Ala F. Nassar

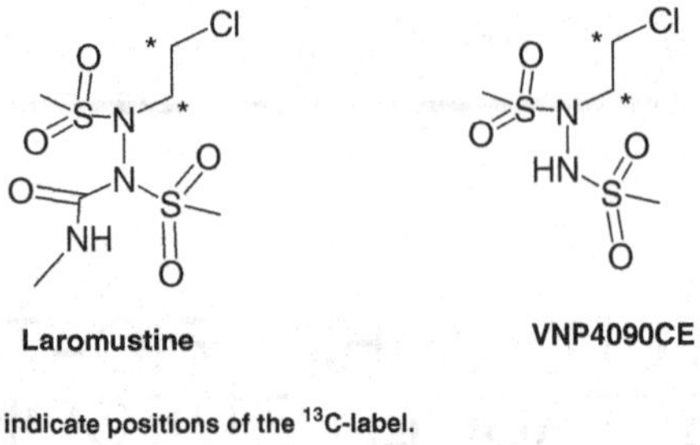

Laromustine **VNP4090CE**

***Asterisk* indicate positions of the ^{13}C-label.**

Scheme 6.1 Chemical structures of laromustine and VNP4090CE.

complexity of matrix and isobaric interferences. Some of the most effective mass spectrometers are time-of-flight MS, Fourier transform ion cyclotron resonance–mass spectrometry (FTICR-MS), and Oribtrap. Recent advances in FTICR-MS have made it possible to routinely generate mass spectra with parts per million mass accuracy and ultrahigh resolving power. Hydrogen-deuterium (H-D) has become common practice for online analysis of metabolite identification of complex matrices (Nassar, 2003). The combination of modern MS with H-D experiments has become ideal for the analysis of structure confirmation. To assess the decomposition/metabolite products of laromustine, a combination of these techniques was used.

In this chapter, we will discuss the structural elucidation of metabolite/decomposition products of laromustine incubated with human liver microsomes (HLMs). We found that laromustine undergoes extensive metabolism and decomposition in *in vitro* systems, and the profiles of metabolite/decomposition products were similar in rat, dog, monkey, and HLMs (Nassar et al., 2010). Most of the identified structures resulted from the chemical decomposition of laromustine and were not P450-mediated (Nassar et al., 2010). The metabolic/decomposition pathway for the formation of metabolite/decomposition products was proposed and investigated.

6.2 MATERIALS AND METHODS

6.2.1 Materials

HLMs, pooled gender (lot number 0610304), were obtained from a commercial source (Xenotech LLC, Lenexa, KS] and were stored at –80°C until use. Magnesium chloride ($MgCl_2$), potassium phosphate, glucose-6-phosphate (G6P), glucose-6-phosphate dehydrogenase (G6PDH), nicotinamide adenine dinucleotide phosphate (NADP), and ethylenediaminetetraacetic acid (EDTA) were obtained from Sigma-Aldrich Corporation (St. Louis, MO). Also, the following were purchased from Sigma-Aldrich: acetic acid 99%, ammonium acetate, formic acid, and phosphoric acid. Acetonitrile and methanol were purchased from ACROS (Morris Plains, NJ). High-performance liquid chromatography (HPLC) grade water was generated from a MilliQ PF plus system from Millipore (Molsheim, France). Deuterium oxide was purchased from Cambridge Isotope Laboratories (Andover, MA). Potassium phosphate monobasic and potassium phosphate dibasic were obtained from Fisher Chemical (Pittsburgh, PA). All other reagents and solvents were of analytical grade. The analytical column was a Prodigy C18, (250 × 2.0 mm, 5 μm), obtained from Phenomenex Inc. (Torrance, CA).

6.2.2 Test Articles

Laromustine Lot number 04-12-0085b, Vion Pharmaceuticals, Inc. (New Haven, CT). [^{14}C]-laromustine, [chloroethyl-1/2-^{14}C], single label, Lot number 645-015-0587, Moravek Biochemical (Brea, CA). [^{13}C]-laromustine, [chloroethyl-$^{13}C_2$] Lot PR16480, Cambridge Isotope Laboratories (Andover, MA).

6.2.3 Microsomal Incubations

Metabolite/decomposition components formed from test articles were generated using mixed pooled HLMs with incubation periods from 0 to 60 min. Microsomal incubations were performed in the presence of a nicotinamide adenine dinucleotide phosphate (NAPDH)-generating system composed of $MgCl_2$ (3 mM), NADP (1 mM), G6P (5 mM), EDTA (1 mM), and G6PDH (1 Unit/mL) in potassium phosphate buffer (50 mM); all concentrations are relative to the final incubation volume. The pH of the final incubation mixture was approximately 7.4. Final protein concentrations were 1.0 mg/mL. Incubations were conducted at 37 ± 1°C with samples taken between 0 and 60 min. The reaction was quenched by the addition of three volumes of 0.05% acetic acid in acetonitrile. The suspension was then vortexed for 1 min and centrifuged at 13,000 rpm for 10 min. The supernatants were collected and then concentrated under nitrogen at room temperature to approximately 0.5 mL. The samples then were loaded onto an LC column for LC-MS^n analysis.

6.2.4 Buffer Incubations

Decomposition components formed from test articles were generated using phosphate buffer (50 mM) at pH 7.4. Incubations were conducted at 37 ± 1°C with samples taken between 0 and 60 min. The reaction was quenched by the addition of three volumes of 0.05% acetic acid in acetonitrile. The reaction was then vortexed for 1 min and then concentrated under nitrogen at room temperature to approximately 0.5 mL. The samples then were loaded onto an LC column for LC-MS^n analysis.

6.2.5 Liquid Chromatography–Mass Spectrometry (LC-MS) Methods

LC-MS and LC-MS^n experiments were carried out by coupling a Surveyor Liquid Chromatography system to a Finnigan linear ion trap mass spectrometer, LTQ (Thermo Quest, CA). The LTQ ion trap mass spectrometer was equipped with an electrospray ionization source (ESI). For this study, the instrument was operated in both ESI positive and negative ion modes. For MS^n experiments, the normalized collision energy used was 50 and the collision gas was helium. Chromatographic separation was achieved at 40°C using a Prodigy C18 column (250 × 2.0 mm, 5 µm) and a linear gradient using increasing proportions of 0.1% formic acid in methanol and aqueous ammonium acetate. The HPLC gradient program was executed over a period of 60.0 min with water containing 5.0 mM ammonium acetate (mobile Phase A) and methanol/0.1% formic acid (mobile Phase B). Both solvents were degassed online. The gradient program was conducted as follows:

Time, min	Solvent A%	Solvent B%	Flow rate, uL/min	Event
0.0	98	2	150	Initial
10.0	95	5	150	Linear gradient
30.0	85	15	250	Linear gradient
40.0	75	25	250	Linear gradient
48.0	2	98	250	Linear gradient
52.0	2	98	250	Hold
53.0	98	2	250	Linear gradient
60.0	98	2	250	Equilibration

6.2.6 H-D Exchange

H-D experiments were used to identify the potential labile protons in the decomposition products. For H-D experiments, 5 mM ammonium acetate in deuterium oxide was made with 5 M ammonium acetate (1 mL) transferred to prepackaged deuterium oxide (1 L), mobile phase A. Formic acid (0.1%) in methanol, (mobile phase B solution) was made by transferring 1 mL of formic acid to a 1000 mL volumetric flask and brought to volume with methanol.

6.2.7 Tools

The tools used for metabolite/decomposition components elucidation are full-scan LC-MS positive and negative modes, multistage MS (MSn) positive and negative modes, N-rule, RDBE (ring double bond equivalents), isotopic patterns, exact mass measurements, H-D experiments, stable isotope (^{13}C-labeled laromustine), and NMR. NMR was used only to confirm C-3 and C-4b.

6.3 RESULTS AND DISCUSSION

6.3.1 LC-MS Analysis

The fragmentation ions of the initial MS^n experiments using LTQ on laromustine suggested that laromustine undergoes rearrangement. The MS signature of chlorine in laromustine was observed and verified in the fragmentation ions. Exact mass measurement using LTQ-Orbitraps and FTICR-MS to confirm the rearrangements were performed. The unique MS signature apparently confirmed the formation of *m/z* 143 and 107. Also, the fragmentation ions were further confirmed using H-D exchange and stable isotope experiments. The product-ion spectra of *m/z* 325 from sequential MS^n experiments showed the fragment ions of *m/z* 308, 251, 143, 107, 93, 81, and 63. The fragment ion at *m/z* 308 resulted from the loss of ammonia from *m/z* 325. The fragment ion at *m/z* 251 resulted from cleavage of the C–N bond. The fragment ion at *m/z* 143 resulted from the loss of CH_3SO_2H and N_2 from *m/z* 251. The fragment ion at *m/z* 107 was due to losses of HCl from *m/z* 143. The fragment ion at *m/z* 81 was due to losses of C_2H_3Cl from *m/z* 143 (Nassar et al., 2010). The identification of these novel rearrangements aids us to understand and identify Phase I decomposition/metabolism products of laromustine formed in *in vitro* systems.

6.3.2 Identification and Characterization of Metabolite/Decomposition Products in HLM

Following incubation of laromustine with HLMs for 60 min in the presence of NADPH, eight metabolite/decomposition products along with the parent drug were detected. *In vitro* mass balance and reaction phenotyping experiments are reported elsewhere (Nassar et al., 2010). The major decomposition products were designated C-1 and C-4. Table 6.1 shows the molecular formula, the molecular weights, the retention times, the results from H-D exchange experiment, *N*-rule data, and RDBE data of the metabolite/decomposition products of laromustine formed in HLM incubations. Under the experimental conditions and on the basis of the experimental data presented here, it was possible to identify the products of Phase I reaction. Phase I reaction determination is described in subsequent sections. The retention times of laromustine and metabolite/decomposition products were between 5 and 43 min with adequate separation efficiency. Peak C-2 was not identified because the MS spectrum of C-2 was not well ionized under the conditions utilized in this study. C-6 product was not reported here because it is one of the potential impurities in laromustine drug substance (Vion's internal document). The peak for [^{14}C]-2-chloroethanol overlaps with peak C-4b using the HPLC conditions described in this study, 2-chloroethanol was designated as C-4a (Table 6.1). The total conversion (sum of C-1, C-2, C-3, C-4, and C-7) of laromustine in HLM was 55%, in the presence of NADPH after 60 min; data are shown graphically in Fig. 6.1. Detailed results from experiments with radioactive 14C-laromustine,

TABLE 6.1 Metabolite/Decomposition Products Formed in HLM Incubation

Component (C) Molecular Formula	MW, amu		RT, min	*N*-rule*		RDBE**		H-D***	
	^{12}C	$^{13}C_2$		Expected	Observed	Expected	Observed	Expected	Observed
C-1 $C_3H_8O_3S$	124.2	126.2	5.5	Even or no	No	0	0	1	1
C-3 $C_4H_{10}N_2O_4S_2$	214.3	216.3	9.3	Even or no	Even	1	1	1	1
C-4a**** C_2H_5ClO	80.5	ND	11.5	Even or no	ND	0	ND	1	ND
C-4b $C_4H_{10}N_2O_4S_2$	214.3	216.3	11.5	Even or no	Even	1	1	1	1
C-5 $C_4H_{10}N_2O_4S_2$	214.3	216.3	19.5	Even or no	Even	1	1	0	0
C-7 $C_6H_{14}ClN_3O_6S_2$	323.8	325.8	36.0	Odd	Odd	1	1	2	2
C-8 (VNP4090CE, 90CE) $C_4H_{11}ClN_2O_4S_2$	250.7	252.7	38.1	Even or no	Even	0	0	1	1
Parent drug (laromustine) $C_6H_{14}ClN_3O_5S_2$	307.8	309.8	43.0	Odd	Odd	1	1	1	1

MW, molecular weight.
$^{13}C_2$, stable isotope (^{13}C-labeled laromustine).
*, If a compound has an odd mass, it definitely has an *N* atom in its structure, and contains an odd number of *N* atoms (1, 3, 5, etc.). If a compound has an even mass, it has either no *N* or an even number of *N* atoms in its structure (0, 2, 4 etc.).
**, Ring double bond equivalents (RDBE).
***, The number of labile protons on the molecule.
****, [^{14}C]2-chloroethanol peak (C-4a) overlaps with the component C-4b; 2-chloroethanol peak was detected by GC-MS.
ND, No data available from LC-MS.

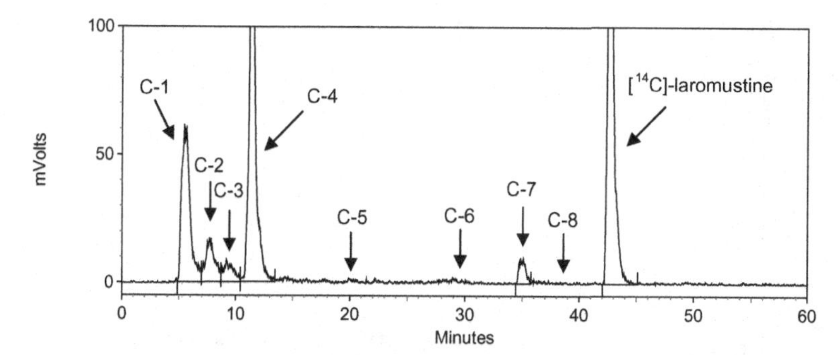

Figure 6.1 LC-Ram chromatogram following incubation of [^{14}C]-laromustine with HLM for 60 min, NADPH.

cross-species liver microsomes, *in vitro* mass balance, and reaction phenotyping are reported elsewhere (Nassar et al., 2010). The profiles of metabolite/decomposition components were similar in rat, dog, monkey, and HLMs, and the relative abundance of individual components between species was not significantly different. *In vitro* studies using HLM have shown that C-7 formation is mediated primarily by CYP3A4/5 and CYP2B6 (Nassar et al., 2010). C-5, C-6, and C-8 do not appear in the radiochromatogram because their formation in the HLM incubation was low.

Identification and Characterization of C-1 The MS spectrum of C-1 was acquired, which revealed an ammonium adducted, protonated molecular ion $[M+NH_4]^+$ at m/z 142. Comparison of the full-scan mass spectrum of undeuterated/deuterated C-1 enabled the determination of the number of labile protons. The full-scan mass spectrum of undeuterated/deuterated C-1 gave ions at m/z 142/147, respectively, which suggested C-1 has only one labile proton. The stable-isotopic experiments gave a full-scan ion at m/z 144, which suggested that the structure of C-1 contains $^{13}C_2$. The product-ion spectra of m/z 142 from sequential MS^n experiments showed the fragment ions of m/z 125, 107, 81, and 79. The fragment ion at m/z 125 resulted from the loss of ammonia from m/z 142. The fragment ion at m/z 107 was due to the loss of water from m/z 125. The fragment ion at m/z 79 resulted from cleavage of the C–S bond. The data in Table 6.1 show that C-1 contains no nitrogen (N) atoms, has only one labile proton, and RDBE is equal to 0, which agrees with the proposed structure (2-(methylsulfonyl)ethanol).

Identification and Characterization of C-3 and C-4b The MS spectra of C-3 and C-4b were acquired, which revealed an ammonium adducted, protonated molecular ion $[M+NH_4]^+$ at m/z 232 (Figs. 6.2a and 3a for components C-3 and C-4b, respectively). The data in Table 6.1 show that C-3 and C-4b contain an even number of nitrogen (N) atoms and RDBE is equal to 1. Comparison of the full-scan mass spectrum of undeuterated/deuterated C-3

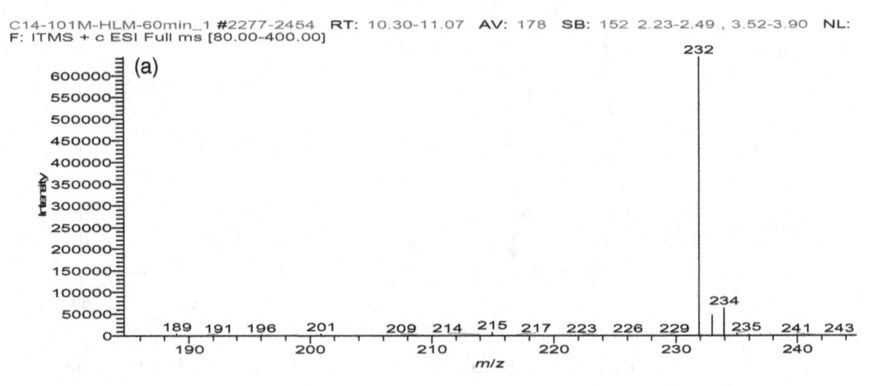

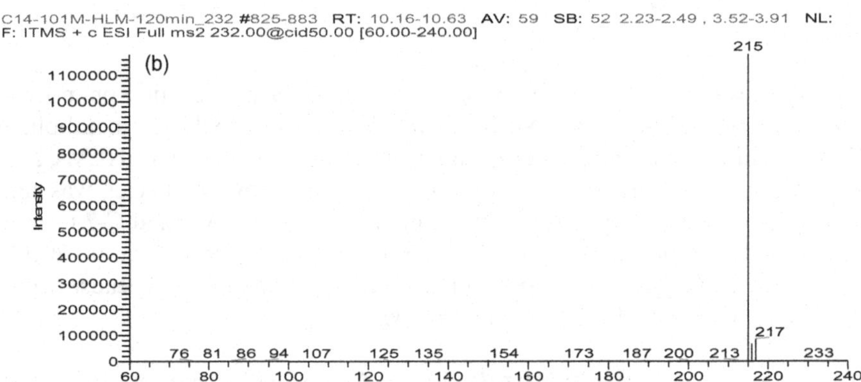

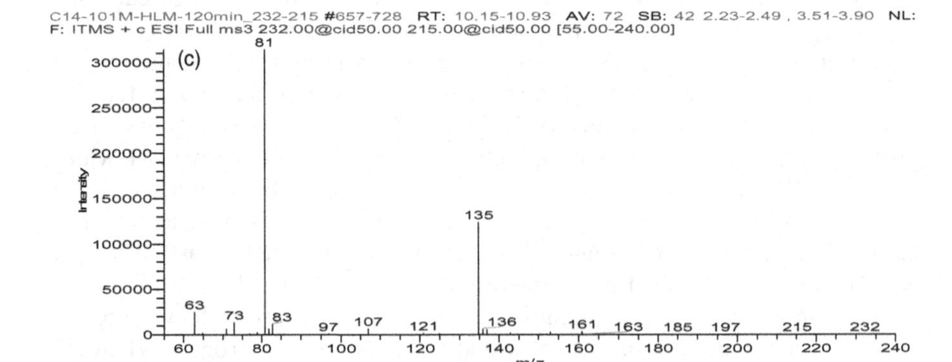

Figure 6.2 Representative LC-MS spectra of $[M+NH_4]^+$, *m/z* 232, C-3 (a) full scan, (b) MS^2, and (c) MS^3.

and C-4b enabled the determination of the number of labile protons. The full-scan mass spectrum of undeuterated/deuterated C-3 and C-4b gave ions at m/z 232/237 respectively, which suggested C-3 and C-4b have one labile proton. The stable-isotopic experiments gave a full-scan ion at m/z 234 for C-3 and C-4b, which suggested that the structure of C-3 and C-4b contains $^{13}C_2$.

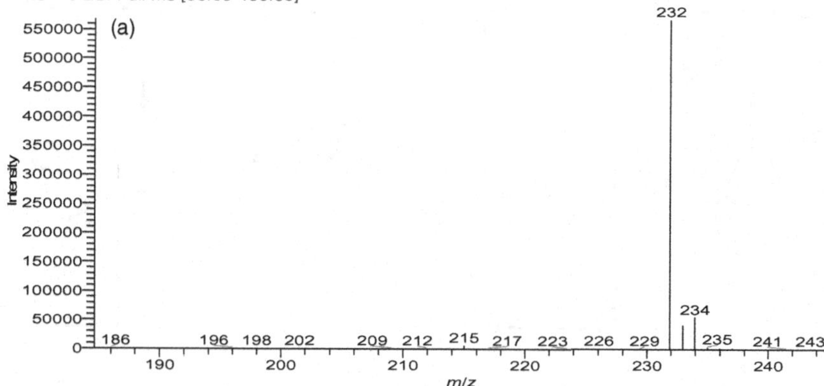

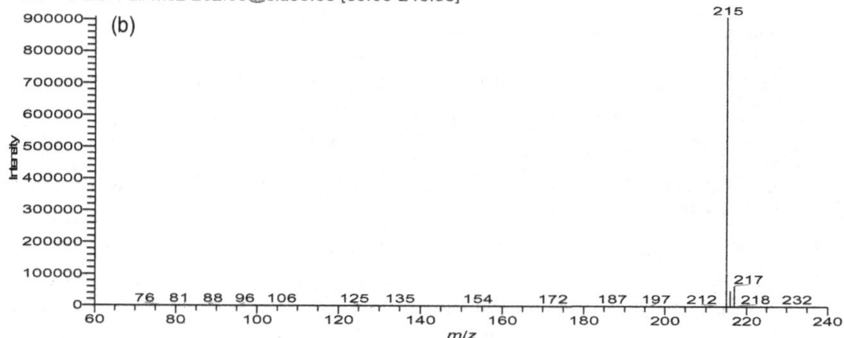

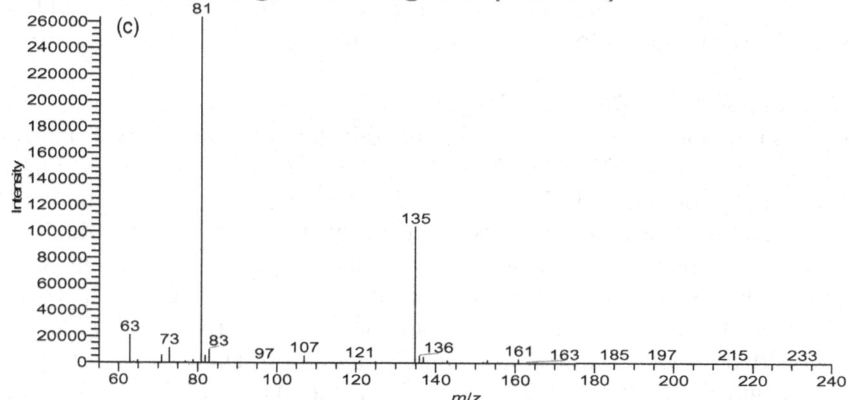

Figure 6.3 Representative LC-MS spectra of $[M+NH_4]^+$, *m/z* 232, C-4b (a) full scan, (b) MS^2, and (c) MS^3.

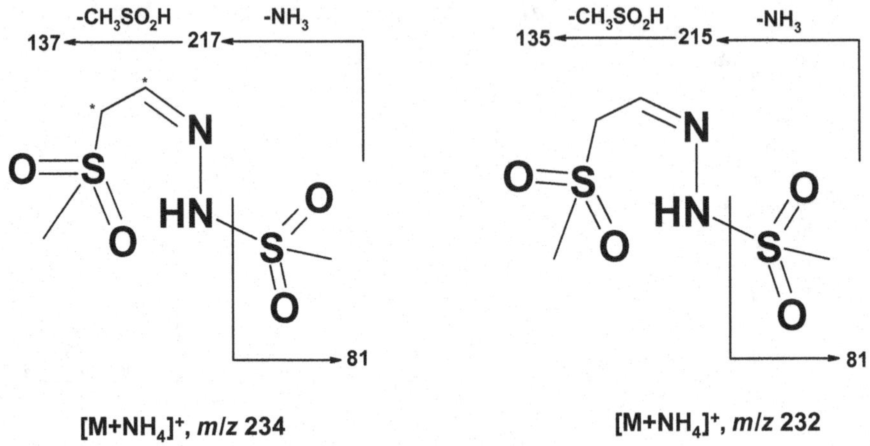

$-CH_3SO_2H$ $-NH_3$ 137 217 81

$[M+NH_4]^+$, *m/z* 234

$-CH_3SO_2H$ $-NH_3$ 135 215 81

$[M+NH_4]^+$, *m/z* 232

Asterisk **indicates position of the ^{13}C-label.**

Figure 6.4 Proposed structure and fragmentation ions for C-3 and C-4b.

Stable-isotopic experiments, in addition to supporting data including N-rule, RDBE, and MS^n, provided evidence to support the structural characterization of C-3 and C-4b. The product-ion spectra of C-3 (Fig. 6.2b,c) and C-4b (Fig. 6.3b,c), from sequential MS^n experiments, showed the fragment ions of m/z 232 to be m/z 215, 135, and 81. The fragment ion at m/z 215 resulted from the loss of ammonia from m/z 232. The fragment ion at m/z 135 was due to losses of CH_3SO_2H moiety from m/z 215. The fragment ion at m/z 81 resulted from cleavage of the N–S bond (Fig. 6.4). The proposed structure for C-3 and C-4b is (E)-N′-(2-(methylsulfonyl)ethylidene)methanesulfonohydrazide (isomer).

Identification and Characterization of C-5 The MS spectrum of C-5 was acquired, which revealed an ammonium adducted, protonated molecular ion $[M+NH_4]^+$ at m/z 232. Comparison of the full-scan mass spectrum of undeuterated/deuterated C-5 enabled the determination of the number of labile protons. The full-scan mass spectrum of undeuterated/deuterated C-5 gave ions at m/z 232/236 respectively, which suggested C-5 has no labile proton. The stable-isotopic experiments gave a full-scan ion at m/z 234, which suggested that the structure of C-5 contains $^{13}C_2$. The product-ion spectra of m/z 232 from sequential MS^n experiments showed fragment ions of m/z 215, 135, 81, and 57. The fragment ion at m/z 215 resulted from the loss of ammonia from m/z 232. The fragment ion at m/z 135 was due to losses of CH_3SO_2H moiety from m/z 215. The fragment ion at m/z 81 resulted from cleavage of the N–S bond (Fig. 6.5). The data in Table 6.1 show that C-5 contains an even number of nitrogen (N) atoms, no labile proton, and RDBE is equal

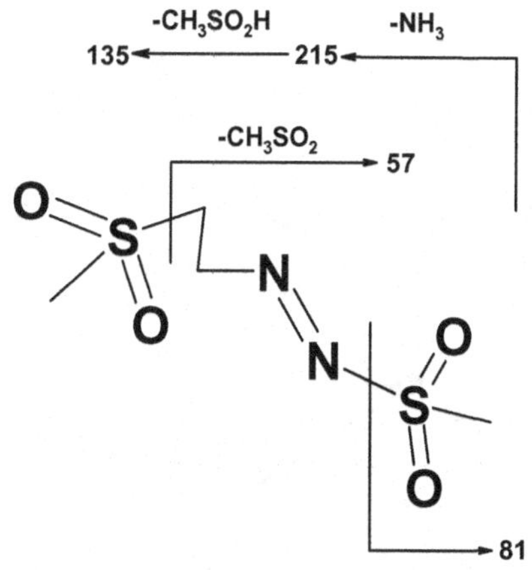

-CH3SO2H -NH3 135 215 -CH3SO2 57 81 [M+NH4]+, m/z 232

Figure 6.5 Proposed structure and fragmentation ions for C-5.

to 1, which agrees with the proposed structure (1-(methylsulfonyl)-2-(2-(methylsulfonyl)ethyl)diazene).

The structural characterizations of the MW 214 components (C-3 and C-4b) were verified using NMR (Vion's internal report). NMR data suggested that peak (C-3) is composed of the free base form with a carbon–nitrogen double bond. The free base (C-3) was found to exist as a mixture of *cis* and *trans* isomers with the *trans* isomer typically more abundant. The unusual observation that the methylene protons adjacent to the double bond are exchangeable in CD_3OD suggests that the double bond can shift in the free base to give a carbon–carbon double bond. In the presence of even weak acids, the free base form is converted to the salt which is present in almost entirely the *trans* form and does not have exchangeable methylene protons, indicating that the double bond does not shift between the carbons in the salt. The salt (C-4b) elutes slightly later than the free base as shown in Fig. 6.1. The third MW 214 peak (C-5) could not be isolated (for NMR analysis) due to its apparent instability. This component has no exchangeable protons, suggesting that the double bond may shift to reside between the nitrogen atoms. Based on the NMR data, C-3 was determined to be the free base and C-4b was determined to be the salt form of *m/z* 232. The proposed structures and fragmentations ions for C-3, C-4b, and C-5 are shown in Figs. 6.4 and 6.5. A proposed mechanism of formation of C-3, C-4b, and C-5 is shown in Fig. 6.6. It has been proposed that laromustine undergoes hydrolysis to form C-8 which then produces C-5 after dehalogenation and rearrangement. C-5 undergoes further rearrangement to form C-3 (free base) or C-4b (salt form).

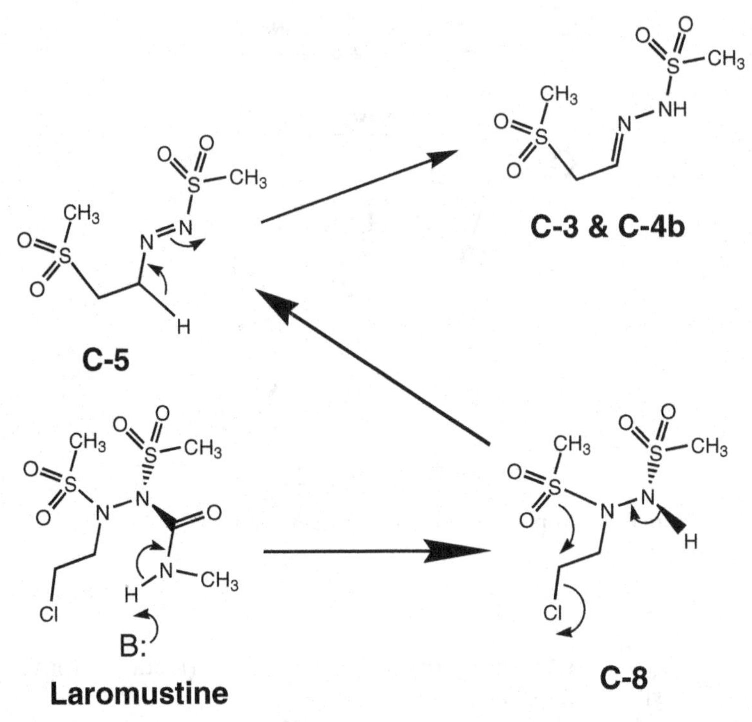

Figure 6.6 Proposed formation mechanism of MW 214 (C-3, C-4b, and C-5).

Identification and Characterization of C-7 This component was detected when laromustine was incubated with HLM in the presence of NADPH, suggesting that this is a metabolite rather than a decomposition product (Fig. 6.1). The fact that C-7 decreased over the incubation time suggests that C-7 continued to metabolize or decompose. The MS spectrum of C-7 was acquired, which revealed an ammonium adducted, protonated molecular ion $[M+NH_4]^+$ at m/z 341, 16 Da more than the parent drug, suggesting the addition of oxygen on laromustine (Omura and Sato, 1964; Schenkman and Kupfer, 1982; Woolf, 1999). The presence of pseudomolecular ions $[M+NH_4]^+$ at m/z 341/343 in a ratio of 3:1 suggested the presence of chlorine on the molecule. The MS signature of chlorine in C-7 was verified by multistage MS. The product-ion spectra of m/z 341 from sequential MS^n experiments showed the fragment ions of m/z 306, 294, 251, 227, 148, 143, and 93. The fragment ion at m/z 306 resulted from the loss of ammonia and water from m/z 341. The fragment ion at m/z 227 was due to the loss of CH_3SO_2 moiety from m/z 306. The fragment ion at m/z 148 was due to the loss of CH_3SO_2 moiety from m/z 227. The fragment ion at m/z 251 resulted from cleavage of the C–N bond. The fragment ion at m/z 143 resulted from the loss of CH_3SO_2H and N_2 moieties from m/z 251. The stable-isotopic experiments gave a full-scan ion at m/z 343. These observa-

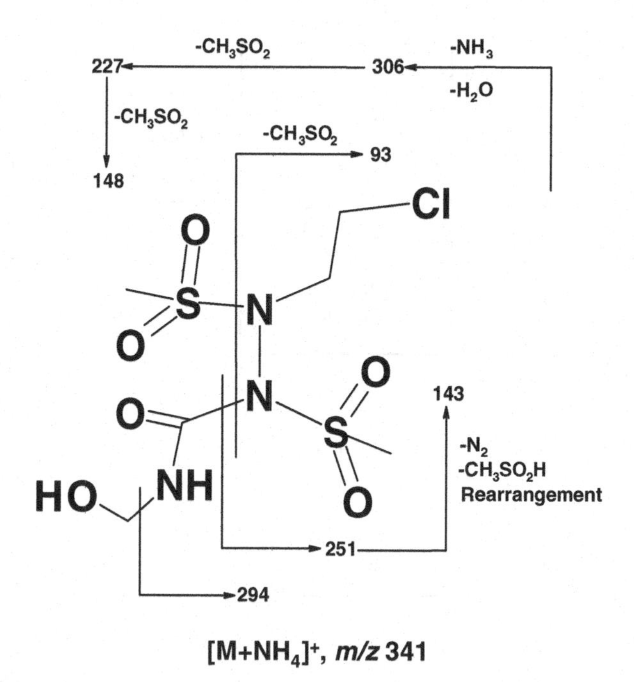

Figure 6.7 Proposed structure and fragmentation ions for C-7.

tions support the proposed structure of C-7. The unique MS signature of chlorine isotope ratios that differs in mass by 2 Da greatly facilitated the structure characterization. The proposed structure and fragmentation is shown in Fig. 6.7.

Identification and Characterization of C-8 (VNP4090CE, 90CE) The C-8 component retention time and fragmentation ions match the reference standard of VNP4090CE. Fig. 6.8 shows the proposed formation of decomposition/metabolite product pathways of laromustine formed in in vitro systems.

6.3.3 Buffer Incubations

Following incubation of laromustine with phosphate buffer (50 mM) at pH 7.4 for 60 min, seven decomposition products along with the parent drug were detected. C-1, C-2, C-3, C-4, C-5, and C-6 were detected and were similar to those detected in HLM.

6.3.4 Summary

The results of this study provide the first analysis of metabolite/decomposition products of laromustine in HLM. H-D exchange , stable isotope (^{13}C-labeled

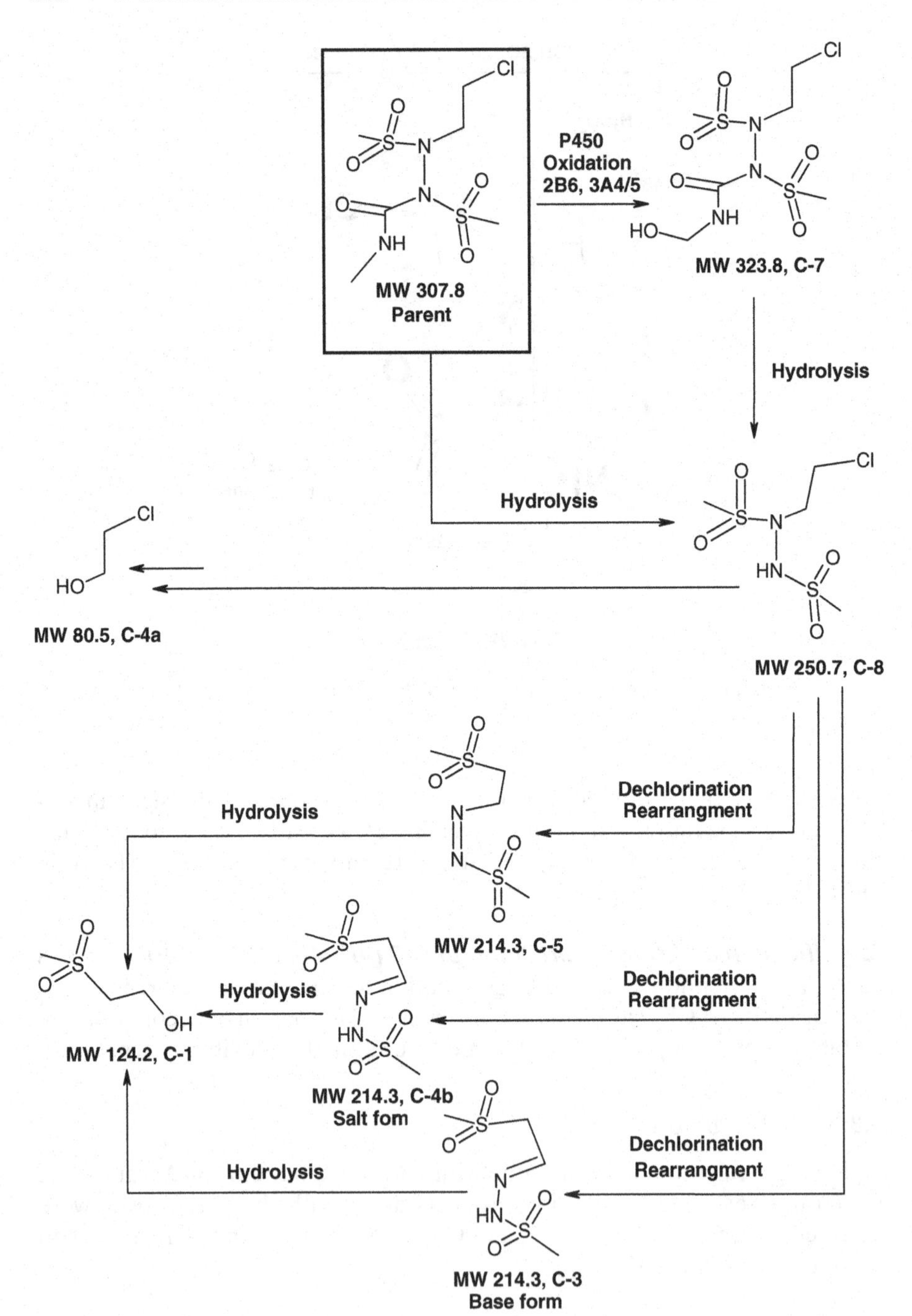

Figure 6.8 Proposed decomposition/metabolite products pathways of laromustine formed in *in vitro*.

laromustine), NMR, and MS^n techniques were applied successfully to identify and characterize the novel metabolite/decomposition components of laromustine. H-D exchange combined with MS is an efficient tool for studying metabolite identification. Because H-D exchange is a low-energy reaction, it can be used to determine the labile protons on the molecule. A combination of these techniques appears promising for maximizing structural information for laromustine. The mechanisms of formation for the decomposition components were proposed. The major decomposition components were not P450-mediated. C-7 was only detected when laromustine was incubated with HLM in the presence of an NADPH-generating system, suggesting that this is a metabolite rather than a decomposition product. Dechlorination and hydrolysis of the methyl isocyanite moiety from laromustine does not require P450 enzymes and occurs at biological pH.

REFERENCES

Baillie TA. 1992. Advances in the application of mass spectrometry to studies of drug metabolism, pharmacokinetics and toxicology. *Int J Mass Spectrom Ion Process* 118/119:289–314.

Baumann RP, Seow HA, Shyam K, Penketh PG, Sartorelli AC. 2005. The antineoplastic efficacy of the prodrug Cloretazine is produced by the synergistic interaction of carbamoylating and alkylating products of its activation. *Oncol Res* 15(6):313–325.

Castro-Perez JM. 2007. Current and future trends in the application of HPLC-MS to metabolite-identification studies. *Drug Discov Today* 12:249–256.

Mutlib AE, Strupczewski JT, Chesson SM. 1995. Application of hyphenated LC/NMR and LC/MS techniques in rapid identification of in vitro and in vivo metabolites of iloperidone. *Drug Metab Dispos* 23:951–964.

Nassar A-E F. 2003. Online hydrogen-deuterium exchange and a tandem quadrupole time-of-flight mass spectrometer coupled with liquid chromatography for metabolite identification in drug metabolism. *J Chromatogr Sci* 41:398–404.

Nassar A-E F, Talaat R. 2004. Strategies for dealing with metabolite elucidation in drug discovery and development. *Drug Discov Today* 9:317–327.

Nassar A-E F, Du J, Belcourt M, Lin X, King I. 2010. *In vitro* profiling and mass balance of the anti-cancer agent laromustine [^{14}C]-VNP40101M by rat, dog, monkey and human liver microsomes. *Open Drug Metab J* 4:1–9.

Omura T, Sato R. 1964. The carbon monoxide binding pigment of liver microsomes: I. Evidence of its hemoprotein nature. *J Biol Chem* 239:2370–2378.

Penketh PG, Shyam K, Baumann RP, Remack JS, Brent TP, Sartorelli AC. 2004. 1,2-Bis(methylsulfonyl)-1-(2-chloroethyl)-2-[(methylamino)carbonyl]hydrazine (VNP40101M): I. Direct inhibition of O6-alkylguanine-DNA alkyltransferase (AGT) by electrophilic species generated by decomposition. *Cancer Chemother Pharmacol* 53(4):279–287.

Pochapsky SS, Pochapsky TC. 2001. Nuclear magnetic resonance as a tool in drug discovery, metabolism and disposition. *Curr Top Med Chem* 5:427–441.

Schenkman J, Kupfer D. 1982. *Hepatic Cytochrome P-450 Monooxygenase System.* New York: Pergamon Press.

Watt AP, Mortishire-Smith RJ, Gerhard U, Thomas SR. 2003. Metabolite identification in drug discovery. *Curr Opin Drug Discov Devel* 1:57–65.

Woolf TF. 1999. *Handbook of Drug Metabolism.* New York: Marcel Dekker.

CHAPTER 7

STRATEGIES FOR THE DETECTION OF REACTIVE INTERMEDIATES IN DRUG DISCOVERY AND DEVELOPMENT

MARK P. GRILLO
Pharmacokinetics and Drug Metabolism, Amgen Inc., San Francisco, CA

7.1 INTRODUCTION

Two major reasons for drug attrition today are preclinical toxicity and human adverse events (Kola and Landis, 2004; Guengerich and MacDonald, 2007). Therefore, it is imperative to be able to understand the chemical and biochemical mechanisms of drug-induced toxicity in order to discover safer drugs and to reduce attrition during drug development. Unfortunately, there are currently no standard criteria used to allow for the prediction of drug toxicity; however a major hypothesis used to explain drug-induced hepatotoxicity is that many toxic drugs become metabolically activated in the liver to reactive metabolites that bind covalently to tissue proteins (Nelson, 2001; Baillie, 2008). If the covalently bound drug–protein adducts represent critical targets important for normal cellular activity, such as mitochondrial proteins involved in energy production and calcium homeostasis (Welch et al., 2005), then it might lead to the loss of mitochondrial function, resulting in cell death and tissue necrosis (Fig. 7.1). For example, recently it was demonstrated that reactive electrophilic intermediates can induce mitochondrion-dependent apoptosis and that covalent modification of specific mitochondrial proteins functions as a triggering event leading to the initiation of apoptosis and subsequent cell death (Wong and Liebler, 2008). Indeed, important cellular targets affected by toxic doses of acetaminophen (APAP), which is a drug that undergoes Phase-I-mediated bioactivation by cytochrome P450 2E1, 1A2, 3A4, and 2A6

Biotransformation and Metabolite Elucidation of Xenobiotics, Edited by Ala F. Nassar

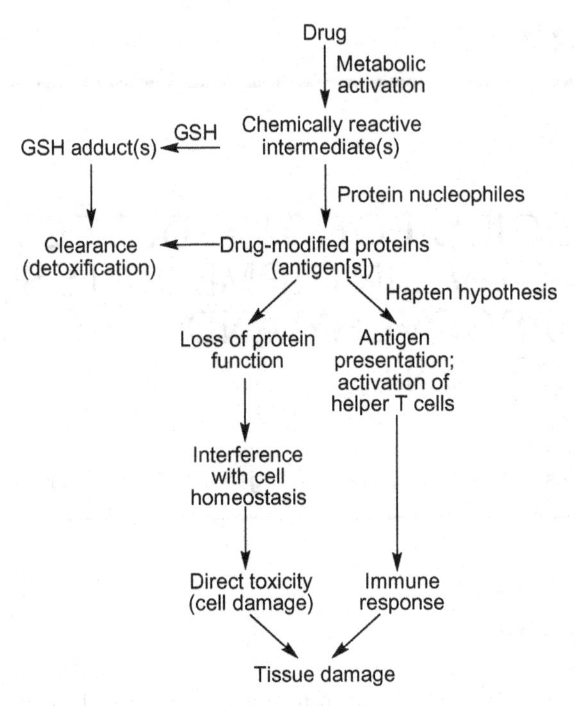

Figure 7.1 Metabolic activation of xenobiotics to chemically reactive products leading to direct toxicity or immune-mediated toxicity.

to a chemically reactive *N*-acetyl-*p*-benzoquinone imine (NAPQI) metabolite (Fig. 7.2), have been proposed to be mitochondrial proteins that result in the loss of energy production, as well as important proteins involved in cellular ion control (Nelson and Pearson, 1990). In studies in mice following high-dose administration of APAP, results from subcellular fractionation techniques showed that APAP treatment produced highest levels of covalent binding to liver mitochondrial proteins in addition to depleting stores of mitochondrial glutathione (γ-glutamylcysteinylglycine, GSH; Tirmenstein and Nelson, 1989). Another example of metabolic activation occurs when carboxylic acid-containing drugs are bioactivated by Phase-II-type metabolism to acyl glucuronide-linked metabolites that can react with tissue protein nucleophiles (Benet et al., 1993). Such protein adducts are proposed to function as antigens that mediate the formation of idiosyncratic allergic reactions to drugs in this chemical class (Boelsterli, 2002; Skonberg et al., 2008).

Based on the pioneering research that occurred in the 1960s and 1970s, and from the collected work and major insights on the metabolic activation of chemicals by membrane-bound cytochrome P450-dependent monooxygenases (Miller and Miller, 1974, 1979, 1981), the covalent binding of chemically reactive drug metabolites to protein was proposed to be associated with target organ toxicity; however, clear mechanistic connections of covalent binding to

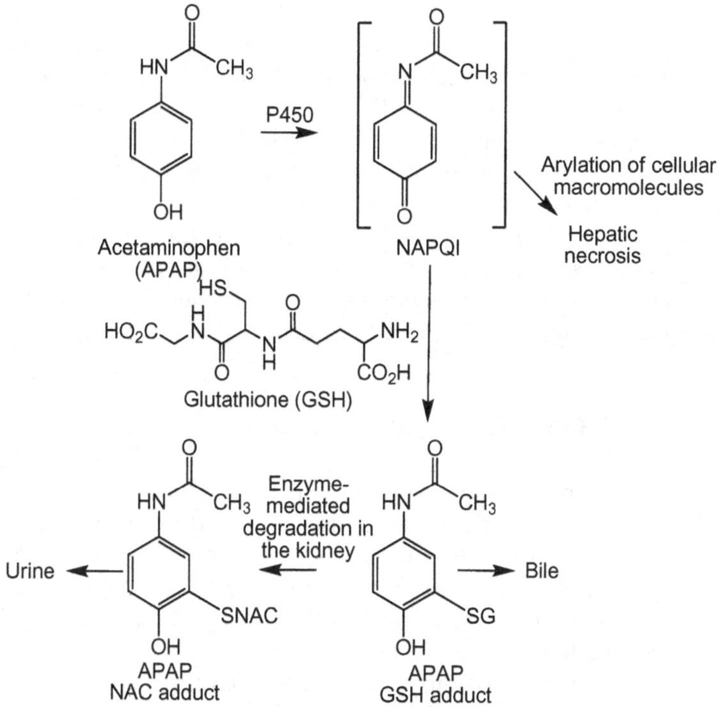

Figure 7.2 Scheme for the metabolic activation of APAP leading to arylation of proteins and GSH-adduct formation.

toxicity still have not yet been demonstrated in most cases (Baillie, 2006, 2008). Molecular and immunochemical approaches have allowed for the identification of specific protein targets of xenobiotic covalent binding for example in studies on the metabolic activation of APAP (Qiu et al., 1998a), benoxaprofen (Qiu et al., 1998b), and diclofenac (Hargus et al., 1995; Jones et al., 2003). Various studies on model compounds like these have shown that protein covalent binding is not arbitrary, but instead that the formation of drug–protein adducts is selective.

Drug–protein adducts also are proposed to mediated allergic reactions that can cause organ failure. A major hypothesis used to explain allergic reactions to drugs is the hapten hypothesis, where a drug or reactive drug metabolite becomes covalently bound to protein forming drug–protein adducts which then are recognized by the immune system as foreign, potentially leading to an immunotoxic reaction (Uetrecht, 2007; Fig. 7.1).

Many hepatotoxicants such as halothane (Satoh et al., 1985), tienilic acid (López-Garcia et al., 1994), felbamate (Dieckhaus et al., 2002), and 3-methylindole (Yan et al., 2007) are believed to cause hepatotoxicity through covalent binding of their corresponding reactive metabolites to tissue protein. Due to the fact that reactive metabolites are chemically unstable, they are in most cases not able to be directly identified and measured. Instead, most studies conducted to assess reactive metabolite formation are performed by measuring captured reactive intermediates in the form of adducts with nucleophilic trapping reagents such as GSH, potassium cyanide (KCN), and semicarbazide. *In vitro* measurement of covalent binding of radiolabeled drug to protein is another valuable approach to measure the extent of reactive metabolite formation. These indirect methods can be used to attempt to correlate reactive metabolite formation with tissue injury by various hepatotoxicants. This chapter deals with current experimental techniques used to identify reactive metabolites and the potential consequences of reactive metabolite formation as it relates to the discovery nontoxic drugs.

7.2 COVALENT BINDING

Results from covalent binding assays can play an invaluable role in the evaluation of drug candidates for continued discovery or development. An important reason for performing covalent binding experiments is to determine the potential exposure of liver tissue to reactive drug metabolites. However, the successful detection of drug covalent binding to protein does not allow for the accurate prediction of potential hepatotoxicity of a drug or drug candidate. To date, it is not known what amount of covalent binding to protein correlates with the risk of a drug being able to cause toxicity (Evans et al., 2004; Baillie, 2006). An *in vitro* experimental approach used to screen for potential drug-induced hepatotoxicity is through the measurement of covalent binding of radioactivity to protein in liver microsomal incubations with radiolabeled (^{14}C, ^{3}H) drug candidates (Pohl and Branchflower, 1981; Evans et al., 2004; Day et al., 2005; Kalgutkar and Soglia, 2005). Other drug–protein adduct detection methods have employed immunoblotting and immunohistochemistry (Kenna et al., 1988). The major impediment for performing covalent binding studies is the availability of radiolabeled drug analogues. However, if the radiolabeled drug is made available, a standardized semi-automated method for the determination of covalent binding of radiolabeled drug equivalents to protein developed by Day et al. (2005) can be used. In their method, Day et al. employed a Brandel cell harvester to be able to isolate drug-adducted protein formed during standardized liver microsome incubations (1 mg protein/mL, 10 μM radiolabeled substrate, 1 mM NADPH, 37°C, 60 min). The method utilizes glass fiber filter paper to capture radioactive precipitated protein, instead of the traditional extraction/centrifuge method (Pohl and Branchflower, 1981). When model compounds [^{14}C]diclofenac, [^{3}H]imipramine and [^{14}C]naphthalene were tested, the levels of covalent binding to human liver microsome (HLM) protein were 57, 127, and 1234 pmol drug equivalents/mg protein, respectively, which correlated well with the traditional method, except with considerable time and labor savings. If covalent binding to protein is detected, then strategic changes to the chemical structure of the drug candidate can be made in order to block or hinder the metabolic activation pathway(s) with the aim of decreasing liver tissue exposure to potentially toxic reactive intermediate(s) (Evans et al., 2004; Doss and Baillie, 2006).

A target (not threshold) covalent binding level of 50 pmol equivalents of drug/mg protein/h of human liver microsomal incubation has been considered as an acceptable amount of irreversible binding to protein where a discovery program might want their candidate drugs to be below (Evans et al., 2004; Evans and Baillie, 2005; Doss and Baillie, 2006). The 50 pmol equivalents of drug residue/mg protein/h level was derived from a literature search for the concentration of covalent binding of radioactive drug equivalents to liver proteins in animals dosed with toxic doses of hepatotoxins, for example, bromobenzene (Monks et al., 1982), isoniazid (Nelson et al., 1978), and APAP (Matthews et al., 1997), where the covalent binding levels were as high as 1000–2000 pmol equivalents/mg liver protein. Thus, the covalent binding target of 50 pmol equivalents of drug/mg protein/h is approximately 20-fold less than that caused by model hepatotoxic compounds. In general, data from covalent binding studies can be of assistance because drug discovery scientists can screen out potentially toxic candidate drugs that form reactive intermediates before progressing to late drug discovery or early drug development.

If covalent binding to protein is detected for a drug candidate, a drug discovery team can put the data into perspective using an assortment of qualifying considerations as proposed by Merck Research Laboratories (Evans et al., 2004; Doss and Baillie, 2006). These qualifying considerations articulate that if a drug is designed for the treatment of a disabling or life-threatening disease, if there is an unmet medical need, or if the duration of use of the drug is acute rather than chronic, then the drug discovery program might decide to progress the drug forward into development. In addition, if the drug is designed for a novel therapeutic target and is to be used only for proof-of-concept purposes, then the compound might still be considered appropriate for progression into target validation studies in humans. However, if developing drugs for pediatric use, the drugs would most likely have to show no evidence for reactive metabolite formation and covalent binding to protein. Another important factor to consider is the clinical dose, since there have been very few drugs removed from the market for toxicity reasons when the daily dose was less than 10 mg (Uetrecht, 1999; Lammert et al., 2008). Thus, if a candidate discovery drug not only possesses covalent binding to protein issues but also has high potency and excellent pharmacokinetic properties, such that the drug is to be given to patients at a low daily dose, then the decision might be to advance the compound further into drug development.

Again, the utility of covalent binding data for predicting toxicities of drug candidates has not yet been shown. Recently, studies on the metabolism-dependent covalent binding of drugs to protein in liver microsomes, supplemented with NADPH, and used to distinguish between hepatotoxic and nonhepatotoxic drugs, were performed on 18 drugs (nine hepatotoxins and nine nonhepatotoxins in humans; Obach et al., 2008). Results from those studies showed that two of the hepatotoxic drugs, namely felbamate and benoxaprofen, did not show covalent binding to protein, whereas paroxetine, the drug that showed the highest covalent binding to protein in liver microsomes, was a nonhepatotoxin. From their covalent binding study results, factors such as the fraction of total metabolism due to covalent binding and the total daily dose led to improved discrimination between hepatotoxic and nonhepatotoxic drugs: however, the liver microsome covalent binding method might still falsely predict some drugs as potentially hepatotoxic with or without consideration of daily dose. When similar *in vitro* covalent binding studies were performed using human hepatocytes, and taking into account estimates of total body burden of covalent binding, an improvement in distinguishing hepatotoxins from nonhepatotoxins over the use of liver microsomes was shown (Bauman et al., 2009).

In another recent study, the covalent binding to protein of 42 radiolabeled drugs in incubations with human hepatocytes was investigated to assess the risk of idiosyncratic drug toxicity (Nakayama et al., 2009). The study results showed that a drug zone classification system categorized as safe, warning/black box warning, or withdrawn, could be used for hepatotoxicity risk assessment when using human hepatocyte covalent binding data and daily dose information. Results indicated that if the daily dose is 10 mg or less, and even if the covalent binding is greater than the target 50 pmol equivalents of drug/mg protein/h, the drug still falls into the safe category, which is in support of the 10 mg nontoxic daily dose proposal by Uetrecht (1999). In general, of the drugs tested by Nakayama et al. (2009), those that were dosed at 20 mg or more/day and also exhibited covalent binding of 50 pmol equivalents of drug/mg protein/h or greater in incubations with human hepatocytes, fell into the warning/black box warning or withdrawn categories. It was proposed that such a zone classification system for the risk assessment of drugs using covalent binding data obtained from studies conducted in human hepatocytes, together with daily dose predictions, could be used for the selection of nonhepatotoxic drug candidates for the development of potentially safer drugs.

In summary, covalent binding studies can be used to provide definitive evidence that a drug or a drug candidate undergoes metabolic activation to chemically reactive intermediates. Whether such data can be used to provide a perspective toward the evaluation of risk/benefit ratios for discovery drugs is not yet known.

7.3 LIQUID CHROMATOGRAPHY–TANDEM MASS SPECTROMETRY (LC/MS/MS) GSH-ADDUCT DETECTION METHODS

In general, the most frequently used method for the detection of reactive drug metabolites involves performing *in vitro* incubations with liver microsomes (usually from rat and human) in the presence of the cofactor for cytochrome P450 function (NADPH) and the nucleophile GSH (10 mM). GSH is a very abundant endogenous sulfur-containing cytoprotective tripeptide (millimolar tissue concentration range). The cysteinyl-sulfhydryl group of GSH is known as a "soft" nucleophile that reacts with soft electrophiles such as carbon in polarized double bonds (e.g., α,β-unsaturated ketones, quinones, and quinone imines) and intermediate "soft-hard" electrophiles such as epoxides, aryl halides, and nitrenium ions (Carlson, 1990). In NADPH-fortified microsomal incubations, GSH reacts with and captures electrophilic metabolites by nucleophilic addition and nucleophilic substitution reactions leading to the formation of thioether-linked GSH adducts. The capturing of reactive metabolites with GSH, followed by structural characterization by LC/MS/MS and NMR, is a very useful strategy because it does not require the synthesis of radioactive drug analogues (Zheng et al., 2007). Alternative strategies that will be discussed in this chapter include the use of an equal mixture of stable isotope-labeled (e.g., ^{13}C, ^{15}N) GSH and unlabeled GSH to trap reactive metabolites for assistance in adduct detection by LC/MS analysis as an "isotope doublet" (Yan and Caldwell, 2004; Mutlib et al., 2005), as well as the use of derivatized GSH for semiquantitative determination of GSH-adduct concentrations (Gan et al., 2005; Soglia et al., 2006). Importantly, such methods employing GSH are not appropriate for trapping and identifying hard electrophilic reactive metabolite species, where a hard nucleophile such as KCN should be used to capture reactive hard electrophile species which include carbocations or iminium ions (Evans et al., 2004; Argoti et al., 2005).

LC/MS/MS methods are routinely used to examine *in vivo* and/or *in vitro* extracts for the presence of GSH adducts using specialized and sensitive scanning techniques. The most often used LC/MS/MS detection techniques for GSH-adduct detection are traditional constant neutral loss (CNL) scanning for 129 Da (Baillie and Davis, 1993) and negative ion tandem LC/MS/MS scanning for precursors of the product ion at *m/z* 272 (Dieckhaus et al., 2005) which are discussed below. Therefore, discovery compounds that are metabolized to reactive intermediates that form GSH adducts can frequently be detected by one or both of these LC/MS/MS screening methodologies which work independent of compound structure, can be identified without a precise understanding of the GSH-adduct chemical structure, and are amenable to high-throughput screening (Baillie and Davis, 1993; Kalgutkar and Soglia, 2005). The promises and challenges of state-of-the-art LC/MS/MS techniques are briefly presented in this chapter from work performed by academic and industry drug metabolism experts.

7.3.1 Neutral Loss Scanning

The first reported LC/MS/MS survey scanning method developed for the detection of GSH adducts was by CNL scanning for 129 Da (Baillie and Davis, 1993). This is a common fragment ion produced by collision-induced

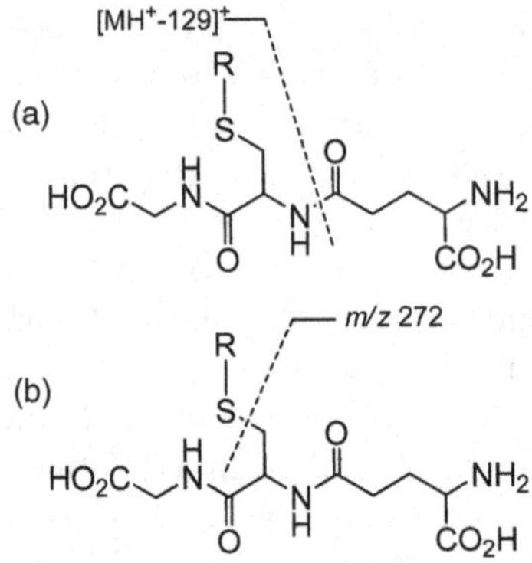

Figure 7.3 Tandem mass spectrometric LC/MS/MS screening for GSH adducts present in biological extracts by (a) constant neutral loss scanning (–129 Da in positive ion scan mode) and (b) precursor ion scanning (*m/z* 272 in negative ion scan mode).

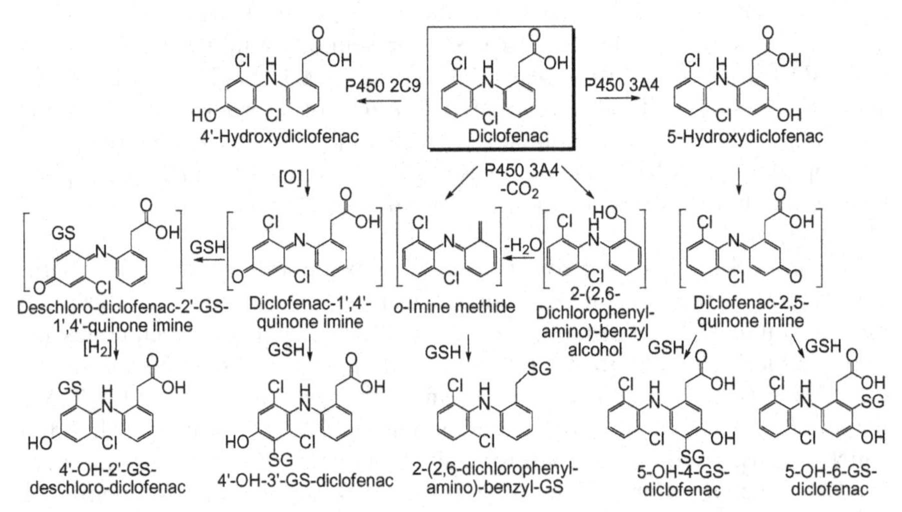

Figure 7.4 Proposed mechanisms in the P450-mediated metabolic activation of diclofenac leading to the formation of GSH adducts (Tang et al., 1999; Yan et al., 2005; Grillo et al., 2008).

dissociation (CID) of GSH adducts in the positive ion mode resulting from the loss elements of pyroglutamic acid from the GSH-adduct protonated molecular ion (Fig. 7.3). One example of the successful use of this technique comes from *in vivo* studies with diclofenac (2-[2-(2,6-dichlorophenyl)amino-phenyl]ethanoic acid; Fig. 7.4), an anti-inflammatory drug used for the treatment of osteoarthritis, rheumatoid arthritis, ankylosing spondylitis, and acute muscle pain (Small, 1989), and which has been shown to cause rare cases of hepatotoxicity (Banks et al., 1995; Boelsterli, 2003; Tang, 2003). Three major

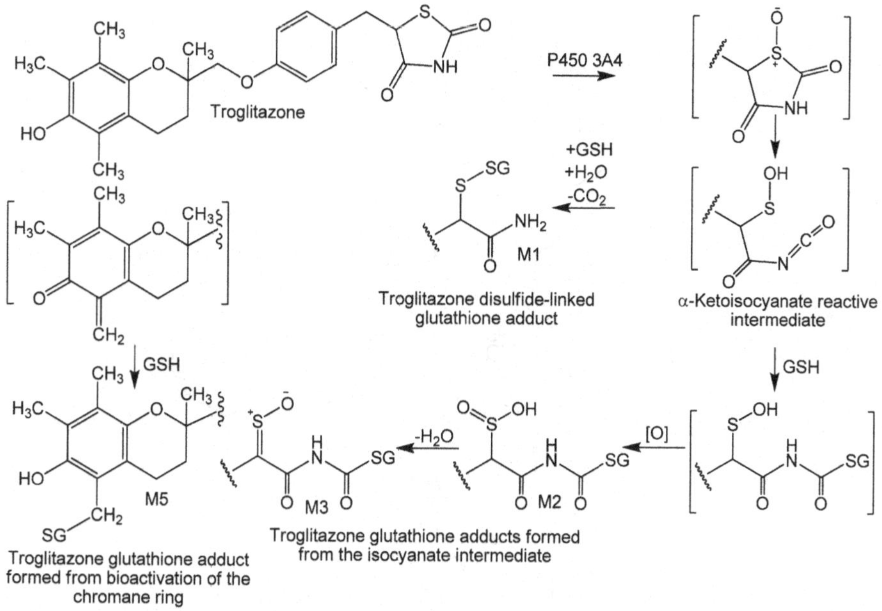

Figure 7.5 Scheme for the metabolic activation of troglitazone leading to the formation of varied GSH adducts. (Adapted from Kassahun et al., 2001.) Not shown in this scheme is the structure of M4 (the quinone derivative of M1).

GSH adducts were identified by LC/MS/MS analysis of bile from diclofenac-dosed rats (200 mg/kg) using the CNL 129 Da detection method. These GSH adducts, namely 4′-hydroxy-3′-gluthionyl-diclofenac, 5-hydroxy-4-glutathionyl-diclofenac, and 5-hydroxy-6-glutathionyl-diclofenac, are formed from cytochrome P450 2C9- and P450 3A4-mediated oxidation of the aromatic rings to 4′- and 5-hydroxylated metabolites, respectively, which undergo subsequent P450-mediated metabolism to electrophilic quinone imine-type metabolites that react with GSH in a Michael-acceptor fashion (Tang et al., 1999; Fig. 7.4).

An additional successful example of the use of survey scanning for CNL of 129 Da comes from studies with troglitazone (Fig. 7.5). Troglitazone is an oral antidiabetic drug that was used for the treatment of Type 2 diabetes until it was taken off the market due to several cases of severe, and sometimes fatal, hepatotoxicity (Isley, 2003). *In vitro* studies in NADPH-fortified preparations of HLMs showed that cytochrome P450 3A4 catalyzes the opening of the troglitazone thiazolidinedione ring system that results in the formation of isothiocyanate- and sulfenic acid electrophilic intermediates (Kassahun et al., 2001; Fig. 7.5). Troglitazone is also metabolized by P450 on the chromane moiety giving rise to a quinone methide reactive metabolite. In one recent report on the metabolic activation of troglitazone (Alvarez-Sánchez et al., 2006), CNL 129 Da scanning of rat and HLM incubation extracts showed the

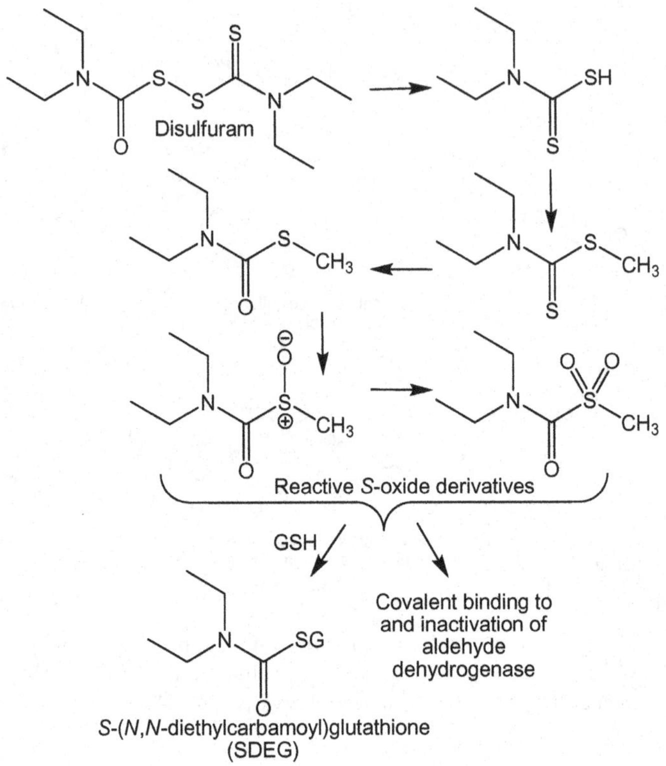

Disulfuram

Reactive *S*-oxide derivatives

GSH

Covalent binding to and inactivation of aldehyde dehydrogenase

S-(*N*,*N*-diethylcarbamoyl)glutathione (SDEG)

Figure 7.6 Scheme for the metabolic activation of disulfuram leading to the inactivation of ALDH and the formation of SDEG. (Adapted from Jin et al., 1994.)

presence of only two predominant GSH adducts. By far, the major adduct detected was due to bioactivation of the chromane ring (M5, Fig. 7.5), and the second adduct (M2) detected by this method was due to ring scission and subsequent reaction of GSH with the isothiocyanate reactive intermediate. As will be discussed below, three other GSH adducts (M1, M3, and M4) characterized by Kassahun et al. (2001) were not detected by the CNL 129 Da loss survey scanning method, which suggested the need for superior LC/MS/MS survey scanning methods to detect GSH adducts of varied structures (Dieckhaus et al., 2005). In addition, LC/MS/MS analysis of *S*-acyl-thioester-linked GSH adducts of carboxylic acid containing drugs under CID conditions has been shown not to produce the CNL 129 Da neutral loss, but instead, these conjugates typically fragment by loss of glutamic acid (147 Da) from the $[M + H]^+$ species (Grillo et al., 2003).

One more example of the successful use of the CNL of 129 Da GSH-adduct survey scanning technique comes from studies with disulfuram [bis(diethylthiocarbamoyl) disulfide, Antabuse; Fig. 7.6], a drug used for the therapy of alcoholism (Jin et al., 1994). Toward an understanding of the mechanism of action leading to the inhibition of liver aldehyde dehydrogenase (ALDH), Jin et al. investigated the formation of reactive metabolites of disulfuram (75 mg/kg) in rats by performing CNL scanning for 129 Da on bile extracts. Results showed the presence of 5 GSH adducts (1 major and 4 minor) in the LC/MS/MS CNL 129 Da chromatogram as common biliary metabolites, where the major GSH adduct was identified as *S*-(*N*,*N*-diethylcarbamoyl) glutathione (SDEG; Fig. 7.6). The identification of SDEG using CNL 129 Da scanning was important because it provided support for the proposal that reactive metabolites of disulfuram (one or both of the *S*-oxide derivatives shown in Fig. 7.6) function as carbamoylating metabolites which may be able to inactivate ALDH by covalent binding to an active site thiol. It is not yet known whether these reactive metabolites of disulfuram contribute to the associated hepatitis which has a considerable mortality risk and where histological signs of immunoallergy are common (Bjornsson et al., 2006).

7.3.2 Negative Ion Tandem LC/MS/MS for Precursors of the *m/z* 272 Production

An alternative LC/MS/MS approach for the detection of GSH adducts by positive ion CNL 129 Da scanning is by negative ion scanning for precursors of the product ion *m/z* 272. The *m/z* 272 ion is a major fragment ion of GSH adducts consisting of the elements of GSH minus H_2S and corresponding to deprotonated γ-glutamyl-dehydroalanyl-glycine produced by CID of $[M-H]^-$ ions of a range (aromatic, benzylic, aliphatic, thioester, etc.) of representative GSH adducts as well as GSH itself (Fig. 7.3; Dieckhaus et al., 2005). Dieckhaus et al. (2005) performed GSH-adduct detection survey scanning studies comparing positive ion CNL scanning for 129 Da to negative ion scanning for parents of *m/z* 272 from a series of *in vivo* and *in vitro* metabolism experiments with APAP, diclofenac, and troglitazone. In summary, results from those experiments showed that scanning for precursors of *m/z* 272 in the negative ion mode revealed the presence of the known GSH adducts and in some instances unmasked additional GSH adducts that had not been previously discovered. For example, during the LC/MS/MS analysis of bile obtained from a rat dosed with troglitazone, results showed that *m/z* 272 negative ion survey scanning led to the detection of two GSH adducts, one adduct with an $[M-H]^-$ ion at *m/z* 745 (consistent with the known reaction of GSH with the reactive quionone methide intermediate formed from metabolic activation of the chromane ring [M5, Fig. 7.5]), and one previously unidentified GSH adduct having an $[M-H]^-$ ion at *m/z* 731. Comparative analyses of the ability of the negative ion scanning for parents of *m/z* 272 method to the positive ion CNL 129 Da scanning method showed the detection of these two adducts, albeit the selectivity and sensitivity of the negative ion scanning technique was markedly superior. Based on these results, the authors proposed that negative ion mode scanning for precursors of *m/z* 272 should provide increased ability for the detection of unknown GSH adducts of candidate drugs present in bile samples

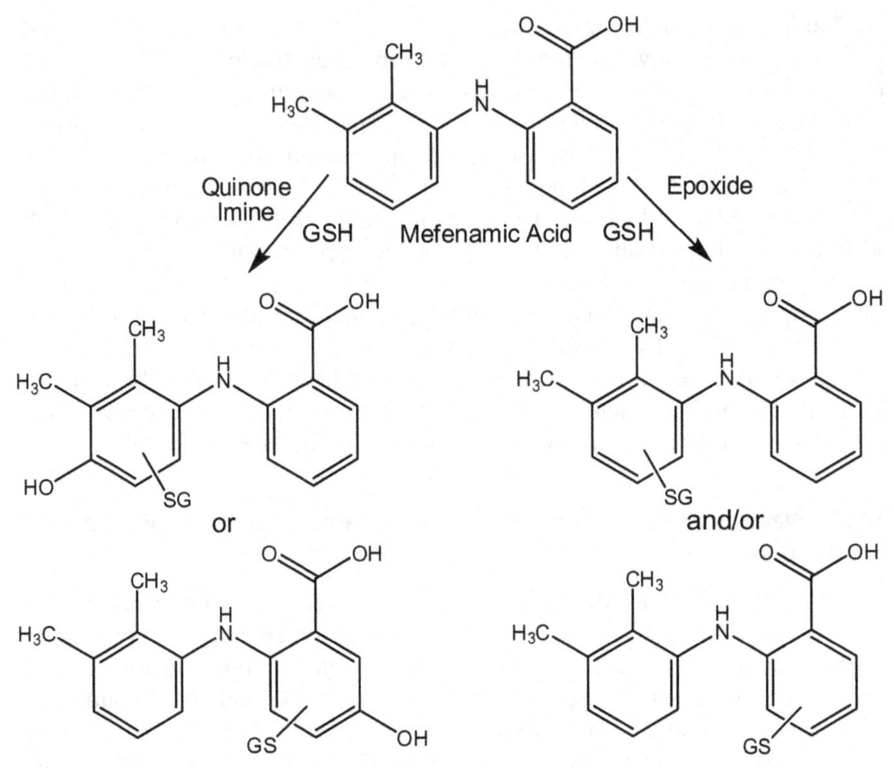

Figure 7.7 Proposed CYP-mediated bioactivation pathways of mefenamic acid occurring in HLM. (Adapted from Zheng et al., 2007.)

of dosed preclinical species. The overall proposed strategy based on these data would be to combine precursor ion scanning of the *m/z* 272 ion in negative ion mode as a survey technique for GSH-adduct detection, with subsequent generation of product ions of the corresponding $[M + H]^+$ ions in the positive ion mode to acquire data for chemical structure determination of the GSH adduct(s). Recently, our laboratory conducted GSH-adduct detection studies with the nonsteroidal anti-inflammatory drug (NSAID) mefenamic acid [(2′,3′)-dimethyl-*N*-phenyl-anthranilic acid]. Mefenamic acid has been shown to cause a rare idiosyncratic hepatotoxicity and nephrotoxicity (Somchit et al., 2004). A proposed mechanism for the development of these toxicities suggests that the drug is metabolized to chemically reactive metabolites that become covalently bound to tissue proteins leading to a potentially toxic immunological response (Fig. 7.7; Zheng et al., 2007). We performed rat bile duct cannulation studies where rats were dosed with mefenamic acid (20 mg/kg) followed by collection of bile over 4 h (unpublished results). Results from GSH adduct LC/MS/MS detection analyses are shown in Figs. 7.8 and 7.9. The positive ion survey scanning CNL 129 Da chromatogram (Fig. 7.8b) showed the

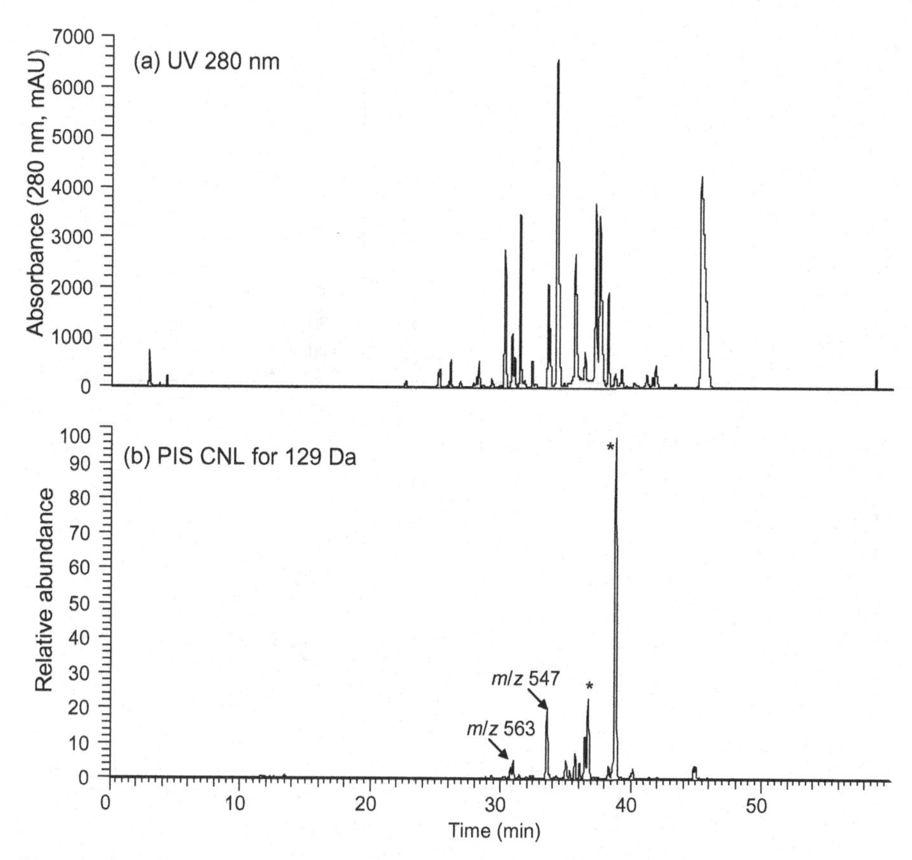

Figure 7.8 Representative (a) UV (280 nm) and (b) positive ion scanning (PIS) for CNL of 129 Da from LC chromatographic analysis of mefenamic acid-dosed rat bile. The total collected bile was treated with an equivalent volume of acetonitrile containing 3% formic acid. The supernatant of the microcentrifuged aliquot was then analyzed by LC/MS/MS using an Agilent 1100 HPLC linked to a Thermo Finnigan Discovery Quantum tandem mass spectrometer and in-line UV detector (280 nm). Chromatographic resolution was achieved with a Shiseido Capcell Pak-AQ C-18 column (4.6 × 250 mm, 3 μm, Tokyo, Japan). Mobile phases consisted of 0.1% aqueous formic acid and acetonitrile with 0.08% formic acid run at a flow rate of 1 mL/min. The solvent gradient was initially held at 0% solvent B for 5 min, and increased linearly to 70% solvent B over another 52 min. (*Denotes endogenous GSH adducts.)

presence of two GSH adducts of mefenamic acid with $[M + H]^+$ ions at *m/z* 547 and *m/z* 563, which are *m/z* values that correspond to the molecular weight of mefenamic acid + GSH − 2H, and mefenamic acid + GSH + atomic oxygen − 2H, respectively. GSH adducts of mefenamic acid with these *m/z* ratios have recently been detected from incubations with HLMs by using hybrid triple quadruple linear ion trap mass spectrometric techniques for screening and identification of GSH-trapped reactive metabolites (Zheng

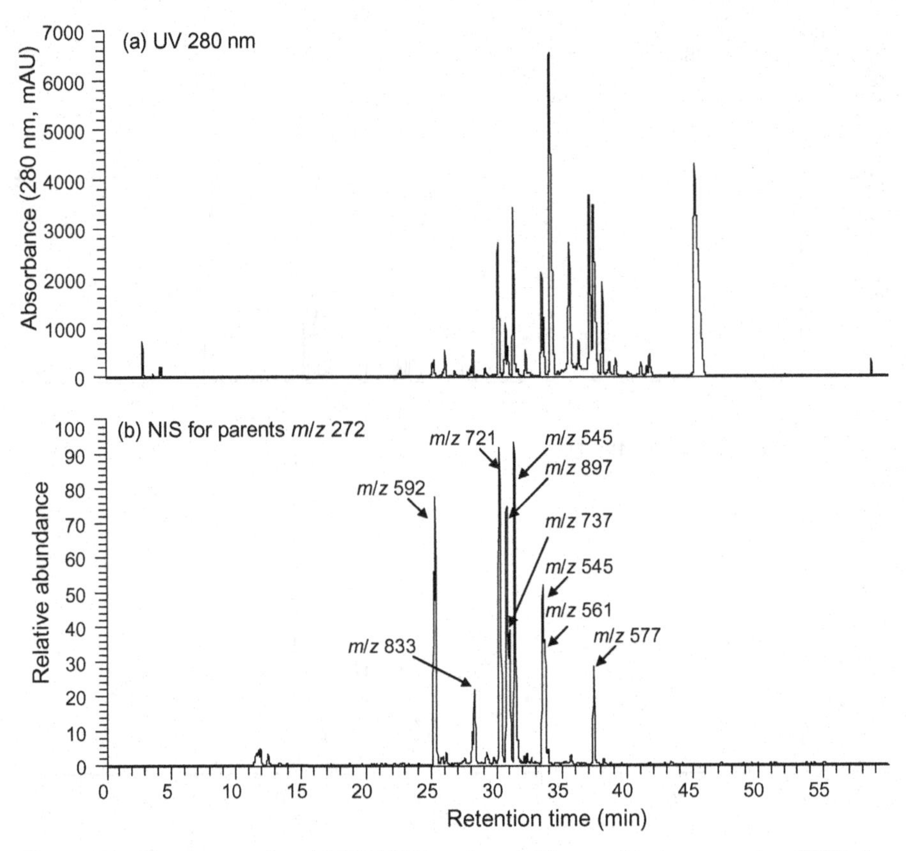

Figure 7.9 Representative (a) UV (280 nm) and (b) negative ion scanning (NIS) for parents of *m/z* 272 from LC/MS/MS chromatographic analysis of mefenamic acid-dosed rat bile (for methods, see legend for Fig. 7.8).

et al., 2007, discussed below). In comparison, the negative ion survey scanning for parents of *m/z* 272 (Fig. 7.9b) led to the detection of many more GSH adducts, some of which represented dual glucuronide/GSH adducts. The negative ion scanning peak at $[M-H]^-$ *m/z* 721, which eluted at ~30 min, corresponds to the molecular weight of the acyl glucuronide conjugate of the $[M + H]^+$ *m/z* 547 GSH adduct discussed above. Therefore, for mefenamic acid rat bile extracts, CNL 129 Da positive ion survey scanning did not allow for the detection of dual GSH-glucuronide mefenamic acid conjugates. The formation of the reactive metabolites leading to GSH adducts with $[M + H]^+$ ions *m/z* 547 and *m/z* 563 has been proposed to be mediated by cytochrome P450 bioactivation of mefenamic to epoxide and quinone imine intermediates, respectively (Zheng et al., 2007; Fig. 7.7). From these preliminary survey scanning studies on the detection of GSH adducts of mefenamic acid, the proposal by Dieckhaus et al. (2005) was confirmed, where the negative ion survey scanning resulted in the detection of GSH adducts that were not detected by positive ion survey scanning for CNL of 129 Da.

7.3.3 Mass Defect Filtering

In a recent study, the use of a mass defect filtering approach for the screening and identification of GSH adducts was performed by acquiring high-resolution LC-MS data with LTQ Orbitrap and Fourier transform ion cyclotron resonance mass spectrometric instruments (Zhu et al., 2007). GSH adducts of APAP, diclofenac, carbamazepine, clozapine, *p*-cresol, 4-ethylphenol, and 3-methylindole (10 μM) formed in incubations with NADPH-fortified HLMs were detected using this method. These hybrid high-resolution accurate mass instruments that were employed are extremely sensitive and have a rapid scan rate and MS^n scanning capacity; however, as opposed to triple quadrupole LC/MS/MS instruments, they may not be able to perform CNL or precursor ion scanning as described above. As a route around this disadvantage, Zhu et al. (2007) adopted a mass defect filtering technique from an approach that they used previously for the comprehensive analysis of oxidative metabolites of various drugs (Zhu et al., 2006). The general strategy for the mass defect filtering method of detection first involves acquiring full-scan LC-MS data using high-resolution mass spectrometry, then selecting GSH-adduct ions by processing the acquired LC-MS data set using mass defect filtering templates (mass defect filters from −0.040 to +0.040 Da around the mass defect of the filter template are capable of selecting common GSH adducts), and then characterizing the structure of the GSH adducts based on *m/z* values and on empirical formulas of the GSH adducts followed by interpretation of the product ion mass spectra through data-dependent analysis. Results from their studies showed that the mass defect filtering approach has many GSH-adduct detection advantages over the CNL 129 Da scanning method, in that it was more sensitive and effective at detecting GSH adducts that do not produce the 129 Da neutral loss as a significant fragment, it was better at detecting GSH adducts present in rat bile, and that the accurate mass spectral data were useful in the determination of GSH-adduct chemical formulas and the elimination of false positives.

7.3.4 Hybrid Triple Quadrupole Linear Ion Trap Mass Spectrometry

Screening and identifying GSH adducts using multiple reaction monitoring (MRM) as the survey scan to trigger the collection of enhanced product ion spectra is a technique recently developed by Zheng et al. (2007) using a triple quadrupole linear mass spectrometer. In their analyses, MRM was employed for survey scanning for more than 100 MRM transitions that are proposed to occur for most GSH adducts. The GSH-adduct $[M + H]^+$ ions that were focused on were those that are usually formed in common bioactivation scenarios that might occur, as mediated by P450 catalysis, in incubations with

HLMs fortified with NADPH. When detected by MRM survey scanning, the acquisition of enhanced product ion spectra of the GSH adduct is activated. Results from their studies using the model compounds APAP, diclofenac, carbamazepine, and from studies with clomipramine and mefenamic acid (chemicals that have structures susceptible to P450-mediated bioactivation), showed that compared to the CNL 129 Da scanning and negative precursor ion *m/z* 272 scanning method, the MRM-based approach gave increased sensitivity and selectivity of GSH-adduct detection, as well as enabled the sensitive acquisition of enhanced informative product ion spectra within the same LC/MS/MS run. Even though this method is more sensitive and selective than CNL 129 Da scanning for GSH adducts with a triple quadrupole, a drawback is that constructing specific MRM acquisition methods for a large amount of samples can be a time-consuming process which may not be amendable to high-throughput screening, particularly for novel types of GSH adducts having unpredictable structures.

Another highly sensitive and efficient method for the detection and characterization of GSH adducts (and for unknown GSH adducts) using a triple quadrupole linear ion trap mass spectrometer employs negative precursor ion *m/z* 272 scanning (discussed above) as the trigger for the acquisition of positive enhanced product ion spectra (Wen et al., 2008). Compared with the major limitation of triple-quadrupole-based survey scanning methods for GSH adducts, where positive ion CNL 129 Da and negative precursor ion *m/z* 272 scanning is a two-step method requiring a second LC/MS/MS run for the acquisition of tandem mass spectra, their "polarity switching" approach allowed for the detection of GSH adducts and the subsequent acquisition of positive ion tandem mass spectra in a single LC/MS/MS run. In their studies, Wen et al. tested this approach by analyzing GSH adducts of APAP, clozapine, diclofenac, imipramine, ticlopidine, and meclofenamic acid that were formed in incubations with NADPH-fortified HLMs. Their results showed superior sensitivity, reliability, and analytical speed compared to corresponding CNL 129 Da scanning analysis. Using this method for validation, they were able to easily detect, for example, the major GSH adducts of diclofenac ($[M + H]^+$ *m/z* 617 [Tang et al., 1999] and $[M + H]^+$ *m/z* 583 [Yan et al., 2005]) formed in incubations with HLMs, as well as previously unreported GSH adducts of diclofenac having $[M + H]^+$ *m/z* values of 567 and 599, representing adducts consisting of diclofenac + GSH + 2O − HCl and diclofenac + GSH − HCl, respectively. Results from their work showed the unprecedented advantage of polarity switching between mass spectrometric detection followed by CID for the rapid screening and characterization of GSH adducts. In agreement with the results from studies conducted by Dieckhaus et al. (2005), the superior selectivity and survey scanning sensitivity allowed by negative precursor ion scanning for *m/z* 272, with subsequent tandem mass spectra acquisition in the positive ion mode, provided GSH-adduct structural information and allowed for efficient identification of reactive intermediates in a single LC/MS/MS run.

7.4 ELECTROCHEMICAL GENERATION OF ELECTROPHILIC DRUG METABOLITES

A recently described and valuable method for generating reactive electrophilic metabolites of drugs *in vitro* is through the use of electrochemistry (Madsen et al., 2007, 2008a,b; Jurva et al., 2008). For example, in studies conducted by Madsen et al. (2008a), the electrochemical oxidation of diclofenac was achieved *in vitro* by oxidizing diclofenac (100 μM) in a solution consisting of 0.1 M potassium phosphate, (pH 7.4)/acetonitrile (3/1) using an ESA 5011 analytical electrochemical cell equipped with a porous graphite working electrode, a Pd counter electrode, and a Pd/H_2 reference electrode with an applied voltage of +1000 mV. The products of the electrochemical reaction were reacted with GSH (5 mM) under physiological conditions leading to the LC/MS/MS detection of two known GSH adducts having $[M + H]^+$ ions at *m/z* 617 produced from the reaction of GSH with the corresponding quinone imine metabolite at the 4- and 6-aromatic carbons (Fig. 7.4). A previously undetected adduct, namely 5-hydroxy-di-GSH adduct ($[M + H]^+$ at *m/z* 922) was also detected. From further electrochemical incubations with the 4′-hydroxy- and 5-hydroxy-diclofenac metabolites (25 μM, +600 mV) and subsequent reactions with GSH, a total of 11 adducts were formed, some of which were di-, tri-, and quatro-linked GSH adducts. After determining the LC/MS/MS characteristics of each of these adducts, the authors were able to develop sensitive LC/MS/MS methods to detect these adducts obtained from diclofenac-dosed rat bile and from diclofenac-treated incubations with rat and HLMs or rat hepatocytes. Their results showed that a number of previously unknown GSH adducts were detected from the sensitive and selective LC/MS/MS analyses of each adduct from these extracts.

Another example of the successful use of electrochemical oxidation comes from studies with troglitazone (Madsen et al., 2008b; Fig. 7.5), where it was shown that electrochemical oxidation of troglitazone led to the generation of a reactive metabolite which was easily trapped by GSH or *N*-acetylcysteine (NAC). Corresponding incubations of troglitazone with rat and HLMs in the presence of either GSH or NAC led to the formation of corresponding conjugates to those produced *in vitro* from electrochemical oxidation. NMR analysis of the purified troglitazone-*S*-(*N*-acetyl)cysteine adduct showed that NAC was linked to the benzylic carbon of the chromane moiety, indicating that the adduct was formed by reaction of an *o*-quinone methide with the nucleophilic thiol of NAC. Similar electrochemical oxidations of rosiglitazone, pioglitazone, and ciglitazone in the presence of GSH were also conducted; however, no GSH conjugates could be detected, even though each of these compounds contains the thiazolidinedione moiety capable of undergoing bioactivation (Fig. 7.5). Conclusions from these electrochemical studies are that only the *o*-quinone methide reactive metabolite of troglitazone was formed during electrochemical oxidation, which was the primary reaction product that was produced in corresponding human and rat liver microsome incubations.

Such results from studies with diclofenac and troglitazone clearly show the value of using electrochemical oxidation together with NMR for small-scale synthesis of metabolites for structural characterization and for collecting data on the potential mechanism(s) involved in the formation of reactive metabolites of drugs or drug candidates.

7.5 USE OF RADIOLABELED, STABLE LABELED, AND DERIVATIZED ANALOGUES OF GSH

Increasingly, there have been attempts to find quantitative techniques for the detection of reactive metabolites formed from non-radiolabeled drug candidates by using derivatives of GSH. Such techniques include the use of radiometric detection of reactive metabolites trapped with [^{3}H]GSH (Mulder and Le, 1998) or [^{35}S]GSH (Hartman et al., 2002; Takakusa et al., 2009) and by the fluorometric detection of reactive metabolites trapped with GSH derivatized with a dansyl group (Gan et al., 2005). These methods can be used to provide quantitative information on GSH adducts formed *in vitro* without the need for radiolabeled drug candidates.

7.5.1 Radiolabeled [^{35}S]GSH

A recent method for identifying and quantifying reactive intermediates which does not require the use of radiolabeled drugs was developed in order to detect and quantify radiolabeled GSH adducts formed in hepatocytes (Hartman et al., 2002). In this method, freshly isolated adherent rat and human hepatocytes were incubated overnight in methionine- and cysteine-free medium and then exposed for 3 h to 100 μM methionine and 10 μCi [^{35}S] methionine, a biosynthetic precursor of GSH, to replenish cellular GSH pools with intracellularly generated [^{35}S]GSH. To test for [^{35}S]GSH-adduct formation using this technique, APAP (Fig. 7.2) was incubated with rat and human adherent hepatocytes for 24 h. LC/radioactivity analysis of incubation extracts showed the presence of radiolabeled APAP-[^{35}S]GSH where adduct formation increased with increasing APAP concentration (0.5 mM–2 mM). For discovery compounds that might form unknown GSH adducts, the suspected GSH adducts identified in this manner would then undergo further characterization by LC/MS/MS techniques.

In a more recent report on the use of [^{35}S]GSH for trapping reactive drug metabolites, [^{35}S]GSH (0.2 mM, 925 kBq/incubation; commercially available from PerkinElmer, Wellesley, MA) and test compounds such as ticlopidine, clozapine, and diclofenac (100 μM) were coincubated with HLMs for 30 min in the presence of an NADPH generating system (Takakusa et al., 2009). Results showed that for these drugs, which are known to form reactive metabolites, the formation of [^{35}S]GSH adducts occurred at a high rate, which was contrary to corresponding experiments with nontoxic drugs (levofloxacin and caffeine). Negative results from their studies where radioactive [^{35}S]GSH adducts were not detected occurred for zomepirac and carbamazepine, which are drugs that have been shown by LC/MS/MS detection to form GSH adducts *in vitro* from the reaction of GSH with reactive epoxide intermediates (Chen et al., 2006; Zhu et al., 2007). In addition, only one [^{35}S]GSH adduct of diclofenac was detected, whereas at least eight diclofenac GSH adducts have been detected from LC/MS/MS analysis of HLM incubation extracts (Madsen et al., 2008a).

Similar to employing [^{35}S]GSH as a trapping nucleophile, another assay for the semiquantitative evaluation of the potential for drugs to undergo metabolic activation was developed using [^{35}S]cysteine as the trapping agent in incubations with NADPH-fortified HLMs (Inoue et al., 2009). Reactive intermediates were captured as [^{35}S]cysteine adducts and were analyzed by LC/radioactivity and LC/MS detection. To validate the assay, drugs known to form reactive intermediates such as clozapine, diclofenac, *R*-(+)-pulegone, and troglitazone were examined for [^{35}S]cysteine-adduct formation. Results showed that the cysteine adducts detected were consistent with the types of GSH adducts reported in the literature for each compound. When they compared the total LC/radiochromatographic peak areas corresponding to the [^{35}S]cysteine adducts of non-radiolabeled test compounds with the extent of covalent binding of radiolabeled compounds to liver microsomal proteins measured under standardized conditions (Day et al., 2005), the method correlated well with the predicted bioactivation potential of each compound tested. Therefore, the trapping assay could be used in place of the standardized covalent binding assay to predict covalent binding to protein *in vitro*. A potential problem with the use of radiolabeled GSH or cysteine analogues for adduct detection is that the chromatographic separation of unreacted radiolabeled GSH or cysteine from the radiolabeled adducts may be difficult, especially during the resolution of di- and tri-linked GSH adducts which can result in insufficient sensitivity. In addition, the use of radiolabeled reagents often requires special facilities, and can be expensive relative to the use of non-radiolabeled GSH derivatives as discussed below.

7.5.2 Stable Isotope-Labeled [$^{13}C_2$,^{15}N-Gly]GSH

Recent investigations on the employment of stable isotope-labeled GSH for the LC/MS/MS detection and characterization of GSH adducts have proven to be very effective. For instance, in metabolic activation experiments conducted with HLMs fortified with NADPH (or an NADPH-regenerating system), incubating test compounds known to undergo P450-mediated reactive metabolite formation with a known mixture of unlabeled and stable isotope-labeled GSH led to the detection of GSH adducts through the LC/MS/MS recognition of an isotopic doublet mass spectrometric signature displaying a known Dalton increase over the naturally occurring adduct (Mutlib et al., 2005; Lim et al., 2008). For these referenced studies, stable labeled-

[$^{13}C_2$,^{15}N-Gly]GSH was used to detect GSH adducts that produced a characteristic pattern of "twin ions" separated by 3 Da in their full scan mass spectra. In the Mutlib et al. (2005) report where they employed this technique, [$^{13}C_2$,^{15}N-Gly]GSH was obtained by a straightforward chemical synthesis starting from [$^{13}C_2$,^{15}N]Gly obtained from Isotec-Sigma Aldrich (Miamisburg, OH) (Mutlib et al., 2005), whereas in the report by Lim et al. (2008), [$^{13}C_2$,^{15}N-Gly]GSH was obtained from Cambridge Isotope Laboratories, Inc (Andover, MA). In the Mutlib et al. (2005) report, the successful detection and characterization of GSH adducts of APAP and flufenamic acid formed in rat and HLM incubations was accomplished using a 1:1 (w:w) mixture of labeled and non-labeled GSH (10 mM) and by using data-dependent scanning on a linear ion trap mass spectrometer where complimentary (full-scan, MS/MS, and MS^3 mass spectra) and confirmatory data were obtained all in one analytical run. It was also determined that a lower amount of the labeled GSH (e.g., 1:2 w/w of labeled/non-labeled GSH) could be used to trap and detect reactive intermediates in order to reduce the cost associated with [$^{13}C_2$,^{15}N-Gly]GSH. In summary, the combination of stable isotope-labeled GSH and linear ion traps can be used effectively in a drug discovery setting to provide for the very sensitive and selective identification of drug candidates susceptible to bioactivation.

Another detection method, developed by Lim et al. (2008), employed an LTQ/Orbitrap tandem mass spectrometer and multiplexed high-resolution accurate mass analysis with isotope pattern triggered data-dependent product ion scanning for the simultaneous detection and structural elucidation of GSH adducts in a single analytical run. In their studies, they were successful when using this technique to detect GSH adducts of APAP, tienilic acid, clozapine, ticlofidine, and mifepristone. The compounds (10 μM) were incubated for 1 h at 37°C with NADPH-fortified HLMs in the presence of a 2:1 ratio of 5 mM GSH and [$^{13}C_2$,^{15}N-Gly]GSH. The GSH adducts were detected by isotope searching of mass defect filtered and control subtracted full-scan accurate mass data using MetWorks software 1.1.0 (Thermo Fisher Scientific, Inc., San Jose, CA). In general, this GSH-adduct detection method has a broad detection potential because the method works independently of the CID properties of GSH adducts which are known to vary widely (Dieckhaus et al., 2005).

In another recently published method, the unbiased high-throughput screening of reactive metabolites using linear ion trap mass spectrometry, polarity switching, and mass tag triggered data-dependent acquisition for GSH-adduct detection was developed (Yan et al., 2008). This method, which was envisioned as an alternative approach for the high-throughput screening and structural characterization of GSH adducts when using a linear ion trap mass spectrometer, proved to be highly efficient. In their studies, a mixture of 0.85:1 (w:w) labeled ([$^{13}C_2$,^{15}N-Gly]GSH) and non-labeled GSH was used. Their methodology consisted of analyzing NADPH-fortified liver microsome extracts from incubations of test compounds (10 μM) with the labeled and non-labeled GSH mixture and then analyzing incubation extracts by LC/MS/MS in both the positive and negative modes (in order to eliminate the potential for response bias in these two modes) via polarity switching all in the same analytical run. As discussed above, the detected GSH adducts would exist as two isotopic partners that would differ in mass by 3.0 Da and would appear at the fixed (0.85:1) intensity ratio to be identified as a highly unique mass spectrometric ion pattern that would trigger a data-dependent tandem mass spectra scan by CID of both the labeled and non-labeled GSH adducts. Proof-of-concept studies were conducted with 4-methylphenol, a compound that is known to undergo P450-mediated metabolic activation in incubations with HLMs to quinone methide and 4-methyl-*o*-benzoquinone reactive intermediates. Results from the use of this technique showed the detection of two adducts, namely 4-(glutathione-*S*-yl-methyl)-hydroquinone and 3-(glutathione-*S*-yl)-5-methyl-*o*-hydroquinone, formed from the reaction of GSH with the quinone methide and 4-methyl-*o*-benzoquinone intermediates, respectively. This mass spectra pattern-based triggering method is useful because it is not likely to cause false negatives. In addition, this approach can provide for wide-range coverage of GSH adducts, regardless of their polarity preferences. Finally, using this technique, Yan et al. (2008) showed that assurance of positive structural identification is obtainable since tandem mass spectra of both labeled and non-labeled GSH adducts, in both positive and negative modes, can be collected for most GSH adducts in one LC/MS/MS run.

7.5.3 Derivatized Analogues of GSH

GSH Ethyl Ester (GSH-EE) GSH-EE (Fig. 7.10) is a GSH-derivatized analog where the glycine carboxylic acid moiety is in the form of the ethyl ester. The use of this of this derivative has been shown to lead to increased sensitivity (~10-fold) toward reactive metabolite detection (Soglia et al., 2004). GSH-EE was recently used in conjunction with hybrid triple quadrupole linear ion trap mass spectrometry to facilitate the identification of varied reactive drug metabolites (Wen and Fitch, 2009). In their studies, Wen and Fitch employed polarity switching, where GSH adducts were detected by negative precursor ion survey scanning for the product anion at *m/z* 300 (an *m/z* ratio corresponding to deprotonated γ-glutamyl-dehydroalanyl-glycine derived from GSH-EE), to trigger the acquisition of enhanced positive product ion scanning in the positive in mode. This method was developed as an approach for screening GSH adducts that would distinguish the drug GSH-EE adducts from endogenous GSH adducts that do not provide the *m/z* 300 product ion in the negative ion survey scanning mode. The method, which was used to assay extracts from HLM incubations with APAP, amodiaquine, carbamazepine, 4-ethylphenol, imipramine, and ticlopidine, showed very sensitive and selective detection of GSH-EE adducts by negative ion precursor ion scanning, and where polarity switching to the positive ion mode for the generation of product ion spectra provided consistent structural information of the varied GSH adducts detected. Another benefit from the use of GSH-EE comes from enhanced chromatographic resolution of the GSH-EE adducts, compared to

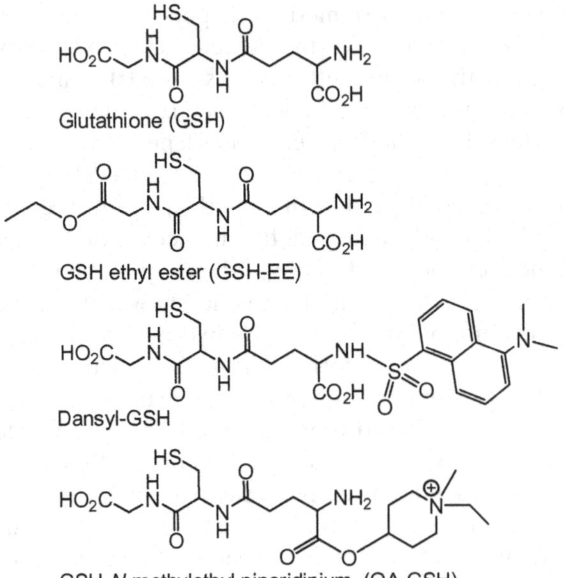

Glutathione (GSH)

GSH ethyl ester (GSH-EE)

Dansyl-GSH

GSH-*N*-methylethyl piperidinium (QA-GSH)

Figure 7.10 Chemical structures of GSH and its derivatized analogs cited in the text.

GSH adducts, due to increased lipophilicity provided by GSH-EE functional group. One potential drawback of using GSH-EE, which should especially be considered when employing rapid LC-gradient elution, is that it might be able to undergo oxidation to the corresponding disulfide which could co-elute with GSH-EE adducts and interfere with their detection.

Dansyl GSH The use of dansyl GSH (Fig. 7.10) as a trapping reagent for the quantitation of reactive metabolite formation was recently developed by Gan et al. (2005). The method utilized a fluorescent GSH analogue, where GSH was derivatized with a dansyl fluorescent tag attached to the free amine moiety of the γ-glutamyl side chain. Dansyl GSH was employed to quantitatively assess metabolic activation of drug candidates that might be occurring in incubations with liver microsomes fortified with NADPH. This alternative method of GSH-adduct quantification was developed to get around problems involved with the use of radiolabel drug analogues (discussed above). The synthesis of dansyl GSH was straightforward and involved derivatization of oxidized GSH with dansyl chloride followed by reduction with dithiothreitol and subsequent purification by preparative HPLC (>99.5% purity; 74% overall yield). In their proof of concept studies, Gan et al. performed standard incubations with HLMs and substrates (50 μM), NADPH (1 mM), in the presence of dansyl GSH (1 mM). From these incubation extracts, reverse phase (C18) HPLC with fluorescence detection was used for the quantification of dansyl-GSH adducts using excitation and emission wavelengths of 340 and 525 nm, respectively. Confirmation of adduct formation was assessed from detection with an in-line ion trap mass spectrometer operating in full scan alternating positive and negative ion modes. In their studies, 14 compounds were tested, 7 of which are known to form GSH adducts (troglitazone, APAP, *R*-(+)-pulegone, diclofenac, clozapine, bromobenzene, and precocene) and 7 that are highly prescribed drugs (omeprazole, lansoprazole, fluoxetine, celecoxib, paroxetine, loratadine, and sertraline) that have not been reported to form GSH adducts and are not known to be toxic via reactive metabolite formation. Their results showed that dansyl-GSH adducts were able to be detected and quantified from incubations of all seven positive control compounds. By contrast, no dansyl-GSH adducts were detected from corresponding incubations with the seven negative control drugs, which is consistent with their safety profile. One notable example where this technique did not work comes from a recent report where the LC/MS/MS detection GSH adducts of paroxetine produced in incubations with HLMs were discovered and which were shown to be NADPH-dependent via the P450-mediated formation of quinone and benzoquinoneimine reactive intermediates (Zhao et al., 2007). A potential explanation for the excellent safety profile of paroxetine could be due to the low once-a-day dosing (20 mg; Uetrecht, 2001), and/or because of the adequate scavenging of the catechol and quinone metabolites by Phase II conjugation and GSH, respectively (Zhao et al., 2007).

One attractive feature of dansyl GSH is that it does not function as a cofactor for glutathione *S*-transferases (GSTs). In general, reactive intermediates that possess sufficient chemical reactivity, such that they do not require GST catalysis to form the corresponding GSH adducts, are most likely to be of greater toxicological concern with respect to the potential for covalent binding to protein nucleophiles. Dansyl GSH and GSH were shown to possess similar rates of chemical reactivity *in vitro* in reactions with chlorodinitrobenzene forming the dansyl-GSH-*S*-yl- and GSH-dinitrobenzene adduct, respectively, in the absence of GST catalysis. In similar reactions of GSH with dichloronitrobenze, a much less reactive electrophile, GST catalysis was necessary for forming the corresponding glutathion-*S*-yl-chloronitrobenzene adduct. However, dansyl GSH in these incubations with GST did not form the danysl-glutathion-*S*-yl-chloronitrobenzene adduct. The lack of catalysis by GST is believed to be due to dansyl GSH having poorer binding affinity for the GSH binding site of GST on account of its increased bulky structure and blockage of the free amine on the γ-glutamyl side chain of GSH which is known to be important for binding of GSH to GST (Gan et al., 2005). In summary, the dansyl-GSH nucleophilic trapping method is useful for the capturing of thiol-reactive intermediates in that the fluorescent danysl moiety provides for the sensitive detection and quantification of GSH adducts by HPLC fluorescence analysis. This method is important because, unlike other trapping methods, it provides a combination of quantification capability, ease of implementation and use (none of the issues regarding the use of radioactivity), and cost-effectiveness. However, the method does require the HPLC separation of fluorescently labeled dansyl-GSH adducts from unreacted dansyl-GSH (including oxidized disulfide) reagent. By its very nature, this strategy, just like the use of radiolabeled GSH, may find constrained high-throughput screening applicability. Another limitation of this strategy is that it is not applicable for the testing of fluorescent drugs due to potential fluorescence interference. In addition, it is not known if dansyl GSH possesses the same reactivity as GSH with the full range of reactive metabolite structures.

Quaternary Ammonium GSH (QA-GSH) Analogues A recent effective method for the determination of GSH-adduct formation was developed by Soglia et al. (2006) and uses LC/MS/MS and a novel QA-GSH analogue (Fig. 7.10). The method was established for the high-throughput screening and semiquantitative detection of reactive metabolite formation occurring in incubations of drug with HLMs. In their proof-of-concept studies, Soglia et al. performed standard incubations with HLMs and test substrates (10–100 μM) fortified with an NADPH-regenerating system and QA-GSH (1 mM). The chemical synthesis of QA-GSH involved a multistep process that led to >95% purity (the overall yield was not reported) of the derivative where the γ-glutamyl side-chain C-terminus carboxylate was tethered to an *N*-methylethyl piperidinium group resulting in a fixed positive charge. Using the γ-glutamyl side-chain carboxylate, rather than derivatization at the glycine carboxylate, was chosen in order to maximize the distance between the fixed positive charge moiety and the nucleophilic cysteinyl-thiol group.

In their studies, LC/MS/MS detection of the QA-GSH adduct (performed using a triple quadrupole mass spectrometer with an orthogonal electrospray ionization interface) was shown to be selective and highly sensitive. Semiquantitation of QA-GSH adducts was accomplished by employing authentic QA-GSH adduct standards added to samples prior to processing and analysis. The mass spectrometric signal response was determined from analyzing equal concentrations of varied QA-GSH standards and showed that the full-scan mass spectrometric responses only differed by ~3.3-fold, even though the drug moiety of the varied QA-GSH adducts varied significantly. This method was tested using the model compounds APAP, clozapine, and flutamide which are known to form reactive metabolites in incubations with NADPH-fortified HLMs. In the studies with APAP, the QA-GSH internal standard semiquantitation method was compared to the use of authentic APAP-GSH adduct standard curves from corresponding incubations of APAP with GSH and showed that the calculated levels of GSH adduct formed compared very closely. The ability of QA-GSH to serve as a cofactor for GST catalysis in QA-GSH adduct formation was not characterized in their studies; however, the authors proposed that since the QA-GSH structure is changed significantly, where the carboxylate on the γ-glutamyl side chain of GSH is changed from a negatively charged ion to a positively charged amine, the QA-GSH might not serve as a cofactor for GST enzymes. As for the dansyl-GSH derivative, the fact that the QA-GSH adduct might not react with reactive drug metabolites in a GST-catalyzed fashion might not matter since it is believed that electrophilic intermediates are more problematic when their chemical reactivity is such that they do not require GST catalysis for GSH-adduct formation.

In summary, the use of the QA-GSH internal standard methodology for potential high-throughput screening proved to be useful for the detection and semiquantitation of reactive metabolites formed from the metabolism of varied drugs. A possible drawback for this method, as is true for all GSH-type trapping reagents, is that they are only able to trap soft electrophiles. The trapping strategies for hard electrophiles (e.g., aldehydes, iminium ions) with methoxylamine, semicarbazide, or KCN will be discussed below. However, the authors state that a similar method could potentially be developed using QA-*N*-acetyl-lysine to generate fixed charge QA-*N*-acetyl-lysine adducts of hard nucleophiles amenable to facile detection by LC/MS/MS.

7.6 USE OF RADIOLABELED DRUGS FOR GSH-ADDUCT DETECTION

Although the synthesis of radiolabeled [^{14}C]- or [^{3}H]-analogues of discovery candidate compounds is expensive and can be time-consuming, they are often made available for studies on the determination of the major routes of metabolism of late-stage discovery compounds occurring *in vivo* in preclinical species (bile, urine, plasma) and *in vitro*. Recent unpublished observations from our laboratory come from investigations on the metabolism of [^{14}C]diclofenac to GSH adducts in incubations with liver microsomes. Figure 7.11 shows representative reverse-phase gradient LC/radioactivity chromatograms generated (under the same chromatographic conditions as described above for the analysis of GSH adducts of mefenamic acid; Fig. 7.8) from the analysis of extracts from rat and HLMs incubated with [^{14}C]diclofenac (50 μM; 62.0 mCi/mmol), NADPH (1 mM), and GSH (10 mM). The representative radiochromatograms show the presence of at least 15 GSH adducts (Fig. 7.11), many of which were predicted by the electrochemical oxidation studies described above (Madsen et al., 2008a). Studies are ongoing to determine the identities of these GSH adducts. However, one of the easily detected adducts that was not detected in any other screening studies conducted to date was the GSH adduct (labeled peak 15) eluting at 34.5 min which was fully characterized by LC/MS/MS and NMR and determined to be 2-(2,6-dichloro-phenylamino)benzyl-*S*-thioether GSH (DPAB-SG; Fig. 7.4). DPAB-SG was proposed to be formed from the P450 3A4-mediated oxidative decarboxylation of diclofenac resulting in the formation of an intermediate benzylic carbon-centered radical that may either recombine with hydroxyl radical (forming a benzyl alcohol intermediate which loses water to produce a reactive *o*-imine methide) or undergo further oxidation leading directly to the *o*-imine-methide intermediate that reacts with GSH (Grillo et al., 2008; Teffera et al., 2008). From our LC/MS/MS analysis of the corresponding extracts (using the same chromatography method as

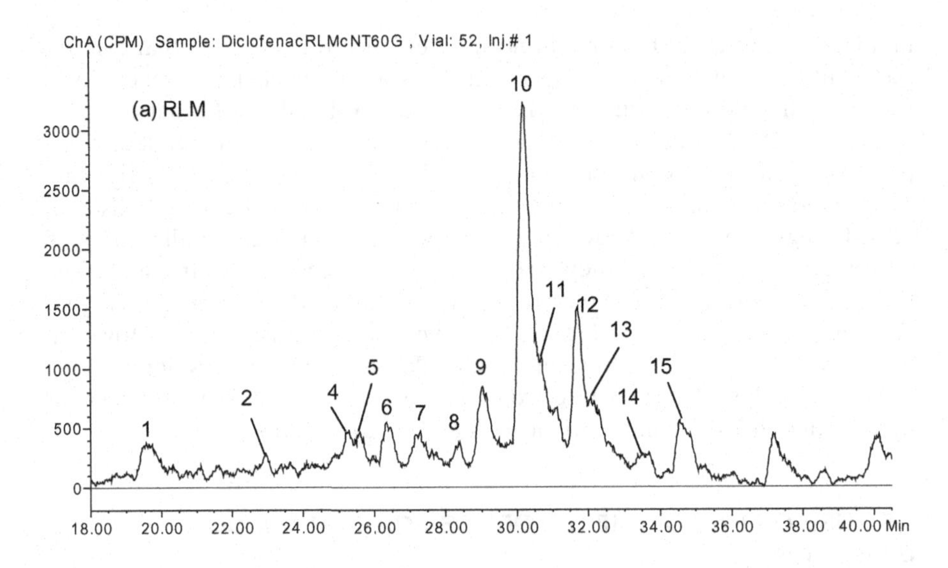

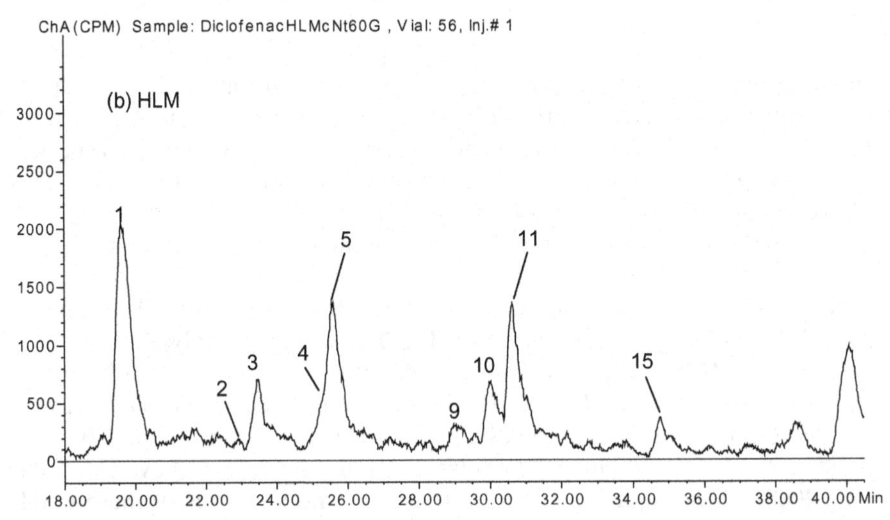

Figure 7.11 Representative reverse-phase gradient LC/radioactivity chromatograms of extract from (a) rat (rat liver microsomes [RLM], 1 mg protein/mL) or human liver microsomes (HLM, 1 mg protein/mL) incubated with [^{14}C]diclofenac (50 μM), NADPH (1 mM), and GSH (10 mM) for 60 min.

described above for LC/radioactivity analysis), the $[M + H]^+$ *m/z* for some of these peaks were determined to be peak 1 (*m/z* 1159), peak 3 (*m/z* 922), peak 5 (*m/z* 888), peak 9 (*m/z* 583), peak 10 (*m/z* 617), peak 11 (*m/z* 617), and peak 15 (*m/z* 557). Using both the positive ion CNL 129 Da and the negative ion precursor scanning for *m/z* 272 GSH-adduct monitoring techniques only allowed for the detection of peaks 10 and 11 ($[M + H]^+$ *m/z* 617; data not

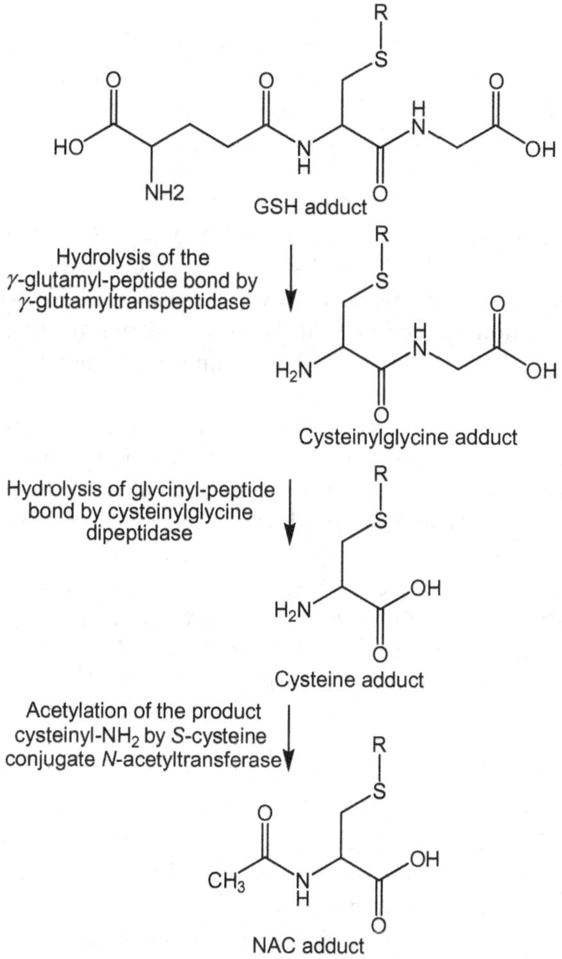

GSH adduct

Hydrolysis of the
γ-glutamyl-peptide bond by
γ-glutamyltranspeptidase

Cysteinylglycine adduct

Hydrolysis of glycinyl-peptide
bond by cysteinylglycine
dipeptidase

Cysteine adduct

Acetylation of the product
cysteinyl-NH_2 by *S*-cysteine
conjugate *N*-acetyltransferase

NAC adduct

Figure 7.12 Enzyme-mediated degradation of GSH adducts in the kidney leading to the formation of mercapturic acid conjugates (*N*-acetylcysteine [NAC] adducts).

shown). Importantly, results from studies with diclofenac clearly demonstrate that more work is required to discover superior methods for the LC/MS/MS detection of GSH adducts present in biological extracts.

7.7 NAC

NAC adducts, also commonly known as mercapturic acid adducts, are *S*-conjugated *N*-acetyl-L-cysteine derivatives that are formed from the conjugation of reactive electrophilic intermediates with GSH usually in the liver (Fig. 7.12). GSH adducts can be transported out of the hepatocyte into the plasma where they undergo "interorgan transport" to the kidney. Once in the kidney, GSH adducts are broken down first by γ-glutatamyl transpeptidase (γ-GT; EC 2.3.2.2), the only known enzyme to cleave GSH and GSH adducts appreciably, which functions by hydrolyzing these conjugates at the γ-glutamyl bond (Tate and Meister, 1985) leading to the formation of cysteinylglycine-*S*-linked adducts (cysteinylglycine adducts). Following this γ-GT step, cysteinylglycine adducts undergo further hydrolysis by cysteinylglycine dipeptidase (EC 3.4.13.6), leading to the formation of the corresponding L-cysteine-*S* adducts (Jones et al., 1979). These L-cysteine-*S* adducts then serve as a substrates for *S*-cysteine conjugate *N*-acetyltransferase (EC 2.3.1.80; Aigner et al., 1996), which functions to acetylate the cysteinyl-amine residue of these conjugates (via the cofactor acetyl-CoA) affording NAC adducts prior to excretion in urine.

NAC adducts can be used as biomarkers for drugs or environmental contaminants that are biotransformed to chemically reactive metabolites *in vivo* in preclinical species and in humans (van Welie et al., 1992; Scholz et al., 2005; Ding et al., 2009). Because NAC is a lower molecular weight compound with fewer bonds, an advantage of using NAC versus GSH as a trapping reagent occurs when NAC adducts are analyzed by LC/MS/MS CID because they are able to generate more product ions from cleavages of the drug moiety, therefore providing important structural information with regard to determining the site of bioactivation on the chemical (Jian et al., 2009).

7.7.1 Lumiracoxib

A successful example of using reactive metabolite trapping with NAC comes from studies with lumiracoxib, a COX-2 selective inhibitor NSAID recently removed from the market for toxicity reasons (Li et al., 2008; Fig. 7.13). Due to its chemical structure containing a 2′-chloro-6′-fluorophenyl-amino group, the exposed 4′-position on the aromatic ring was predicted to undergo metabolic activation by cytochrome P450 to a reactive quinone imine intermediate, similar to the mechanism of cytochrome P450-mediated bioactivation of diclofenac (Tang et al., 1999; Fig. 7.4). In addition, because reports suggested that the drug may cause hepatotoxicity in patients (Laine et al., 2008); lumiracoxib was studied *in vitro* in incubations with rat and HLMs to examine potential bioactivation to reactive intermediates that may be trapped by NAC. In their studies, lumiracoxib (50 μM) was incubated with liver microsomes, NADPH, and NAC (5 mM), or with human hepatocytes in suspension for 60 min at 37°C. Extracts from these incubations were analyzed by LC/MS/MS on an ion trap mass spectrometer in the positive ion mode. Results from these studies, which included NMR analysis of purified NAC adducts, showed the identification of two NAC adducts, namely 3′-NAC-4′-hydroxy lumiracoxib and 4′-hydroxy-6′-NAC-desflouro lumiracoxib, from each of the incubation systems tested (Fig. 7.13). Inhibition studies with anti-CYP2C9 antibody provided strong evidence of CYP2C9 mediating the bioactivation of the drug.

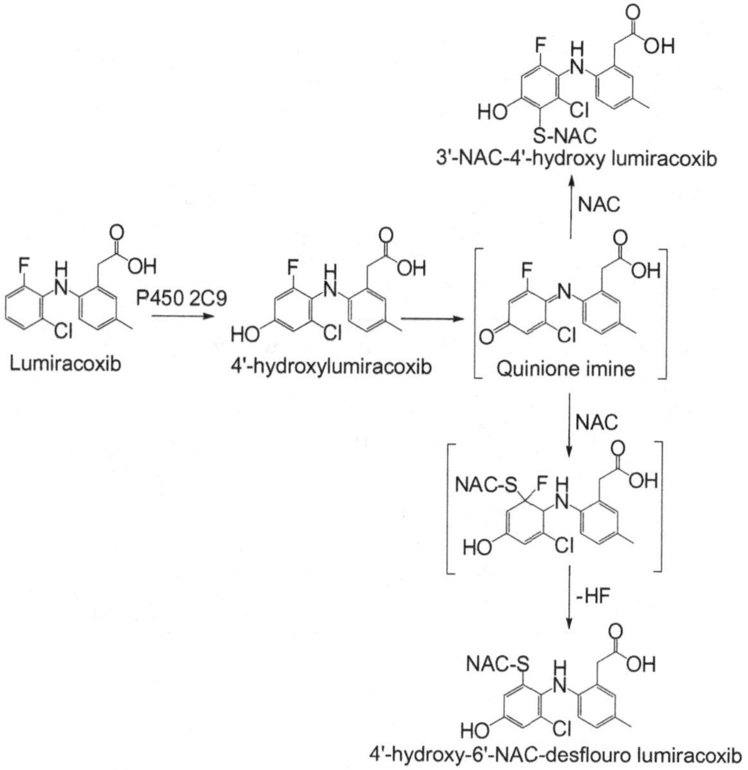

Lumiracoxib → (P450 2C9) → 4'-hydroxylumiracoxib → Quinione imine → (NAC) → 3'-NAC-4'-hydroxy lumiracoxib; Quinione imine → (NAC) → intermediate → (-HF) → 4'-hydroxy-6'-NAC-desflouro lumiracoxib

Figure 7.13 Proposed mechanism for the P450 2C9-mediated metabolic activation of lumiracoxib leading to the formation of NAC adducts.

The ability to use NAC to trap reactive quinone imine metabolites of lumiracoxib in incubations with human hepatocytes suggested that the same reactive metabolite formation might also be occurring in humans dosed with the drug. These mechanistic *in vitro* studies employing NAC as a trapping nucleophile are very helpful toward understanding a potential relationship between the metabolism of lumiracoxib to reactive intermediates and subsequent liver injury. Corresponding GSH-adduct detection studies in HLMs using an ion trap LC/MS/MS in data-dependent scanning mode showed the presence of two major GSH adducts, one of which was 4′-hydroxy-6′-GSH-desflouro lumiracoxib (the GSH adduct which corresponds to 4′-hydroxy-6′-NAC-desflouro lumiracoxib discussed above). The second major GSH adduct was formed from P450-mediated oxidation of lumiracoxib to an imino methide intermediate that reacts with GSH at the benzylic position. Many minor GSH adducts of the drug were also detected and included di- and tri-GSH adducts. The authors concluded that the detection of NAC and GSH adducts of lumiracoxib provided evidence that a reactive quinone imine intermediate of the drug

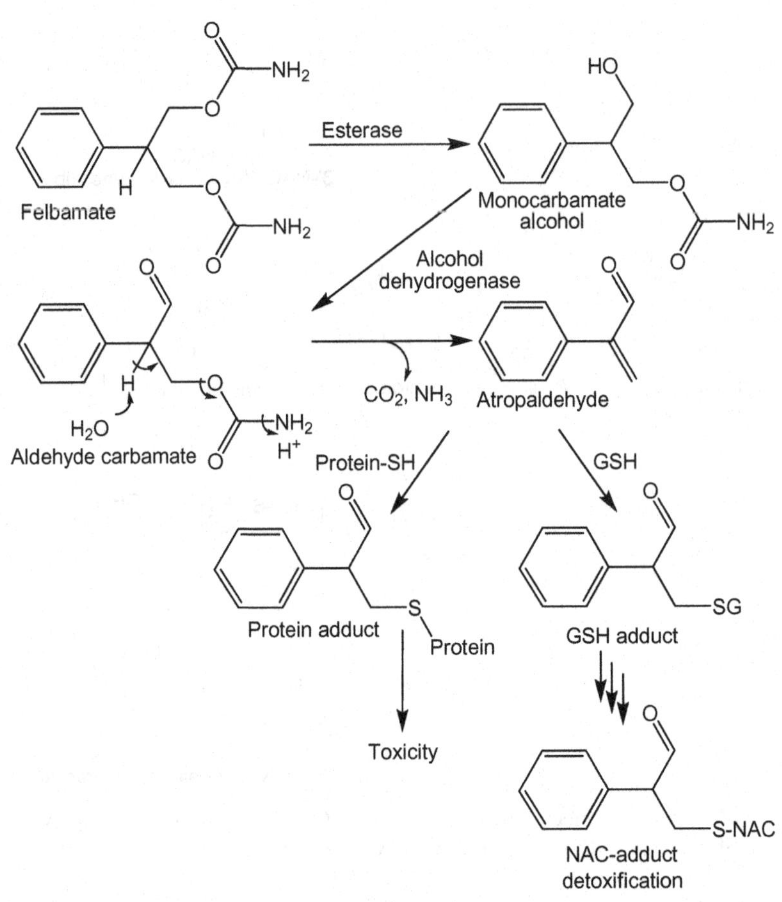

Figure 7.14 Proposed mechanism for the metabolic activation of felbamate to the reactive intermediate, atropaldehyde, that either binds covalently with protein leading to potential toxicity, or with GSH leading to elimination of the NAC adduct in urine (Dieckhaus et al., 2002.)

formed *in vivo* might lead to GSH-depletion, oxidative stress, covalent binding to protein, and to potential hepatotoxicity.

7.7.2 Felbamate

Another example of successful NAC-adduct detection comes from studies with felbamate (2-phenyl-1,3-propanediol dicarbamate; Fig. 7.14). Felbamate is a broad spectrum anticonvulsant drug that was approved by the Food and Drug Administration and introduced into the market in 1993; however, the use of felbamate was highly restricted when black box warnings were given after only 1 year on the market. These black box warnings occurred because there were 33 cases of aplastic anemia and 18 cases of hepatic failure out of 120,000 patients receiving the medication (Pellock, 1999). A proposed mechanism to explain the hepatotoxicity was that reactive metabolites of felbamate may have been formed in the livers of these patients (Dieckhaus et al., 2002; Kapetanovic et al., 2002). The metabolic activation of felbamate occurs by esterase-mediated formation of a monocarbamate alcohol intermediate followed by alcohol dehydrogenase-mediated oxidation to an aldehyde carbamate intermediate which then releases CO_2 and ammonia, giving rise to the reactive metabolite, atropaldehyde (2-phenylpropenal; Fig. 7.14). Atropaldehyde is proposed to bind to protein cysteinyl sulfhydryls forming covalent protein adducts that are immunogenic, leading to hepatotoxic allergic reactions (Dieckhaus et al., 2000). Evidence in support of this proposal comes from the LC/MS identification of reduced (*N*-acetyl-*S*-(2-phenylpropan-3-ol)-L-cysteine) and oxidized (*N*-acetyl-*S*-(2-phenyl-3-propanoic acid)-L-cysteine) mercapturic acids of atropaldehyde in the urine of felbamate-dosed rats and humans (Thompson et al., 1997, 1999). These findings were important because they established the proposal that the reactive atropaldehyde intermediate is formed in felbamate-dosed humans, which undergoes GSH-adduct formation, and finally is excreted in urine in the form of a NAC adduct. Therefore, the search for a NAC adduct in urine of dosed patients assisted in potentially understanding the toxic mechanism that might be occurring during the therapeutic use of felbamate.

7.7.3 Diclofenac

A rapid, sensitive, and specific detection method for NAC adducts of unknown structure was recently developed by Scholz et al. (2005). The method employed a linear ion trap LC/MS/MS with negative ion mode CNL scanning for 129 Da for the detection of NAC adducts which then would trigger the collection of product ion spectra for structural determination (Fig. 7.15). Using this method, both urine extracts and extracts from liver microsomal incubations fortified with NADPH and NAC were examined. In their studies, varied NAC-adduct standards were analyzed for characteristic product ions to determine specific CNL fragment, followed by the generation of theoretical MRM transitions using the observed CNL fragments for comparison to the

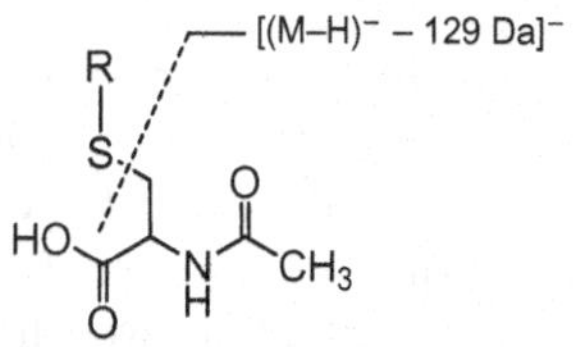

Figure 7.15 Negative ion LC/MS/MS screening for NAC adducts by CNL scanning for 129 Da.

information-dependent CNL method. Their results showed that for a range of 16 authentic NAC-adduct standards tested, negative ion mode LC/MS/MS product ion spectra produced by CID of the $[M-H]^-$ ions gave the common CNL of 129 Da as major fragment consisting of the elements of NAC minus H_2S (*N*-acetyl-dehydroalanine; Fig. 7.15). By contrast, CID of the $[M+H]^+$ ions in the positive ion mode produced neither a common fragment nor common CNL. In order to test the suitability of this technique in the detection of reactive intermediates formed *in vitro*, diclofenac was selected as a model substrate to scan for NAC adducts produced in rat and HLM incubations in the presence of NAC (10 mM). Results showed the detection of two $[M-H]^-$ ions both with *m/z* 471, which corresponded to two regioisomeric NAC adducts of hydroxy-glutathionyl-diclofenac. In summary, the method proved to be capable of the detection of NAC adducts from complex matrices. The two information-dependent methods employing either CNL or theoretical MRM resulted in similar sensitivity of NAC-adduct screening detection when analyzing urine and microsomal incubation extracts.

7.7.4 Quadrupole-Linear Ion Trap Mass Spectrometry and Polarity Switching

A recently reported method for the identification and characterization of NAC adducts of drugs was conducted using a hybrid triple quadrupole-linear ion trap LC/MS (Q-Trap; Jian et al., 2009). The method was developed to include the detection of NAC adducts excreted in urine or formed *in vitro* in incubations with liver microsomes. The method used polarity switching, where CNL scanning (using the survey scanning for CNL 129 Da) or MRM (multiple transitions corresponding to the loss of 129 Da from the $[M–H]^-$ ion to a product ion) were performed in the negative ion mode which would trigger information-dependent enhanced product ion scanning of detected NAC adducts in the positive ion mode. Again, as was the case for the analysis of GSH adducts, the polarity switching ability of the hybrid Q-trap instrument made possible the acquisition of both total ion chromatograms of the CNL scanning (or MRM) and the total ion chromatogram for the enhanced product ion scan in one LC/MS/MS run. To test this method, NAC adducts of APAP were searched for in urine extracts from APAP-dosed (500 mg) volunteers. Results showed one major NAC adduct, namely 3-NAC-APAP, and two minor NAC adducts, tentatively assigned as 3-hydroxy-5-NAC-APAP and 3-methoxy-5-NAC-APAP. The method also proved to be effective for the analysis of NAC adducts formed *in vitro* in incubations of the test compounds (30 μM) diclofenac and clozapine with rat liver microsomes, where two and seven NAC adducts, respectively, were readily detected and where the product ion spectra provided structurally informative fragment ions. The MRM-enhanced product ion scanning method showed high sensitivity; however, it required a compound-dependent acquisition protocol. The authors propose that this method, which is applicable for high-throughput analysis, might be useful in target analysis of low concentrations of urinary NAC adducts during early drug development studies.

7.8 CYSTEINE- AND CYSTEINYLGLYCINE-*N*-ACYL-AMIDES

Carboxylic acid-containing drugs can be metabolized to reactive acyl glucuronides and/or acyl-CoA thioester metabolites that are capable of transacylating protein nucleophiles and GSH (Benet et al., 1993; Shore et al., 1995; Boelsterli, 2002; Sidenius et al., 2004; Skonberg et al., 2008). Transacylation of GSH results in the formation of *S*-acyl-GSH thioesters that are often excreted and detected in bile (Grillo and Benet, 2002; Grillo et al., 2003; Grillo and Hua, 2003, 2008). *S*-acyl-GSH thioesters that are not excreted unchanged in bile would be predicted, as discussed above (Fig. 7.12), to undergo sequential enzyme catalyzed degradation steps to the *S*-acyl-linked mercapturic acid conjugate (*S*-acyl-NAC) prior to excretion from the kidney into the urine. However, there has only been one report on the excretion of an *S*-acyl-NAC conjugate of a carboxylic acid drug in urine. In that report, clofibryl-*S*-acyl-NAC was detected in the urine of clofibric acid-dosed rats (Stogniew and Fenselau, 1982). However, in another report, LC/MS analysis of urine extracts of clofibric acid or clofibric acid-*S*-acyl-GSH-dosed rats showed that clofibryl-*S*-acyl-NAC was not detected, or was barely detectable, respectively (Grillo and Benet, 2001). A report by Tate (1975) offered an explanation for this finding by showing that γ-GT, the enzyme that catalyzes the first step in the degradation of GSH adducts to NAC adducts, interacts with *S*-acyl-GSH yielding *N*-acyl-cysteinylglycine-amide-linked conjugates. The degradation of the *S*-acyl-GSH adducts by γ-GT, via cleavage of the γ-glutamyl group, leads to *S*-acyl-cysteinylglycine products that quickly and quantitatively convert to the *N*-acyl-dipeptide conjugates (Fig. 7.16). Therefore, the rearrangement reaction that occurs *in vitro* for these *S*-acyl-GSH-thioester derivatives via the transpeptidase may also occur *in vivo* for *S*-acyl-GSH adducts formed from reactions of GSH with drug acyl glucuronides and/or *S*-acyl-CoA thioesters. Thus, the rearrangement of the *S*-acyl-cysteinylglycine intermediates to the *N*-acyl-cysteinylglycine derivative would preclude its further degradation to the corresponding mercapturate and explain the lack of its detection. Indeed, analysis of urine extracts from clofibric acid-dosed rats also resulted in the detection of clofibryl-*N*-acyl-cysteine and cysteinylglycine conjugates as their symmetrical disulfide dimers (Grillo and Benet, 2001). Another recent report showed the excretion in bile of *S*-methyl-*N*-acyl-cysteine and *N*-acyl-cysteinylglycine disulfide conjugates of three aromatic carboxylic acid prostaglandin I_2-preferring receptor antagonist discovery analogs (Fitch et al., 2004). From examples such as these, *S*-acyl-linked-NAC adducts are not predicted to be the important adducts of interest surveyed for in urine extracts as

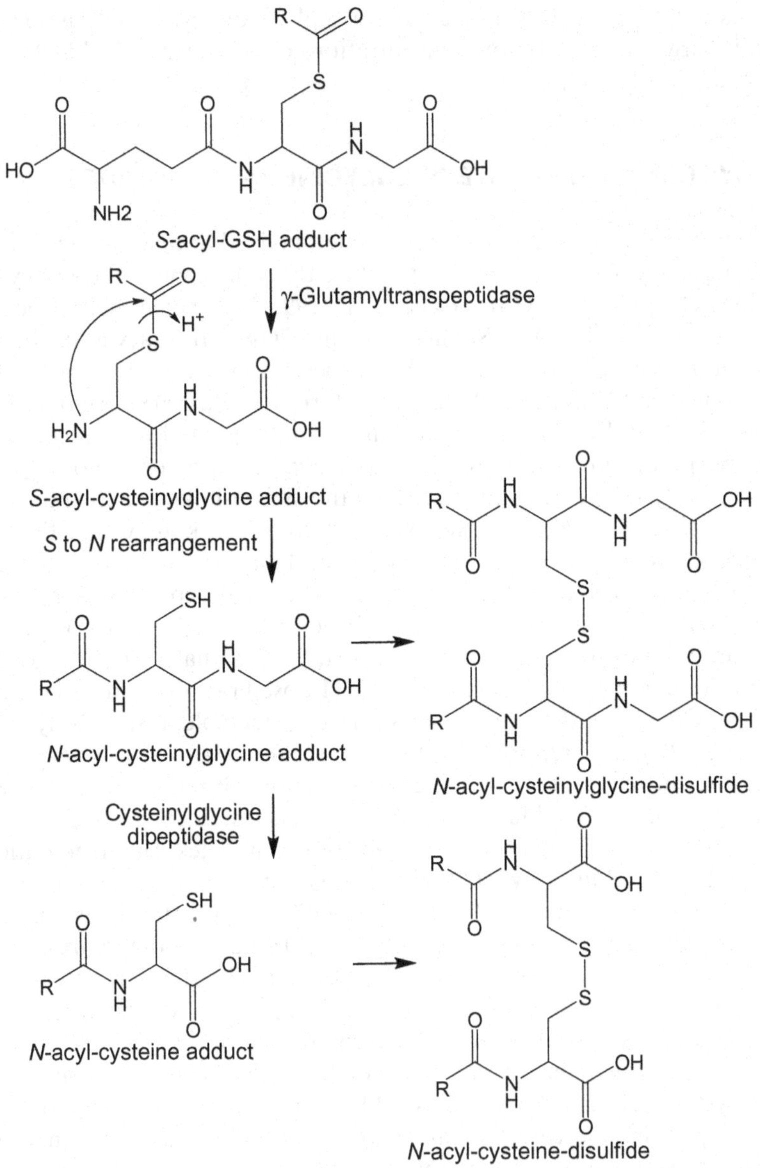

S-acyl-GSH adduct
γ-Glutamyltranspeptidase
S-acyl-cysteinylglycine adduct
S to N rearrangement
N-acyl-cysteinylglycine adduct
N-acyl-cysteinylglycine-disulfide
Cysteinylglycine dipeptidase
N-acyl-cysteine adduct
N-acyl-cysteine-disulfide

Figure 7.16 Enzyme-mediated degradation of *S*-acyl-GSH adducts in the kidney or liver leading to the formation of *N*-acyl-cysteinylglycine and *N*-acyl-cysteine adducts and their respective disulfide products.

markers of GSH-adduct formation occurring *in vivo*. More importantly, it is proposed that drug-*N*-acyl-cysteine or *N*-acyl-cysteinylglycine derivatives should be used as markers of GSH-adduct formation occurring *in vivo* when studying urine extracts from preclinical species or patients dosed with carboxylic acid-containing drugs that undergo bioactivation by acyl glucuronidation and/or *S*-acyl-CoA formation.

7.9 TRAPPING REAGENTS FOR HARD ELECTROPHILES

Whereas soft electrophiles (e.g., epoxides and benzoquinones) are usually uncharged and are strongly polarizable species that react with soft nucleophiles (strongly polarizable, relatively weak bases such as thiols found in cysteinyl residues of proteins and GSH), hard electrophiles (Lewis acids) are sometimes charged, hardly polarizable electrophilic species (e.g., aldehydes, iminium ions, and alkyl carbonium ions) that react with hard nucleophiles (hardly polarizable strong bases; Lewis bases such as amines, hydroxide anion, conjugate bases of alcohols, oxygen in purines and pyrimidines of nucleic acids, and phosphate oxygens in DNA; Carlson, 1990). Therefore, in general, soft electrophiles react with soft nucleophiles, whereas hard electrophiles react with hard nucleophiles. Intermediate soft-hard electrophiles such as aryl carbonium ions, aryl halides, benzylic carbonium ions, nitrenium ions, and the carbon of epoxides can react with both soft and hard nucleophiles. Intermediate soft-hard nucleophiles including nitrogen in primary and secondary amino groups of proteins, nitrogen in amino groups in purine bases in DNA, and CN^- anion can react with both soft and hard electrophiles.

7.9.1 Cyanide Trapping of Iminium Ion Reactive Intermediates

Because GSH is a soft nucleophile, employing it for trapping reactive metabolites is most often unsuccessful for detecting hard electrophilic intermediates. For example, iminium ions and aldehydes are hard electrophiles which are much less likely to react with GSH. A suitable reagent for the trapping and identification of intermediate to hard electrophilic intermediates is KCN, which has been used, for example, in capturing reactive carbocation or iminium ion intermediates (Evans et al., 2004; Argoti et al., 2005). Drugs containing secondary or tertiary alicyclic amine moieties in their structure are numerous, and their biotransformation to chemically reactive iminium ions has been investigated where KCN, $Na^{14}CN$ in radiometric assays, and stable-labeled $K^{13}C^{15}N$ have been employed.

KCN One of the first reports of KCN successfully being used as a trapping reagent of iminium ions was in metabolism studies with nicotine (Murphy, 1973). In these studies, extracts from incubations containing nicotine (1 mM), rabbit liver microsomes, an NADPH-generating system, and KCN (5 mM) showed the formation of 5′-cyanonicotine. This CN adduct is known to be produced from P450-mediated metabolism to a nicotine-$\Delta^{1'(5')}$ iminium ion that reacts with CN^- ion. In other studies, the production of an iminium ion as an intermediate in the demethylation of nicotine was reported where *N′*-

cyanomethylnornicotine was detected in incubation extracts of rabbit liver microsomes with nicotine and NaCN (Nguyen et al., 1976). Increasing numbers of reports on the use of cyanide for the trapping iminium ion-type reactive intermediates, which have been considered as potentially toxic intermediates, have been shown for a large number of structurally varied alicyclic amine-containing drugs (Gorrod and Aislaitner, 1994).

Radiolabeled $Na^{14}CN$ or $K^{14}CN$ The metabolism of a range of alicyclic amine-containing drugs has been characterized with the use of radiolabeled $Na^{14}CN$ in radiometric assays where the ^{14}CN adducts were extracted and quantified by scintillation counting (Gorrod et al., 1991). In these novel experiments, radiolabeled cyanide was used to trap reactive iminium ion intermediates. However, although this method is successful in detecting CN adducts, no structural information can be obtained. A quantitative high-throughput trapping assay employing $K^{14}CN$ and using non-radiolabeled compounds incubated with liver microsomes (Meneses-Lorente et al., 2006) proved to be specific, sensitive, and robust in the assessment of bioactivation potential of varied drugs susceptible to iminium ion formation. In this assay, the level of ^{14}CN-adduct formation in the incubation was determined by liquid scintillation counting after unreacted $^{14}CN^-$ was removed from the incubation extract by solid-phase extraction. An attractive feature of this method was that many of the steps were performed using a liquid-handling workstation, where minimal manual manipulation was required.

Stable-Labeled $K^{13}C^{15}N$ A recent report on the use of a relatively high-throughput LC/MS/MS method for the detection of CN adducts formed liver microsomal incubations employed both unlabeled and stable isotope-labeled KCN (Argoti et al., 2005) for trapping reactive iminium ion intermediates. In those studies, nicotine (100 µM) was used as a positive control compound in rat and HLM incubations with 1 mM KCN (or $K^{13}C^{15}N$) fortified with NADPH. Importantly, acetonitrile containing 5% ammonium hydroxide was used to quench incubations in order to prevent the formation of potentially toxic HCN gas. In their studies, an Applied Biosystems/MDS Sciex API-4000 triple quadrupole mass spectrometer was used to analyze extracts from separate incubations, one with KCN and the other with $K^{13}C^{15}N$, in order to authenticate the formation of the detected CN adducts. Extracts from the incubations were analyzed for CN and $K^{13}C^{15}N$ adducts by LC/MS/MS analysis employing CNL scanning for 27 or 29 Da, which is due to the loss of HCN or $HK^{13}C^{15}N$, respectively. Another tandem mass spectrometric run was made to acquire product ion spectra of the detected CN adducts in order to determine the structures of the trapped iminium ion intermediates. Results showed that nicotine was metabolized to two regioisomeric CN adducts resulting from CN addition to each of the two α-carbon atoms next to the endocyclic nitrogen atom, which was consistent with a published report on the bioactivation of nicotine

(Kalgutkar et al., 2002). In addition to the detection of CN adducts, parallel incubations of nicotine with the soft nucleophile GSH followed by a CNL scanning of 129 Da by LC/MS/MS showed no detection of GSH adducts, thus confirming that the iminium ion intermediates were trapped specifically by the hard CN nucleophile. This LC/MS/MS screening method was also successful in the detection of CN-trapped iminium ions produced from corresponding incubations with a range of alicyclic amine-containing compounds. In summary, results from the studies by Argoti et al. showed that LC/MS/MS screening for CN-trapped iminium ion intermediates is a very effective strategy for detecting those metabolites having the potential to form the reactive iminium ion intermediates. The method is valuable because it does not require the use of radioactivity, and also because the high-throughput LC/MS/MS screen is rapid compared to LC/radiometric assays where it is necessary to resolve unreacted $^{14}CN^-$ from the ^{14}CN adducts. Another considerable benefit over radiometric CN-adduct assays is the ability to characterize the structures of the CN adducts by tandem mass spectrometry. A drawback of this LC/MS/MS method is that it does not provide for quantitative information on the CN adducts formed which can be readily obtained by LC/radiometric ^{14}CN-adduct methods (discussed above).

In another CN-adduct trapping technique, it was shown that by employing a mixture containing equal amounts of unlabeled and isotopically labeled CN in liver microsomal incubations, CN adducts were able to be detected using LC/MS/MS CNL scanning of 27 and 29 Da (Kalgutkar et al., 2002). In those studies, 1 mM KCN (CN and $^{13}C^{15}N$, 1:1) was used as the trapping reagent, where the detection of CN adducts of a piperidine-containing drug candidate by LC/MS was recognized by the existence of isotopic "doublets" separated in mass by 2 Da. Finally, positive ion tandem mass spectra of these adducts were characterized by a dominant neutral loss of 27/29 Da (HCN/$H^{13}C^{15}N$]).

7.9.2 Aldehyde Trapping

Both semicarbazide and methoxylamine have been shown to be useful reagents for the trapping of chemically reactive aldehydes via reaction with nucleophilic amino groups leading to the formation of semicarbazone- and oxime-type Schiff-base-linked adducts, respectively. These adducts are relatively stable and can be detected by various chromatographic methods as discussed below.

Methoxylamine A remarkable example of the use of methoxylamine to trap a reactive aldehyde metabolite comes from mechanistic *in vitro* studies with abacavir (Walsh et al., 2002). Abacavir is a reverse transcriptase inhibitor used for the treatment of HIV-1 infection, where hypersensitivity reactions to the drug occur in a small percentage of patients potentially mediated by immune system recognition of a drug–protein adduct. An important route of metabolism of abacavir occurs by oxidation of the primary alcohol to a

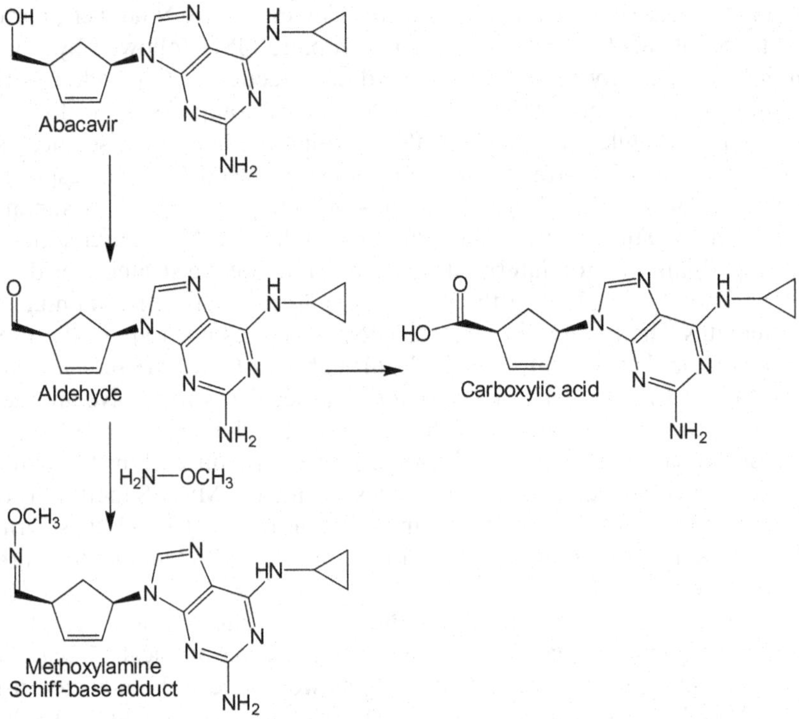

Figure 7.17 Scheme for the metabolism of abacavir and the methoxylamine trapping of the reactive aldehyde intermediate.

carboxylic acid via a reactive aldehyde intermediate (Fig. 7.17). This route of metabolism was shown *in vitro* to be mediated by specific human isoforms of cytosolic alcohol dehydrogenase (ADH). In their studies, [^{14}C]abacavir (10 μM, 0.5 μCi/mL) was incubated with specific ADH isozymes (αα and γ2γ2) fortified with NAD^+ and 10 mM methoxylamine. The aldehyde intermediate was trapped during these incubations as an oxime derivative which was readily detected by LC/radiometric analysis. The derivatized product was more lipophilic (later eluting) than abacavir, and LC/MS/MS analysis of the $[M + H]^+$ ion at *m/z* 314 (^{12}C-isotope) provided a tandem mass spectrum consistent with a methoxylamine-adduct structure (Fig. 7.17). The significance of these observations toward understanding the mechanism of hypersensitivity reactions observed with the use of abacavir remains to be determined. However, the understood polymorphisms of the ADH3 gene encoding for the γ2γ2 isozyme, having differing allele frequencies in different populations, constitutes a genetic intersubject variable directly linked to abacavir bioactivation, and according to the authors, would be a rational target for genotyping studies of hypersensitivity incidence in addition to the current pre-therapy genetic testing for the HLA-B*5701 allele (Hetherington et al., 2002).

Other successful aldehyde metabolite trapping studies employing methoxylamine and LC/MS/MS analysis have been reported. For example, in mechanistic studies with the antimalarial drug amodiaquine (Johansson et al., 2009), aldehyde metabolites of amodiaquine and desethylamodiaquine were identified as major products formed in incubations with recombinantly expressed human cytochromes P450s CYP1A1 and CYP1B1. In another example, liver microsomal metabolism studies with the furan-containing 5-lipoxygenase inhibitor, L-739,010, showed strong evidence for furan bioactivation through the identification of *O*-methoxylamine adducts of a ring-opened aldehyde moiety of the drug (Zhang et al., 1996).

A potential problem with the use of methoxylamine as a trapping reagent is due to its ability to cause reversible inhibition of P450s with $IC_{(50)}$ values of 0.53 mM for CYP1A2, 4.12 mM for CYP2C9, 2.04 mM for CYP2C19, 9.72 mM for CYP2D6, and 1.26 and >10 mM for CYP3A4/5 (Zhang et al., 2009). In that same report, GSH and CN KCN were shown not to inhibit these enzymes at concentrations up to 10 mM.

Semicarbazide Reactive aldehyde-containing metabolites or intermediates are able to be trapped in liver microsomal incubations containing semicarbazide leading to the formation of stable semicarbazone derivatives (Ravindranath et al., 1984). An example of the successful use of semicarbazide as a trapping reagent comes from mechanistic studies with the 5-lipoxygenase inhibitors L-746,530 and L-739,010 (Chauret et al., 1995). It was shown that metabolic activation of these compounds in incubations with rhesus monkey liver microsomes (species with the highest reactive metabolite formation for these compounds) by P450 3A-mediated oxidation of the dioxabicyclo moiety led to an exocyclic primary aldehyde that was efficiently trapped by coincubation with semicarbazide (6 mM) leading to the detection of two isomeric adducts (Fig. 7.18). These semicarbazide adducts were identified and their structures were partially characterized by both positive and negative ion capillary HPLC/continuous flow-liquid secondary ion mass spectrometry on a JEOL-HX110A mass spectrometer. Results from these mechanistic trapping experiments led to the synthesis of a hindered gem-dimethyl analog of L-739,010 (where the dioxabicyclo moiety was blocked from potential *O*-dealkylation that would result in reactive aldehyde formation) which, when tested for metabolic activation *in vitro*, showed significantly less covalent binding to liver microsomal protein cross-species (85–95% reduction).

In one last example, a semicarbazide adduct of the NSAID suprofen (Fig. 7.19) was detected in incubations with human recombinant P450 2C9 (O'Donnell, et al., 2003). Suprofen was used in the United States until several cases of nephrotoxicity occurred, which led to the drug being withdrawn from the market (Hart, 1987; Fung et al., 2001). Suprofen contains a thiophene moiety that is bioactivated to thiophene *S*-oxide, thiophene epoxide, and γ-thioketo-α,β-unsaturated aldehyde reactive intermediates (Fig. 7.19). When suprofen (100 μM) was incubated with recombinant P450 2C9 in the

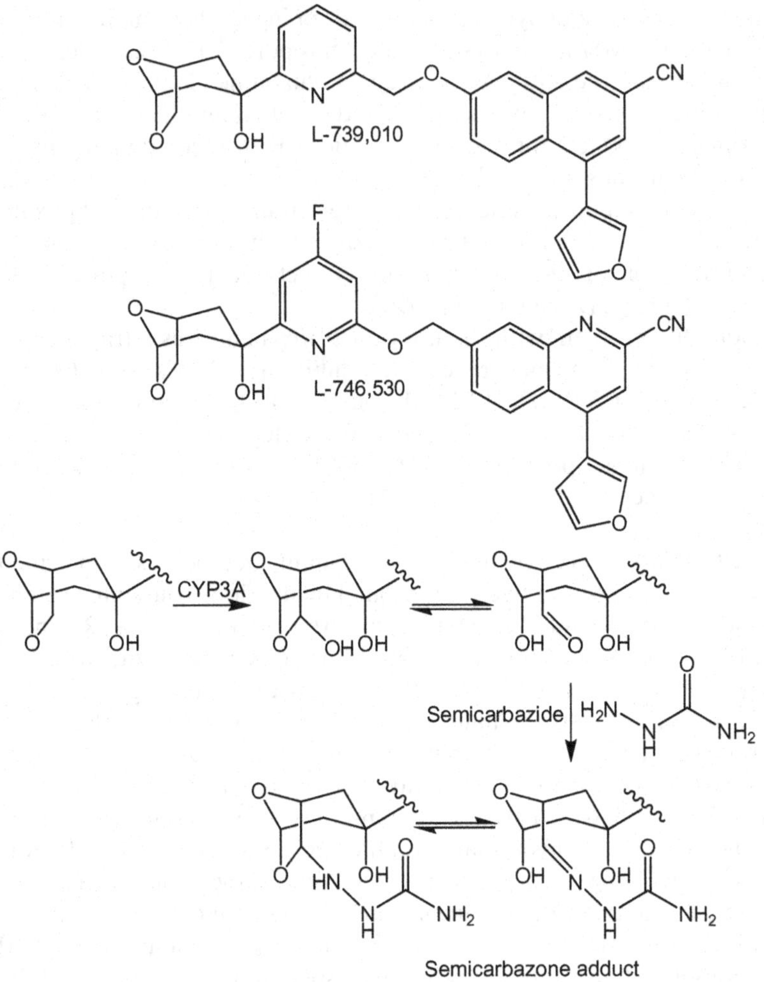

Figure 7.18 Scheme for the metabolism of L-739,010 and L-746,530 by P450-mediated hydroxylation leading to the reactive aldehyde on the dioxabicyclo moiety which is trapped by semicarbazide. (Adapted from Chauret et al., 1995.)

presence of NADPH and semicarbazide (1 mM), the inactivation of the enzyme was not prevented, but one semicarbazide adduct, namely α-methyl-4-(3-pyridazinylcarbonyl)benzene acetic acid, was detected (indirectly) using an LC-MS/qTOF mass spectrometer. The trapping and LC/MS detection of a subsequently formed pyridazine adduct provided the proposal that the γ-thioketo-α,β-unsaturated aldehyde reactive intermediate is to be considered the most likely reactive species leading to mechanism-based inactivation of P450 2C9 (O'Donnell, et al., 2003; Fig. 7.19).

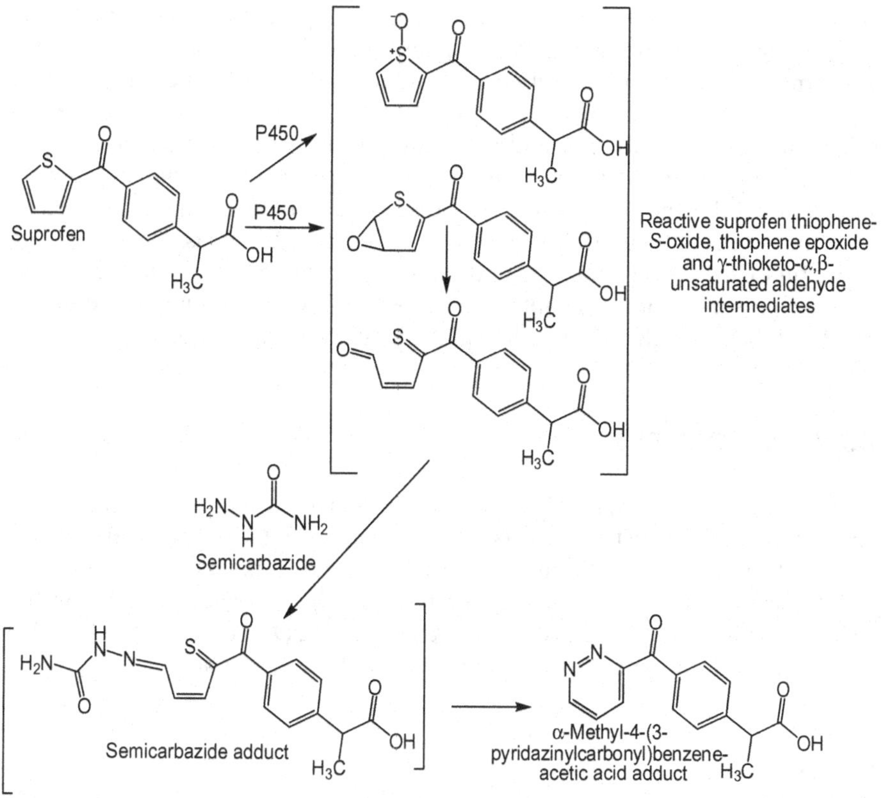

Figure 7.19 Scheme for the metabolic activation of suprofen to electrophilic thiophene *S*-oxide-, thiophene epoxide-, and γ-thioketo-α,β-unsaturated aldehyde-type intermediates. (Adapted from O'Donnell, et al., 2003.)

7.10 CONCLUSIONS

In summary, the current trend in modern drug metabolism has been to use increasingly sophisticated methods for the detection of reactive intermediates of drugs or drug candidates in order to address the growing need for the discovery and development of safer drugs. Improved drug safety is needed, since it is known that most drugs discovered today fail due to toxicity reasons (Kola and Landis, 2004; Guengerich and MacDonald, 2007). Although drug metabolism scientists have made excellent progress in the ability to detect reactive drug metabolites and in determining the probable biochemical mechanisms that mediate their formation, we still do not understand the pathways of subsequent organ toxicity. The toxic mechanisms are complex within cellular systems, and this creates an important challenge for drug metabolism scientists and toxicologists toward the discovery of nontoxic drugs. The speed at which novel reactive intermediates are being discovered is increasing over time largely because of the increasing speed of development of superior LC/MS/MS instruments having higher and higher selectivity and sensitivity. What is going more slowly is the rate of understanding of the biochemical and cellular mechanisms of drug toxicity. In terms of trying to understand these mechanisms, it seems that nature is a lot more complicated than we may have hoped. But arguably, that is a large part of what makes mechanistic drug metabolism so exciting. Like the other sciences, the science of modern drug metabolism sometimes seems to be driven more by tools than by ideas. Hopefully, the powerful LC/MS/MS techniques and capabilities that are available today, such as those presented in this chapter, and those that are continuing to be developed, will be more than just technical skills that are fun to exercise.

ACKNOWLEDGMENTS

MPG would like to thank Dr. Christian Skonberg (Department of Pharmaceutics and Analytical Chemistry, University of Copenhagen, Denmark) and Howard Horng (Department of Biopharmaceutical Science, University of California, San Francisco) for constructive criticism and helpful suggestions during the writing of this chapter, and also to Amgen scientists, Dr. Tim Carlson (PKDM) and Dr. Kathila Rajapaksa (Toxicology) for helpful discussions.

REFERENCES

Aigner A, Jäger M, Pasternack R, Weber P, Wienke D, Wolf S. 1996. Purification and characterization of cysteine-*S*-conjugate *N*-acetyltransferase from pig kidney. *Biochem J* 317:213–218.

Alvarez-Sanchez R, Montavon F, Hartung T, Pähler A. 2006. Thiazolidinedione bioactivation: A comparison of the bioactivation potentials of troglitazone, rosiglitazone, and pioglitazone using stable isotope-labeled analogues and liquid chromatography tandem mass spectrometry. *Chem Res Toxicol* 19:1106–1116.

Argoti D, Liang L, Conteh A, Chen L, Bershas D, Yu CP, Vouros P, Yang E. 2005. Cyanide trapping of iminium ion reactive intermediates followed by detection and structure identification using liquid chromatography-tandem mass spectrometry (LC-MS/MS). *Chem Res Toxicol* 18:1537–1544.

Baillie TA. 2006. Future of toxicology-metabolic activation and drug design: Challenges and opportunities in chemical toxicology. *Chem Res Toxicol* 19:889–893.

Baillie TA. 2008. Metabolism and toxicity of drugs. Two decades of progress in industrial drug metabolism. *Chem Res Toxicol* 21:129–137.

Baillie TA, Davis MR. 1993. Mass spectrometry in the analysis of glutathione conjugates. *Biol Mass Spectrom* 22:319–325.

Banks AT, Zimmerman HJ, Ishak KG, Harter JG. 1995. Diclofenac-associated hepatotoxicity: Analysis of 180 cases reported to the Food and Drug Administration as adverse reactions. *Hepatology* 22:820–827.

Bauman JN, Kelly JM, Tripathy S, Zhao SX, Lam WW, Kalgutkar AS, Obach RS. 2009. Can in vitro metabolism-dependent covalent binding data distinguish hepatotoxic from nonhepatotoxic drugs? An analysis using human hepatocytes and liver S-9 fraction. *Chem Res Toxicol* 22:332–340.

Benet LZ, Spahn-Langguth H, Iwakawa S, Volland C, Misuma T, Mayer S, Mutschler E, Lin ET. 1993. Predictability of covalent binding of acidic drugs in man. *Life Sci* 53:PL141–PL146.

Bjornsson E, Nordlinder H, Olsson R. 2006. Clinical characteristics and prognostic markers in disulfuram-induced liver injury. *J Hepatol* 44:791–797.

Boelsterli UA. 2002. Xenobiotic acyl glucuronides and acyl-CoA thioesters as protein-reactive metabolites with the potential to cause idiosyncratic drug reactions. *Curr Drug Metab* 3:439–450.

Boelsterli UA. 2003. Diclofenac-induced liver injury: A paradigm of idiosyncratic drug toxicity. *Toxicol Appl Pharmacol* 192:307–322.

Carlson RM. 1990. Assessment of the propensity for covalent binding of electrophiles to biological substrates. *Environ Health Perspect* 87:227–232.

Chauret N, Nicoll-Griffith D, Friesen R, Li C, Trimble L, Dubé D, Fortin R, Girard Y, Yergey J. 1995. Microsomal metabolism of the 5-lipoxygenase inhibitors L-746,530 and L-739,010 to reactive intermediates that covalently bind to protein: The role of the 6,8-dioxabicyclo[3.2.1]octanyl moiety. *Drug Metab Dispos* 23:1325–1334.

Chen Q, Doss GA, Tung EC, Liu W, Tang YS, Braun MP, Didolka V, Strauss JR, Wang RW, Stearns RA, Evans DC, Baillie TA, Tang W. 2006. Evidence for the bioactivation of zomepirac and tolmetin by an oxidative pathway: Identification of glutathione adducts in vitro in human liver microsomes and in vivo in rats. *Drug Metab Dispos* 34:145–151.

Day SH, Mao A, White R, Schulz-Utermoehl T, Miller R, Beconi MG. 2005. A semi-automated method for measuring the potential for protein covalent binding in drug discovery. *J Pharmacol Toxicol Methods* 52:278–285.

Dieckhaus CM, Miller TA, Sofia RD, Macdonald TL. 2000. A mechanistic approach to understanding species differences in felbamate bioactivation: Relevance to drug induced idiosyncratic reactions. *Drug Metab Dispos* 28:814–822.

Dieckhaus CM, Thompson CD, Roller SG, Macdonald TL. 2002. Mechanisms of idiosyncratic drug reactions: The case of felbamate. *Chem Biol Interact* 142:99–117.

Dieckhaus CM, Fernández-Metzler CL, King R, Krolikowski PH, Baillie TA. 2005. Negative ion tandem mass spectrometry for the detection of glutathione conjugates. *Chem Res Toxicol* 18:630–638.

Ding YS, Blount BC, Valentin-Blasini L, Applewhite HS, Xia Y, Watson CH, Ashley DL. 2009. Simultaneous determination of six mercapturic acid metabolites of volatile organic compounds in human urine. *Chem Res Toxicol* 22:1018–1025.

Doss GA, Baillie TA. 2006. Addressing metabolic activation as an integral component of drug design. *Drug Metab Rev* 38:641–649.

Evans DC, Baillie TA. 2005. Minimizing the potential for metabolic activation as an integral part of drug design. *Curr Opin Drug Discov Devel* 8:44–50.

Evans DC, Watt AP, Nicoll-Griffith DA, Baillie TA. 2004. Drug-protein adducts: An industry perspective on minimizing the potential for drug bioactivation in drug discovery and development. *Chem Res Toxicol* 17:3–16.

Fitch WL, Berry PW, Tu Y, Tabatabaei A, Lowrie L, Lopez-Tapia F, Liu Y, Nitzan D, Masjedizadeh MR, Varadarajan A. 2004. Identification of glutathione-derived metabolites from an IP receptor antagonist. *Drug Metab Dispos* 32:1482–1490.

Fung M, Thornton A, Mybeck K, Wu JH, Hornbuckle K, Muniz E. 2001. Evaluation of the characteristics of safety withdrawal of prescription drugs from world wide pharmaceutical markets, 1960 to 1999. *Drug Inform J* 35:293–317.

Gan J, Harper TW, Hsueh MM, Qu Q, Humphreys WG. 2005. Dansyl glutathione as a trapping agent for the quantitative estimation and identification of reactive metabolites. *Chem Res Toxicol* 18:896–903.

Gorrod JW, Aislaitner G. 1994. The metabolism of alicyclic amines to reactive iminium ion intermediates. *Eur J Drug Metab Pharmacokinet* 19:209–217.

Gorrod JW, Whittlesea CM, Lam SP. 1991. Trapping of reactive intermediates by incorporation of ^{14}C-sodium cyanide during microsomal oxidation. *Adv Exp Med Biol* 283:657–664.

Grillo MP, Benet LZ. 2001. Interaction of gamma-glutamyltranspeptidase with clofibryl-*S*-acyl-glutathione in vitro and in vivo in rat. *Chem Res Toxicol* 14: 1033–1040.

Grillo MP, Benet LZ. 2002. Studies on the reactivity of clofibryl-S-acyl-CoA thioester with glutathione in vitro. *Drug Metab Dispos* 30:55–62.

Grillo MP, Hua F. 2003. Identification of zomepirac-*S*-acyl-glutathione in vitro in incubations with rat hepatocytes and in vivo in rat bile. *Drug Metab Dispos* 31:1429–1436.

Grillo MP, Hua F. 2008. Enantioselective formation of ibuprofen-S-acyl-glutathione in vitro in incubations of ibuprofen with rat hepatocytes. *Chem Res Toxicol* 21:1749–1759.

Grillo MP, Knutson CG, Sanders PE, Waldon DJ, Hua F, Ware JA. 2003. Studies on the chemical reactivity of diclofenac acyl glucuronide with glutathione: identification of diclofenac-*S*-acyl-glutathione in rat bile. *Drug Metab Dispos* 31:1327–1336.

Grillo MP, Ma J, Teffera Y, Waldon DJ. 2008. A novel bioactivation pathway for diclofenac initiated by P450-mediated oxidative decarboxylation. *Drug Metab Dispos* 36:1740–1744.

Guengerich FP, MacDonald JS. 2007. Applying mechanisms of chemical toxicity to predict drug safety. *Chem Res Toxicol* 20:344–369.

Hargus SJ, Martin BM, George JW, Pohl LR. 1995. Covalent modification of rat liver dipeptidyl peptidase IV (CD26) by the nonsteroidal anti-inflammatory drug diclofenac. *Chem Res Toxicol* 8:993–996.

Hart D, Ward M, Lifschitz MD. 1987. Suprofen-related nephrotoxicity. A distinct clinical syndrome. *Ann Intern Med* 106:235–238.

Hartman, NR, Cysyk RL, Bruneau-Wack C, Thenot JP, Parker RJ, Strong JM. 2002. Production of intracellular ^{35}Sglutathione by rat and human hepatocytes for the quantification of xenobiotics reactive intermediates. *Chem Biol Interact* 142:43–55.

Hetherington S, Hughes AR, Mosteller M, Shortino D, Baker KL, Spreen W, Lai E, Davies K, Handley A, Dow DJ, Fling ME, Stocum M, Bowman C, Thurmond LM, Roses AD. 2002. Genetic variations in HLA-B region and hypersensitivity reactions to abacavir. *Lancet* 359:1121–1122.

Inoue K, Shibata Y, Takahashi H, Ohe T, Chiba M, Ishii YA. 2009. A trapping method for semi-quantitative assessment of reactive metabolite formation using [^{35}S]cysteine and [^{14}C]cyanide. *Drug Metab Pharmacokinet* 24:245–254.

Isley WL. 2003. Hepatotoxicity of thiazolidinediones. *Expert Opin Drug Saf* 2:581–586.

Jian W, Yao M, Zhang D, Zhu M. 2009. Rapid detection and characterization of in vitro and urinary *N*-acetyl-L-cysteine conjugates using quadrupole-linear ion trap mass spectrometry and polarity switching. *Chem Res Toxicol* 22:1246–1255.

Jin L, Davis MR, Hu P, Baillie TA. 1994. Identification of novel glutathione conjugates of disulfiram and diethyldithiocarbamate in rat bile by liquid chromatography-tandem mass spectrometry. Evidence for metabolic activation of disulfiram in vivo. *Chem Res Toxicol* 7:526–533.

Johansson T, Jurva U, Grönberg G, Weidolf L, Masimirembwa C. 2009. Novel metabolites of amodiaquine formed by CYP1A1 and CYP1B1: Structure elucidation using electrochemistry, mass spectrometry, and NMR. *Drug Metab Dispos* 37:571–579.

Jones DP, Moldeus P, Stead AH, Ormstad K, Jörnvall H, Orrenius S. 1979. Metabolism of glutathione and a glutathione conjugate by isolated kidney cells. *J Biol Chem* 254:2787–2792.

Jones JA, Kaphalia L, Treinen-Moslen M, Liebler DC. 2003. Proteomic characterization of metabolites, protein adducts, and biliary proteins in rats exposed to 1,1-dichloroethylene or diclofenac. *Chem Res Toxicol* 16:1306–1317.

Jurva U, Holmén A, Grönberg G, Masimirembwa C, Weidolf L. 2008. Electrochemical generation of electrophilic drug metabolites: Characterization of amodiaquine quinoneimine and cysteinyl conjugates by MS, IR, and NMR. *Chem Res Toxicol* 21:928–935.

Kalgutkar AS, Soglia JR. 2005. Minimising the potential for metabolic activation in drug discovery. *Expert Opin Drug Metab Toxicol* 1:91–142.

Kalgutkar AS, Dalvie DK, O'Donnell JP, Taylor TJ, Sahakian DC. 2002. On the diversity of oxidative bioactivation reactions on nitrogen-containing xenobiotics. *Curr Drug Metab* 3:379–424.

Kapetanovic IM, Torchin CD, Strong JM, Yonekawa WD, Lu C, Li AP, Dieckhaus CM, Santos WL, Macdonald TL, Sofia RD, Kupferberg HJ. 2002. Reactivity of atropaldehyde, a felbamate metabolite in human liver tissue in vitro. *Chem Biol Interact* 142:119–134.

Kassahun K, Pearson PG, Tang W, McIntosh I, Leung K, Elmore C, Dean D, Wang R, Doss G, Baillie TA. 2001. Studies on the metabolism of troglitazone to reactive intermediates in vitro and in vivo. Evidence for novel biotransformation pathways involving quinone methide formation and thiazolidinedione ring scission. *Chem Res Toxicol* 14:62–70.

Kenna JG, Satoh H, Christ DD, Pohl LR. 1988. Metabolic basis for a drug hypersensitivity: Antibodies in sera from patients with halothane hepatitis recognize liver neoantigens that contain the trifluoroacetyl group derived from halothane. *J Pharmacol Exp Ther* 245:1103–1109.

Kola I, Landis J. 2004. Can the pharmaceutical industry reduce attrition rates? *Nat Rev Drug Discov* 3:711–715.

Laine L, White WB, Rostom A, Hochberg M. 2008. COX-2 selective inhibitors in the treatment of osteoarthritis. *Semin Arthritis Rheum* 38:165–187.

Lammert C, Einarsson S, Saha C, Niklasson A, Bjornsson E, Chalasani N. 2008. Relationship between daily dose of oral medications and idiosyncratic drug-induced liver injury: Search for signals. *Hepatology* 47:2003–2009.

López-Garcia MP, Dansette PM, Mansuy D. 1994. Thiophene derivatives as new mechanism-based inhibitors of cytochromes P-450: Inactivation of yeast-expressed human liver cytochrome P-450 2C9 by tienilic acid. *Biochemistry* 33:166–175.

Li Y, Slatter JG, Zhang Z, Li Y, Doss GA, Braun MP, Stearns RA, Dean DC, Baillie TA, Tang W. 2008. In vitro metabolic activation of lumiracoxib in rat and human liver preparations. *Drug Metab Dispos* 36:469–473.

Lim H-K, Chen J, Cook K, Sensenhauser C, Silva J, Evans D. 2008. A generic method to detect electrophilic intermediates using isotope pattern triggered data-dependent high-resolution accurate mass spectrometry. *Rapid Commun Mass Spectrom* 22: 1295–1311.

Madsen KG, Olsen J, Skonberg C, Hansen SH, Jurva U. 2007. Development and evaluation of an electrochemical method for studying reactive phase-I metabolites: Correlation to in vitro drug metabolism. *Chem Res Toxicol* 20:821–831.

Madsen KG, Skonberg C, Jurva U, Cornett C, Hansen SH, Johansen TN, Olsen J. 2008a. Bioactivation of diclofenac in vitro and in vivo: correlation to electrochemical studies. *Chem Res Toxicol* 21:1107–1119.

Madsen KG, Grönberg G, Skonberg C, Jurva U, Hansen SH, Olsen J. 2008b. Electrochemical oxidation of troglitazone: identification and characterization of the major reactive metabolite in liver microsomes. *Chem Res Toxicol* 21:2035–2041.

Matthews AM, Hinson JA, Roberts DW, Pumford NR. 1997. Comparison of covalent binding of acetaminophen and the regioisomer 3′-hydroxyacetanilide to mouse liver protein. *Toxicol Lett* 90:77–82.

Meneses-Lorente G, Sakatis MZ, Schulz-Utermoehl T, De Nardi C, Watt AP. 2006. Quantitative high-throughput trapping assay as a measurement of potential for bioactivation. *Anal Biochem* 351:266–272.

Miller EC, Miller JA. 1974. Biochemical mechanisms of chemical carcinogenesis. In *The Molecular Biology of Cancer*, ed. Busch H, 377–402. New York: Academic Press.

Miller EC, Miller JA. 1979. Milestones in chemical carcinogenesis. *Oncology* 6:445–456.

Miller EC, Miller JA. 1981. Reactive metabolites as key intermediates in pharmacologic and toxicologic responses: Examples from chemical carcinogenesis. *Adv Exp Med Biol* 136:1–21.

Monks TJ, Hinson JA, Gillette JR. 1982. Bromobenzene and *p*-bromophenol toxicity and covalent binding in vivo. *Life Sci* 30:841–848.

Mulder GJ, Le CT. 1998. A rapid, simple in vitro screening test, using [^{3}H]glutathione and L-[^{35}S]cysteine as trapping agents, to detect reactive intermediates of xenobiotics. *Toxic In Vitro* 2:225–230.

Murphy PJ. 1973. Enzymatic oxidation of nicotine to nicotine 1′(5′) iminium ion. A newly discovered intermediate in the metabolism of nicotine. *J Biol Chem* 248:2796–2800.

Mutlib A, Lam W, Atherton J, Chen H, Galatsis P, Stolle W. 2005. Application of stable isotope labeled glutathione and rapid scanning mass spectrometers in detecting and characterizing reactive metabolites. *Rapid Commun Mass Spectrom* 19:3482–3492.

Nakayama S, Atsumi R, Takakusa H, Kobayashi Y, Kurihara A, Nagai Y, Nakai D, Okazaki O. 2009. A zone classification system for risk assessment of idiosyncratic drug toxicity using daily dose and covalent binding. *Drug Metab Dispos* 37: 1970–1977.

Nelson SD. 2001. Molecular mechanisms of adverse drug reactions. *Curr Ther Res* 62:885–899.

Nelson SD, Pearson PG. 1990. Covalent and noncovalent interactions in acute lethal cell injury caused by chemicals. *Ann Rev Pharmacol Toxicol* 30:169–195.

Nelson SD, Mitchell JR, Snodgrass WR, Timbrell JA. 1978. Hepatotoxicity and metabolism of iproniazid and isopropylhydrazine. *J Pharmacol Exp Ther* 206:574–585.

Nguyen TL, Gruenke LD, Castagnoli N Jr. 1976. Metabolic *N*-demethylation of nicotine. Trapping of a reactive iminium species with cyanide ion. *J Med Chem* 19: 1168–1169.

Obach RS, Kalgutkar AS, Soglia JR, Zhao SX. 2008. Can in vitro metabolism-dependent covalent binding data in liver microsomes distinguish hepatotoxic from nonhepatotoxic drugs? An analysis of 18 drugs with consideration of intrinsic clearance and daily dose. *Chem Res Toxicol* 21:1814–1822.

O'Donnell JP, Dalvie DK, Kalgutkar AS, Obach RS. 2003. Mechanism-based inactivation of human recombinant P450 2C9 by the nonsteroidal anti-inflammatory drug suprofen. *Drug Metab Dispos* 31:1369–1377.

Pellock JM. 1999. Felbamate. *Epilepsia* 40:57–62.

Pohl LR, Branchflower RV. 1981. Covalent binding of electrophilic metabolites to macromolecules. *Methods Enzymol* 77:43–50.

Qiu Y, Benet LZ, Burlingame AL. 1998a. Identification of the hepatic protein targets of reactive metabolites of acetaminophen in vivo in mice using two-dimensional gel electrophoresis and mass spectrometry. *J Biol Chem* 273:17940–17953.

Qiu Y, Burlingame AL, Benet LZ. 1998b. Mechanisms for covalent binding of benoxaprofen glucuronide to human serum albumin: Studies by tandem mass spectrometry. *Drug Metab Dispos* 26:246–256.

Ravindranath V, Burka LT, Boyd MR. 1984. Reactive metabolites from the bioactivation of toxic methylfurans. *Science* 224:884–886.

Satoh H, Fukuda Y, Anderson DK, Ferrans VJ, Gillette JR, Pohl LR. 1985. Immunological studies on the mechanism of halothane-induced hepatotoxicity: immunohistochemical evidence of trifluoroacetylated hepatocytes. *J Pharmacol Exp Ther* 233:857–862.

Scholz K, Dekant W, Völkel W, Pähler A. 2005. Rapid detection and identification of *N*-acetyl-L-cysteine thioethers using constant neutral loss and theoretical multiple reaction monitoring combined with enhanced product-ion scans on a linear ion trap mass spectrometer. *J Am Soc Mass Spectrom* 16:1976–1984.

Shore LJ, Fenselau C, King AR, Dickinson RG. 1995. Characterization and formation of the glutathione conjugate of clofibric acid. *Drug Metab Dispos* 23:119–123.

Sidenius JM, Skonberg C, Olsen J, Hansen SH. 2004. In vitro reactivity of carboxylic acid-CoA thioesters with glutathione. *Chem Res Toxicol* 17:75–81.

Skonberg C, Olsen J, Madsen KG, Hansen SH, Grillo MP. 2008. Metabolic activation of carboxylic acids. *Expert Opin Drug Metab Toxicol* 4:425–438.

Small RE. 1989. Diclofenac sodium. *Clin Pharm* 8:545–558.

Soglia JR, Harriman SP, Zhao S, Barberia J, Cole MJ, Boyd JG, Contillo LG. 2004. The development of a higher throughput reactive intermediate screening assay incorporating micro-bore liquid chromatography-micro-electrospray ionization-tandem mass spectrometry and glutathione ethyl ester as an in vitro conjugating agent. *J Pharm Biomed Anal* 36:105–116.

Soglia JR, Contillo LG, Kalgutkar AS, Zhao S, Hop CE, Boyd JG, Cole MJ. 2006. A semiquantitative method for the determination of reactive metabolite conjugate levels in vitro utilizing liquid chromatography-tandem mass spectrometry and novel quaternary ammonium glutathione analogues. *Chem Res Toxicol* 19: 480–490.

Somchit N, Sanat F, Gan EH, Shahrin IA, Zuraini A. 2004. Liver injury induced by the non-steroidal anti-inflammatory drug mefenamic acid. *Singapore Med J* 45:530–532.

Stogniew M, Fenselau C. 1982. Electrophilic reactions of acyl-linked glucuronides. Formation of clofibrate mercapturate in humans. *Drug Metab Dispos* 10:609–613.

Takakusa H, Masumoto H, Makino C, Okazaki O, Sudo K. 2009. Quantitative assessment of reactive metabolite formation using ^{35}S-labeled glutathione. *Drug Metab Pharmacokinet* 24:100–107.

Tang W. 2003. The metabolism of diclofenac—Enzymology and toxicology perspectives. *Curr Drug Metab* 4:319–329.

Tang W, Stearns RA, Bandiera SM, Zhang Y, Raab C, Braun MP, Dean DC, Pang J, Leung KH, Doss GA, Strauss JR, Kwei GY, Rushmore TH, Chiu SH, Baillie TA. 1999. Studies on cytochrome P-450-mediated bioactivation of diclofenac in rats and in human hepatocytes: Identification of glutathione conjugated metabolites. *Drug Metab Dispos* 27:365–372.

Tate SS. 1975. Interaction of gamma-glutamyl transpeptidase with *S*-acyl derivatives of glutathione. *FEBS Lett* 54:319–322.

Tate SS, Meister A. 1985. γ-Glutamyl transpeptidase from kidney. *Methods Enzymol* 13:400–421.

Teffera Y, Waldon DJ, Colletti AE, Albrecht BK, Zhao Z. 2008. Identification of a novel glutathione conjugate of diclofenac by LTQ-orbitrap. *Drug Metab Lett* 2:35–40.

Thompson CD, Gulden PH, Macdonald TL. 1997. Identification of modified atropaldehyde mercapturic acids in rat and human urine after felbamate administration. *Chem Res Toxicol* 10:457–462.

Thompson CD, Barthen MT, Hopper DW, Miller TA, Quigg M, Hudspeth C, Montouris G, Marsh L, Perhach JL, Sofia RD, Macdonald TL. 1999. Quantification in patient urine samples of felbamate and three metabolites: Acid carbamate and two mercapturic acids. *Epilepsia* 40:769–776.

Tirmenstein MA, Nelson SD. 1989. Subcellular binding and effects on calcium homeostasis produced by acetaminophen and a nonhepatotoxic regioisomer, 3′-hydroxyacetanilide, in mouse liver. *J Biol Chem* 264:9814–9819.

Uetrecht JP. 1999. New concepts in immunology relevant to idiosyncratic drug reactions: The "danger hypothesis" and innate immune system. *Chem Res Toxicol* 12:387–395.

Uetrecht J. 2001. Prediction of a new drug's potential to cause idiosyncratic reactions. *Curr Opin Drug Discov Devel* 4:55–59.

Uetrecht J. 2007. Idiosyncratic drug reactions: Current understanding. *Annu Rev Pharmacol Toxicol* 47:513–539.

van Welie RT, van Dijck RG, Vermeulen NP, van Sitter NJ. 1992. Mercapturic acids, protein adducts, and DNA adducts as biomarkers of electrophilic chemicals. *Crit Rev Toxicol* 22:271–306.

Walsh JS, Reese MJ, Thurmond LM. 2002. The metabolic activation of abacavir by human liver cytosol and expressed human alcohol dehydrogenase isozymes. *Chem Biol Interact* 142:135–154.

Welch KD, Wen B, Goodlett DR, Yi EC, Lee H, Reilly TP, Nelson SD, Pohl LR. 2005. Proteomic identification of potential susceptibility factors in drug-induced liver disease. *Chem Res Toxicol* 18:924–933.

Wen B, Fitch WL. 2009. Screening and characterization of reactive metabolites using glutathione ethyl ester in combination with Q-trap mass spectrometry. *J Mass Spectrom* 44:90–100.

Wen B, Ma L, Nelson SD, Zhu M. 2008. High-throughput screening and characterization of reactive metabolites using polarity switching of hybrid triple quadrupole linear ion trap mass spectrometry. *Anal Chem* 80:1788–1799.

Wong HL, Liebler DC. 2008. Mitochondrial protein targets of thiol-reactive electrophiles. *Chem Res Toxicol* 21:96–804.

Yan Z, Caldwell GW. 2004. Stable-isotope trapping and high-throughput screenings of reactive metabolites using the isotope MS signature. *Anal Chem* 76:6835–6847.

Yan Z, Li J, Huebert N, Caldwell GW, Du Y, Zhong H. 2005. Detection of a novel reactive metabolite of diclofenac: Evidence for CYP2C9-mediated bioactivation via arene oxides. *Drug Metab Dispos* 33:706–713.

Yan Z, Easterwood LM, Maher N, Torres R, Huebert N, Yost GS. 2007. Metabolism and bioactivation of 3-methylindole by human liver microsomes. *Chem Res Toxicol* 20:140–148.

Yan Z, Caldwell GW, Maher N. 2008. Unbiased high-throughput screening of reactive metabolites on the linear ion trap mass spectrometer using polarity switch and mass tag triggered data-dependent acquisition. *Anal Chem* 80:6410–6422.

Zhang KE, Naue JA, Arison B, Vyas KP. 1996. Microsomal metabolism of the 5-lipoxygenase inhibitor L-739010: Evidence for furan bioactivation. *Chem Res Toxicol* 9:547–554.

Zhang C, Wong S, Delarosa EM, Kenny JR, Halladay JS, Hop CE, Khojasteh-Bakht SC. 2009. Inhibitory properties of trapping agents: Glutathione, potassium cyanide, and methoxylamine, against major human cytochrome p450 isoforms. *Drug Metab Lett* 3:125–129.

Zhao SX, Dalvie DK, Kelly JM, Soglia JR, Frederick KS, Smith EB, Obach RS, Kalgutkar AS. 2007. NADPH-dependent covalent binding of [^{3}H]paroxetine to human liver microsomes and S-9 fractions: Identification of an electrophilic quinone metabolite of paroxetine. *Chem Res Toxicol* 20:1649–1657.

Zheng J, Ma L, Xin B, Olah T, Humphreys WG, Zhu M. 2007. Screening and identification of GSH-trapped reactive metabolites using hybrid triple quadruple linear ion trap mass spectrometry. *Chem Res Toxicol* 20:757–766.

Zhu M, Ma L, Zhang D, Ray K, Zhao W, Humphreys WG, Skiles G, Sanders M, Zhang H. 2006. Detection and characterization of metabolites in biological matrices using mass defect filtering of liquid chromatography/high resolution mass spectrometry data. *Drug Metab Dispos* 34:1722–1733.

Zhu M, Ma L, Zhang H, Humphreys WG. 2007. Detection and structural characterization of glutathione-trapped reactive metabolites using liquid chromatography-high-resolution mass spectrometry and mass defect filtering. *Anal Chem* 79:8333–8341.

CHAPTER 8

SAFETY TESTING OF DRUG METABOLITES: MIST GUIDANCE IMPACT ON THE PRACTICE OF INDUSTRIAL DRUG METABOLISM*

J. GREG SLATTER
Pharmacokinetics and Drug Metabolism, Amgen Inc., Seattle, WA

8.1 INTRODUCTION

> The area of metabolites and safety testing is one frequently shrouded in dogma and almost defying some of the basic principles of ligand interaction and pharmacology.
>
> —Smith and Obach (2009, pp. 267–279)

This chapter discusses the potential impact of the 2008 Center for Drug Research and Evaluation (CDER) Guidance "Safety Testing of Drug Metabolites" (STDM, 2008[1]) on the practice of industrial drug metabolism. At the center of the important problem of how to assess the safety of human drug metabolites are the issues of clinical relevance and development resource allocation. The challenge to an absorption, distribution, metabolism, and excretion (ADME) practitioner is to develop a plan that helps to accurately define the actual contribution of metabolite(s) to patient risk versus benefit. The plan should do so in an unambiguous, cost-effective, and regulatory-compliant fashion. Many aspects of the final guidance are a step forward in defining the case-by-case approach required to address this

*Presented at The 12th Annual Land O'Lakes Conference on Drug Metabolism and Applied Pharmacokinetics, Merrimac WI, September 16, 2009.

[1] Also known as the Metabolites in Safety Testing or "MIST" guidance.

Biotransformation and Metabolite Elucidation of Xenobiotics, Edited by Ala F. Nassar

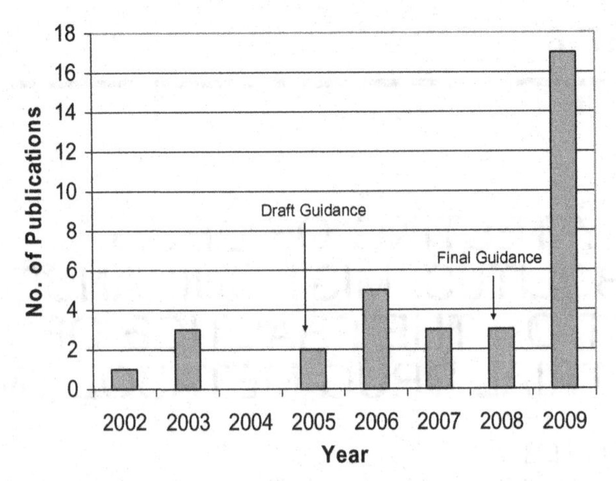

Figure 8.1 Mist-related publication growth.

complex issue and assure patient safety. Other aspects, including relatively minor word changes from the draft guidance, such as the shift from 10% of drug-related materials (DRMs) to 10% of parent drug, have the potential to increase scrutiny of human metabolites that have a negligible chance of contributing to drug toxicity. That is, unless these are treated as case-by-case exceptions, as advocated in the guidance and in a subsequent opinion article (Atrakchi, 2009). Over the next 5–10 years, actual case-by-case regulatory interactions, training of regulatory reviewers in core ADME technologies, and harmonization with MIST-related language in the International Conference on Harmonization (ICH) M3 (R2) and ICH S9 guidance (ICH, 2008, 2009) are likely to define how the 2008 STDM (MIST) guidance is implemented, and how well it serves patients by promoting the development of safe and affordable drug products.

8.2 SCOPE OF THE REVIEW

The chapter encompasses the history of MIST guidance development and implementation from 1999 to July 2009 and will focus on strategic and tactical implications for ADME practitioners. Publication growth related to MIST over this time period is summarized in Fig. 8.1. The chapter will only briefly touch on safety-related strategy or the drug safety activities that would be triggered by the elucidation of a disproportionate metabolite.

8.3 HISTORY OF THE MIST GUIDANCE

8.3.1 From the Beginning to a Draft Guidance

In 1999, technical groups within the industry trade organization Pharmaceutical Research and Manufacturers of America (PhRMA) examined the issue of drug metabolites as potential mediators of drug toxicity and convened a joint FDA/PhRMA workshop in November 2000 to review their findings. A multidisciplinary position paper "Drug Metabolites in Safety Testing" (later MIST) was subsequently published (Baillie et al., 2002). The goal of the workshop and position paper were to define practical and scientifically based approaches to the use of metabolite data to address contemporary issues in the safety evaluation of drug candidates and develop points to consider for ongoing dialogue between industry and regulators. Key definitions of a "major," "important active," or "structural alert" (potentially reactive) metabolite were put forward. These defined metabolites, when quantified across humans and preclinical species, enabled the definition of a "unique" human metabolite. Separate toxicology assessment was considered appropriate in the rare instance when a unique human metabolite was also major, important, and active/reactive (Baillie et al., 2002). In a prescient fashion, Baillie et al. (2002) pointed out that modern liquid chromatography–mass spectrometry (LC-MS) methods now enable the quantitation of extremely low concentrations of drug metabolites, but that "quantifying metabolites that account for only a small percentage of the total administered dose rarely will result in data that can be used in the interpretation of drug safety." As such, industry settled on "in the absence of other scientific considerations a major metabolite... accounts for 25% or more of exposure to circulating drug-related material." A case-by-case approach driven by good scientific judgment was advocated, and many of the elements suggested in the position paper persisted into the draft and final guidance that followed in 2005 (CDER, 2005) and 2008 (CDER, 2008), respectively. The Baillie et al. (2002) paper drew a response from the FDA on the lack of precedent for the 25% value, citing veterinary, EPA, and ICH impurities guidance precedents for more conservative values of 10% of drug residue, 5% of administered compound, or >0.1% of drug substance (Hastings et al., 2003). This was based on concern over potential toxicities of "minor" metabolites such as those of halothane and aflatoxin B1. The CDER scientists came out against an arbitrary quantitative trigger for defining unique human metabolites, favoring instead a case-by-case approach to deciding the nature and extent of any required toxicology and carcinogenicity studies. After an answer from the authors of the MIST document, it was clear that a guidance would be in the works at the FDA (Baillie et al., 2003). Before the draft guidance arrived, Smith and Obach (2005) launched the first pun of the MIST dialogue ("Seeing through the MIST") and parsed some of the complexities of "percentage of drug-related materials" approaches to MIST. They suggested a set of criteria to be used with absolute metabolite abundance as an alternative to the relative "25% of plasma radiochemical AUC" threshold. This was based primarily on the premise that high-dose drugs are most often likely to cause metabolite-derived toxicity. The additional factors were metabolite structure, physical chemistry, and type of toxicity (A, B, C, or D).[2] As experienced ADME practitioners, they were able to highlight the practical limitations of conventional human [^{14}C] ADME studies on which MIST decision making would be predicated.

Human ADME data are typically expressed as "percentage of drug-related materials," but do not usually consider the relevance of overall metabolite abundance to the pharmacology of the administered drug. For example, consider the relative risk of a 10% of dose metabolite from a 1 g dose of acetaminophen (100 mg of metabolite) versus a 0.125 mg dose of triazolam (12.5 µg of metabolite). In the Smith and Obach (2005) algorithms, human metabolite abundance (excreta > 10 mg/day) or overall concentration (plasma>1 µM) drove decision trees that were parsed by structural similarity, pharmacology/toxicology, protein binding, and dose-mass to define when metabolite monitoring might be scientifically justified.

In the end, the draft guidance defined metabolites of interest as "unique" or "major" and set a human threshold of "greater than 10% of drug-related material (administered dose or systemic exposure, whichever is less)" advocating additional attention in compounds with narrow therapeutic index, significant, irreversible, or unmonitorable toxicity or significantly diverse metabolite profiles between humans and nonclinical species (CDER, 2005). In a perspective article, Davis-Bruno and Atrakchi (2006) correctly asserted that unique human metabolites exceeding this threshold would be rare and introduced the concept of "disproportionate" metabolites, where the same metabolite may be present but preclinical exposure is not comparable to human therapeutic exposure (Davis-Bruno and Atrakchi, 2006; Huang, 2007). Overall, the draft guidance acknowledged the complexities of defining metabolites as unique, major, or disproportionate, and advocated acquiring necessary species comparison data early in development, with a case-by-case assessment facilitated by early consultation with the FDA.

8.3.2 Draft Guidance to Final Guidance: 2005–2008

The publication of the draft guidance (CDER, 2005) triggered a wave of industry and academic responses that highlighted the complexity of the issue and diversity of opinion (Humphreys and Unger, 2006; Prueksaritanont et al.,

[2] Metabolite-derived toxicity was parsed as dose-dependent and pharmacologically mediated through on- or off-target activity at a receptor or enzyme (types A1 and A2, respectively), versus type B (idiosyncratic) toxicity that is not dose-dependent and possibly related to metabolite reactivity, versus type C, toxicity related to drug or metabolite reactivity (acetaminophen), or type D, a delayed type B or C toxicity such as carcinogenesis (Park et al., 1998).

2006; Smith and Obach, 2006; Naito et al., 2007). This period of MIST history was summarized by Luffer-Atlas (2008) and Guengerich (2006). The key issues and solutions raised prior to the final guidance are summarized as follows:

Contribution of Stable Metabolites to Drug Toxicity Humphreys and Unger (2006) pointed out that examples of off-target adverse events not encountered for the parent drug and mediated by stable (as opposed to reactive) drug metabolites are very few. Chemotype similarity of oxidized metabolites to parent drug makes similar on-target and off-target effects more likely. Similar off-target effects could be mediated by crosstalk among similar receptors or enzyme families (such as in the CNS or kinome). Only rarely were there literature examples of distinct off-target toxicity, usually arising after significant metabolite structural change, as in hydrolytic cleavage or conjugation. Off-target toxicities included phospholipidosis, bile salt export pump (BSEP) inhibition, CYP2C inhibition, and solubility-related kidney effects. They also estimated that the synthesis and assay validation costs for a metabolite were $100,000 (2006 U.S. dollars) and that analytical compromises in parent drug dynamic range and sensitivity are trade-offs for the analysis of parent and multiple metabolites. They concluded with resource-sparing arguments for a flexible, tiered approach to stable metabolite characterization across species, based on the low likelihood of a significant departure from parent pharmacology and generally lower exposure relative to parent.

Role of Reactive Metabolites and Dose-Mass in Drug Toxicity Reactive metabolites are difficult or impossible to measure directly and have had a long-standing "guilt by association" link to adverse events. High dose-mass is a common theme among drugs that generate reactive metabolites and also have idiosyncratic or exposure-related human toxicities. In a sequel to "Seeing through the MIST," Smith and Obach (2006) revisited the percentage-versus-abundance issue with a paper that, surprisingly, had no MIST-related puns, although the reasons for this MIST opportunity remain a MISTery. Human and preclinical radiolabel mass balance studies were highlighted as the traditional and definitive source for cross-species metabolite comparisons that drive subsequent decisions on metabolite assessment. In a retrospective assessment of 24 drugs that were withdrawn from the market due to safety problems, 14 were deemed to be mechanistically related to metabolites, with 11 of 14 presumed to relate to chemical reactivity. In the opinion of the authors, of the 14 metabolism-related drug withdrawals from the market, only one case (flosequinan) would have benefited from risk assessment of the metabolite. Bromfenac, nefazodone, tolcapone, and trovafloxacin might have possibly benefited. They concluded that these market withdrawals would not have manifested as failure of the toxicology species to generate human metabolites. Specific caveats related to abundance (dose-mass), chemical structure, toxicity mechanism, absolute plasma concentrations, and absolute mass of metabolites recovered in excreta were proposed as the case-by-case factors that might identify the rare disproportionate metabolite that is actually worthy of some level of toxicological scrutiny.

Japanese Industry Perspective Naito et al. (2007) reviewed the Japanese Pharmaceutical Manufacturers' (JPMA) perspective on the U.S. FDA draft MIST guidance. They said that no guidance exists in Japan and proposed a case-by-case approach. A survey of 123 Japanese new drug approvals in the period 2000–2006, discerned that 48 (39%) had investigated metabolite safety. Of these 48 metabolite investigations, 98% were single dose administrations, 15% had repeated dose administrations, and 31% had genotoxicity assessment. The reasons given for the metabolite investigations were most commonly that it was a major and/or unique human metabolite or had a genotoxicity structure alert. In no instance was the metabolite toxicity greater than the parent drug. The 10% of DRM (or radioactivity) threshold in the draft guidance was especially problematic to the JPMA, since human ADME studies are needed to define the metabolite exposure and, although common, these studies are not required in Japan.

8.3.3 After the Final Guidance (February 2008)

Positive Aspects of the MIST Guidance Despite controversy over what is rational and necessary to discern a disproportionate metabolite and when to test metabolite safety, there are many elements of the guidance that reflect the shared mission of industry and regulators to protect patients. Notable examples are recommendation for case-by-case assessment and early FDA consultation; flexibility for drug candidates addressing serious or life-threatening illness or unmet medical need; need for adequate exposure in only one animal test species; exclusions for Phase II conjugates; and recommendation to accrue cross-species metabolism data as early as possible.

Other Aspects of the Final Guidance The final guidance MIST paradigm is shown in Fig. 8.2 along with modifications to illustrate the workstream codependencies that arise when metabolite quantitation activities are conducted. The low "10% of parent" threshold for metabolite selection will activate diverse groups across preclinical and clinical development to quantify metabolites in plasma and determine whether metabolite safety studies are needed.

Extensive Parent Metabolism and the "10% of Parent" Threshold "Human metabolites that can raise a safety concern are those formed at greater than 10% of parent drug systemic exposure at steady state (CDER, 2008)." This was the most significant departure from the draft guidance and was a surprise to most stakeholders in the pharmaceutical industry. Normalization of metabolite exposure to parent drug exposure shifted most of the burden of MIST compliance to ADME practitioners because it increased the analytical scru-

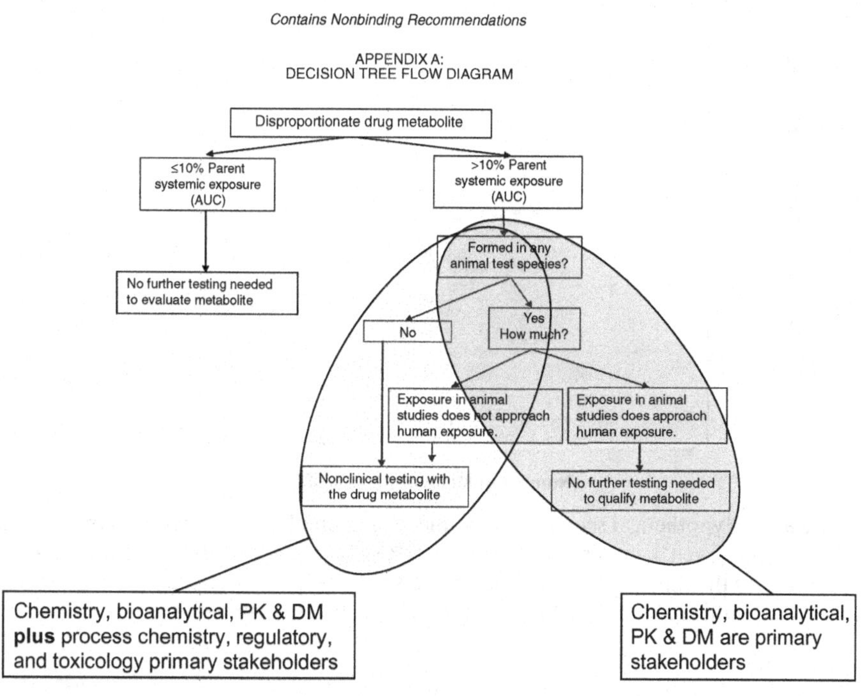

Figure 8.2 Adapted from the MIST Guidance decision tree (CDER, 2008), modified with workload codependencies related to metabolite and internal standard synthesis, bioanalysis, regulatory interactions, and safety studies on a disproportionate metabolite.

tiny needed to rule out metabolites that would never (realistically) contribute to adverse events. When human plasma radiochromatograms reveal any parent metabolism, parent area under the plasma concentration versus time curve (AUC) will fall to zero as the relative abundance and number of metabolites increases. As such, metabolite/parent (M/P) AUC quickly exceeds 10% of parent and can approach infinity (Smith and Obach, 2009). In Fig. 8.3, a hypothetical scenario shows how an increasing number of metabolites that are 9% of DRM each evokes an exponential increase in M/P ratio. The percent of parent threshold therefore becomes increasingly impractical as the extent of metabolism increases. Unless case-by-case approaches advocated in the guidance are endorsed, it has high cost implications related to metabolite synthesis and bioanalytical qualification. Consider the example shown in a recent MIST review of an extensively metabolized Lilly drug candidate (table 1 in Luffer-Atlas, 2008). Under the 10% of DRM conditions in place at the time of writing, metabolites M2, M3, and M4 accounted for 9%, 5%, and 8% of total radioactive materials and were therefore below the threshold for further scrutiny under the draft guidance. When the 10% of parent requirement in the final guidance is applied to the same data, these metabolites come in at 38%, 55%, and 81% of parent drug radiochemical AUC.

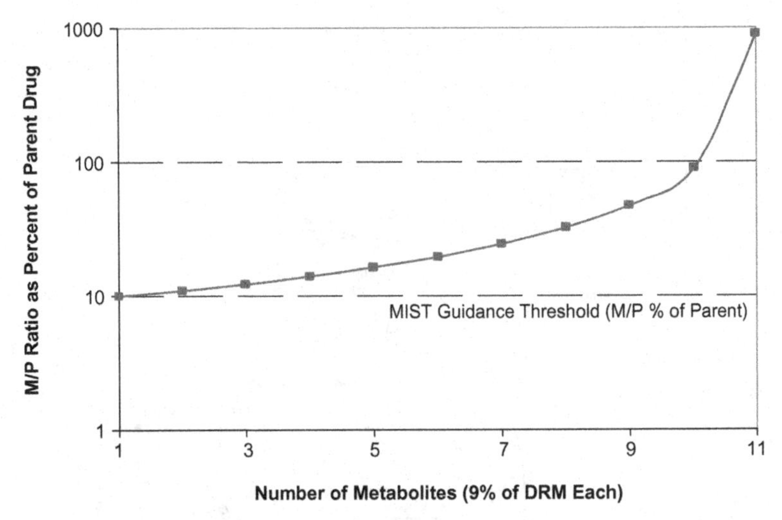

Figure 8.3 Hypothetical scenario for an increasing number of metabolites that are all 9% of DRM. Even 1 metabolite at 9% of DRM exceeds the "10% of parent AUC" threshold. As the number of minor metabolites increases, the parent drug denominator falls away, artificially inflating the apparent abundance of each minor metabolite.

Leclercq et al. (2009) conducted a retrospective analysis on 12 structurally diverse J&J drug candidates. They determined that the number of metabolites requiring additional scrutiny under final guidance language nearly doubled to 28 from 15, although 7 of the additional 13 metabolites were glucuronides. Most metabolites were covered by at least one toxicology species. In two instances, rabbit data were needed to cover two N-glucuronides.

Smith and Obach (2009) surveyed 15 compounds from the literature that had enough public data to discern how the percent of parent threshold affects the number of metabolites that may require analytical scrutiny. They also defined the number of times metabolites could be deemed unique or disproportionate based on animal data. Parent drug comprised an average of 44% (range 0–95% of circulating DRM) with an average of 5.5 circulating metabolites (range 1–15). They concluded, based on the *verbatim* guidance, that one in three new chemical entities will require quantitative bioanalysis of *at least* one metabolite in humans and in animals at steady state (SS). By comparison, under the 10% of DRM threshold, about one in six compounds would have one metabolite requiring scrutiny. It is fair to say that the subtle change in wording in the final guidance has the potential to markedly increase the scrutiny of inconsequential metabolites.

Normalizing to parent drug exposure was rationalized in a post guidance opinion paper as based on the toxicokinetic notion that parent drug is the benchmark for drug exposure assessments (Atrakchi, 2009). Importantly, although rationalizing the choice of parent drug as benchmark, it was acknowledged that "There are, however, circumstances when such comparison is inappropriate and other metrics should be used... A case-by-case approach should be exercised when information provided in the guidance does not apply" ... "in these cases it is more appropriate to make the comparison relative to total exposure or total administered dose" (Atrakchi, 2009, pp. 1218–1219). Based on this opinion, most ADME practitioners would prefer that any metabolite greater than 10% of parent but less than 10% of DRM does not warrant further scrutiny.

Metabolite AUCs versus Standard ADME Development Practices Radiometric quantitation of human radiolabeled metabolites has many analytical caveats but is still the most reliable way to observe exposure to "all" human metabolites. However, conventional human radiolabel studies are rarely done with multiple radiolabel administrations to SS, and also are not routinely done before there is evidence of human pharmacological activity. Even before human ADME data are available, a preliminary call must be made based on *in vitro* metabolism and other preclinical data as to whether metabolite(s) will be disproportionate in humans at SS, how much to accelerate human radiolabel ADME studies, and what metabolites to make and monitor in single ascending dose (SAD) and multiple ascending dose (MAD) first in human (FIH) studies. When the single dose human ADME data become available, the data may still trigger labor-intensive attempts to define clinical and preclinical SS exposures to multiple metabolites.

The SS requirement was rationalized in the post guidance opinion paper as based on the potential for accumulation or differences in protein binding that alter the disposition of metabolite relative to parent. It was acknowledged that "this may be inappropriate and impractical and the case-by-case approach as stated in the guidance should be employed" (Atrakchi, 2009, p. 1219). The opinion paper suggested that modeling of the single dose metabolite data may be an appropriate alternative (Atrakchi, 2009). This is an attractive option since analytical qualification of metabolite accumulation at SS could be limited to cases when the half-life of radioactivity or a metabolite markedly exceeds that of parent drug.

Separate Administration of Metabolites in Safety Studies The physicochemical characteristics of separately dosed metabolites can result in different ADME characteristics, potentially limiting the relevance of toxicological observations in metabolite safety studies (Prueksaritanont et al., 2006; Smith and Obach, 2006; Pang, 2009; Powley et al., 2009). For example, low bioavailability may result in high gut exposure but preclude the attainment of adequate plasma exposure of the preclinical species. A switch to parenteral administration to boost exposure may incur spurious toxicity that is not relevant to an oral drug. An absorbed metabolite may have higher hepatic extraction and biliary excretion than parent drug, may compete for metabolism or transport, or not be subject to renal tubular reabsorption that prolonged exposure to the parent drug. Different protein binding, chemical instability, or different tissue distribution and metabolism complete the list of at least four potential sites (intestine, liver, kidney, and plasma) and four potential mechanisms (ADME) that can act singly or in consort to limit preclinical species exposure to a separately administered metabolite, even at high doses. As such, separate administration of the metabolite does not assure preclinical exposure at SS and may result in spurious toxicity.

Smith and Obach (2009) reviewed species differences in drug metabolism enzymology that potentially create MIST issues and in turn could complicate separate administration of a metabolite.

Pang (2009) used physiologically based pharmacokinetic models of intestine, liver, and kidney to show that the kinetics of a preformed (separately administered) metabolite will not mimic the kinetics, and therefore, the toxicity of the metabolite formed *in vivo* after parent drug administration. Theoretically, a combination of certain conditions may render parent and metabolite disposition similar, for example, when drug and metabolite are not transported, have flow-limited distribution, and metabolism enzymes are accessible in a well-mixed compartment.

Powley et al. (2009) surveyed the pros and cons of options for *in vivo* studies on drug metabolites, and included a new option as an alternative to separate administration of the metabolite. Safety studies could be done on parent drug in mouse strains that are genetically engineered to express the human P450s that are responsible for the disproportionate metabolite. Although this may be expensive and has risks related to a lack of historical toxicology data on the mouse strain, given the cost (estimated at about $1M for a separate toxicology program) and questionable relevance of direct administration of the metabolite to rats, it may warrant consideration as a last resort, although this may not have ever been done before.

8.3.4 Post CDER Final Guidance Discussion

The CDER (2008) final guidance resulted in a burst of papers and presentations (Fig. 8.1). In a timely special issue of *Chemical Research in Toxicology* in Spring 2009 (Guengerich, 2009), the MIST guidance was discussed from a variety of academic and industrial ADME and toxicology perspectives (Anderson et al., 2009; Baillie, 2009; Espina et al., 2009; Guengerich, 2009; Leclercq et al., 2009; Smith and Obach, 2009; Vishwanathan et al., 2009). Similarly, an issue of *Chemico-Biological Interactions* covered metabolism-based toxicity in drug development (Hanzlik et al., 2009; Lauer et al., 2009; Li, 2009a,b; Ma and Zhu, 2009; Pang, 2009; Smith et al., 2009; Vickers, 2009). The position paper discussed elsewhere in this chapter appeared shortly afterward in 2009, rationalizing guidance wording and emphasizing case-by-case approaches to various aspects of the guidance (Atrakchi, 2009).

Retrospectives and Alternate Paradigms After the final guidance (CDER, 2008), the discussion of the importance of dose and metabolite abundance was revived with a literature review relating mechanism of toxicity to whether toxicity is plasma concentration and/or duration dependent (Smith and Obach, 2009; Smith et al., 2009). Class A and class D toxicities are most related to free plasma concentrations of drug or metabolite, whereas class B and class C toxicities (idiosyncratic and reactive metabolite direct effect, respectively) are more likely to be related to metabolites only detectable in excreta as conjugates (except acyl glucuronides). Drug withdrawals from 1980 to the present were classified as A1 (5 examples), A2 (8 examples), and B/C (11 examples). Class B/C withdrawals were generally at a higher dose (100–800 mg) and included six withdrawals for hepatotoxicity. Hence, for Class B/C withdrawals, dose-mass was important, and plasma metabolite data were less likely to be informative. Based on dose–response and causality arguments, these observations were worked into amount-based threshold recommendations for metabolites as follows: A1 > 25% of parent on-target activity; A2 – free plasma concentrations greater than 1 μM; B or C – 10 mg body burden based on the amount of adduct observed in excreta. The authors thought *in vivo* carcinogenicity studies on disproportionate metabolites should not be conducted when treatment duration was not life long, when the metabolite has similar pharmacology to parent or was inactive, and when there was no new genotoxic structure alert relative to parent. *In silico* prediction and *in vitro* pharmacology were suggested as alternatives to *in vivo* testing (Smith and Obach, 2009; Smith et al., 2009).

Isoherranen (2009) used the Metabolism and Transport Drug Interaction Database to determine how often significant cytochrome P450 inhibitors have circulating metabolites that may contribute to cytochrome P450 inhibition. In terms of MIST, drug–drug interactions (DDI) from a metabolite would constitute an off-target effect, regardless of whether or not the parent was an inhibitor or not. Eighty two percent of 129 P450 inhibitors had confirmed circulating metabolites and an additional 11% were likely to have them, based on extensive metabolism. Only 7% did not, and these were drugs that were mainly cleared intact via the kidneys. Only 34 metabolites had been investigated for DDI *in vitro*. As such, metabolites may be confounders of *in vitro–in vivo* correlations or be DDI perpetrators themselves. So when plasma metabolite concentrations warrant and standards are available, *in vitro* assessment of potential for DDI is relatively simple and may be useful, regardless of whether the metabolite is disproportionate.

Anderson et al. (2009) reviewed 12 Lilly molecules to determine how well *in vitro* data predicted major plasma metabolites. *In vitro* data were predictive 41% of the time, underpredicted 35% of the time, and overpredicted 24% of the time. Not surprisingly, multistep metabolism, transformations that are species-specific (e.g., N-glucuronidation), and disproportionate metabolites decreased the likelihood of prediction. Although there is obvious value in early *in vitro* and *in vivo* metabolism studies, human plasma metabolite profiles are difficult to predict, and the definitive answer comes from the human radiolabel ADME study.

Similarly, Dalvie et al. (2009) conducted a retrospective analysis on 48 compounds to determine whether pooled human liver microsomes, liver S9 fraction, and hepatocytes adequately predicted *in vivo* metabolite profiles. Seventy percent success was obtained in predicting primary metabolites; however, secondary metabolites were less reliably predicted. Even with relatively good prediction, the importance of metabolite profiling of plasma samples from SAD and MAD and human ADME studies was emphasized.

8.3.5 Emerging Technologies and Approaches

Reactive Metabolites Reactive metabolites are excluded from consideration in the guidance as they are not consistently observed in plasma and are, in most cases, only measured as excreted adducts. In addition, the consequences of protein adduct formation are not well understood and differ from drug to drug. Companies will continue to avoid developing low potency, high-dose drug candidates with high covalent binding and chemical motifs commonly associated with genotoxicity (Baillie, 2009). The guidance does focus on acyl glucuronides, possibly because some are reactive or subject to acyl migration; most are measurable in plasma; and some may contribute to higher parent drug exposure by enterohepatic recycling of the parent aglycone.

New Approaches to Nonradioactive Metabolite Quantitation Interest of bioanalysts in MIST increased markedly after the final guidance because of bioanalysis timing issues and the need for early semiquantitative methods for plasma metabolite comparisons prior to the availability of synthetic standards, stable-labeled internal standards, or radiolabel. Crystal City III (Viswanathan et al., 2007) and European Bioanalysis Forum (Abbott and Brudny-Kloeppel, 2009) meetings recommended that characterization of major human metabolites should employ a flexible and tiered approach that is comprised of screening, qualified, and then validated assays that defer major bioanalytical resource allocation until later in development. Ultimately, a single validated assay for both parent and the metabolite(s) of interest is useful when human exposure is not adequately covered in preclinical test species, or when metabolite pharmacological activity warrants this level of scrutiny. The single validated assay paradigm assumes that the inclusion of multiple analytes does not compromise validation and application of the assay. With a parent plus metabolite(s) assay available, metabolite exposure can be determined in healthy volunteers early in clinical development. Decisions about the timing of human radiolabel study or when it is appropriate to monitor the metabolite in other clinical or preclinical studies are then handled in a data-driven, case-by-case fashion.

In advance of a synthetic standard, semiquantitative approaches are needed to compare metabolite peaks across species. Approaches include benchmarking mass spectral ionization to mass using UV-diode array detector (DAD), or to a sample containing a known mass of radioactive metabolite, or to an estimated mass of isolated metabolite (Yu et al., 2007; Leclercq et al., 2009).

Leclercq et al. (2009) reviewed recent advances in the separation and identification of metabolites, as well as the development of semiquantitative methods without standards or radioactivity. Ultra performance liquid chromatography combined with high-resolution mass spectrometry, mass defect filtering, and other scanning techniques, when combined with new software-assisted metabolite identification programs, have been used to separate metabolite signals from endogenous interferences. This afforded more detailed structural characterizations of a more complex slate of nonradioactive metabolites than previously possible (Baillie, 2009; Leclercq et al., 2009). Nonetheless, without radioactivity, caveats are always needed, since recovery and stability of metabolites during extraction, concentration, and analysis are never assured, especially with protracted storage.

Vishwanathan et al. (2009) used pooling of plasma samples from human MAD studies to identify major metabolites by LC-MS. When chemical synthesis is not trivial or is not available, biological *in vitro* or *in vivo* synthesis and isolation can be used to quickly generate standards that are purified and quantified by LC/UV/MS and NMR (Espina et al., 2009). The standards are then used to benchmark an LCMS or LCUV-based comparison of pooled plasma AUC across species (Vishwanathan et al., 2009).

Timing of the [^{14}C] Human ADME Study The ADME study provides the definitive answer to what metabolites circulate in humans after a single dose (reviewed by Roffey et al., 2007). Most companies do not conduct conventional human ADME studies until after proof of pharmacological activity in Phase II because of costs related to the risk of candidate attrition, lack of radiodosimetry data from a rat quantitative whole body autoradiography study, and the time required for radiosynthesis and radioformulation. Overall, early *in vitro* and *in vivo* metabolism data can be used to accelerate the human ADME study when a disproportionate metabolite appears to be possible. The strong guidance recommendation to conduct the human *in vivo* metabolism study as early as possible and the requirement to have MIST issues addressed before large-scale trials commence creates a conundrum. For cost reasons, most companies will probably continue to conduct these studies "at risk" after clinical proof of concept, unless *in vitro* data prompt an accelerated approach where accelerator mass spectrometry (AMS) or conventional study designs might be pursued.

AMS AMS is a technology that has the potential to answer, quantitatively and early in development, whether metabolites are disproportionate in humans. Until the final MIST guidance issued, AMS was regarded as expensive and was mainly promoted for microdose bioavailability and exploratory metabolism studies. Recently, AMS vendors such as Xceleron have offered to help companies address the MIST guidance. Clinical and sample analysis strategies that have the potential to reduce the cost of AMS-based bioanalysis and allow human ADME to be done during Phase I are available (Xceleron, 2009). The human [^{14}C] doses typically used (about 200 nCi for AMS vs. 100 μCi) for a conventional study (Dain et al., 1994; Lappin, 2008) are low enough that rat radiodosimetry data are not needed and radioactivity contamination control is relatively easy. Pared-down human ADME study designs in 4–8 healthy volunteers with AMS analysis might enable the guidance recommendation to obtain these data as early as possible in clinical development. Costs depend somewhat on extent of metabolism and ability to separate metabolites by HPLC.

AMS can also be used with nonconventional human ADME study designs that address the MIST guidance steady-state requisites. For example, a single [^{14}C] dose can be given after dosing the cold drug to SS, a design that could be added to a MAD FIH trial. The disadvantage of this design is a change in specific activity of DRMs *in vivo*, such that LCMS and radiometric quantitation may no longer align. Alternately, with such low radiochemical doses, multiple radiochemical administrations could be used to obtain SS data. The risks are an increase in the likelihood of a compounded error in both calculation of administered radioactive dose and recovered radioactivity and uncertainty with respect to how long it takes for metabolites to reach SS.

Plasma pooling strategies, similar to those described by Hop et al. (1998) and Hamilton et al. (1981) can be used to derive total radioactivity, parent drug, and individual metabolite AUCs from pooled serial plasma samples. This significantly decreases the number of samples to analyze, since only one plasma sample per subject undergoes HPLC fractionation for AMS analysis. Paradoxically, the disadvantage of AMS as a tool to elucidate metabolism may be that it is too sensitive. AMS can measure plasma metabolites of potent drugs that are not monitorable by any other means. However, based on miniscule dose-mass and exposure, trace level metabolites have no reasonable possibility of causing adverse events that are distinct from those of parent drug.

Perhaps at the right dose, with good recoveries, efficient extractions (or urinary excretion), and high-quality time-averaged metabolite profiles, there is the possibility of forgoing the definitive human ADME study altogether. This would be a significant cost-saving option that could be enabled with FDA consultation.

ICH M3 (R2): ICH MIST Provisions The ICH guidance on nonclinical safety studies for the conduct of human clinical trials and marketing authorization for pharmaceuticals M3 (Revision 2) Step 4 was issued in June 2009. Step 5 (implementation) occurred in December 2009 (ICH, 2009). This ICH guidance contains new language that addresses the issue of disproportionate metabolites. Pursuant to the points raised in this chapter, it is more aligned with current ADME practice than the final FDA guidance as follows: The disproportionate human metabolite plasma exposure threshold is 10% of total DRMs, and human exposure must be "significantly greater" than the maximum exposure seen in toxicity studies; Metabolite studies should be conducted to support Phase III clinical trials; If the daily administered dose is <10 mg, greater fractions of the DRM are needed to trigger metabolite safety testing; Adducts such as glutathione conjugates do not warrant testing; and finally, nonclinical characterization of metabolites with an identified cause for concern (e.g., a unique human metabolite) should be considered on a case-by-case basis.

ICH S9: MIST for Anticancer Agents The FDA MIST applies to small molecule therapeutics and does not apply to some cancer therapies where a risk–benefit assessment is considered. A more specific guidance for the STDM in cancer therapy was said to be in progress. Related to this, MIST language already exists for anticancer pharmaceuticals in an ICH draft guidance (ICH, 2008). The S9 draft guidance indicates that separate general toxicology evaluation might not be warranted for disproportionate metabolites in patients with late-stage or advanced cancer, where human metabolite safety would have been assessed in Phase I clinical trials; or where parent compound is positive for embryo-fetal toxicity or genotoxicity. The guidance clearly states that unless there is a specific cause for concern, nonclinical testing of the metabolite is not warranted.

8.4 CONCLUSION

The final MIST guidance promotes metabolism and safety investigations to define the potential risk of human exposure to drug metabolites. However, minor changes in wording from the draft guidance, if interpreted *verbatim*, create the potential for a resource burden during early development through Phase II that could increase the scrutiny of metabolites that have a negligible chance of contributing to pharmacology or toxicology. Specifically, the "10% of parent at steady state" concept and lack of caveats for dose-mass increase the number of low concentration metabolites that will require measurement. Single dose metabolite exposure measurements, benchmarked to total radioactivity, with appropriate caveats for radiochemical half-life and dose of parent drug, would focus attention on instances where there is significant human exposure to the metabolite. Some simple changes in wording to harmonize with the ICH M3 (R2) guidance would make the need for separate administration of disproportionate metabolites to toxicology species both rare and more likely to be informative. Alternatively, the case-by-case approach to disproportionate metabolite identification advocated in the guidance should be pursued through frequent consultation with the FDA.

ACKNOWLEDGMENTS

GS acknowledges valuable discussions with the PhRMA Drug Metabolism Technical Group and Amgen PKDM scientists, Dr. Raju Subramanian, Dr. Ben Amore, Mr. Robert Foti, and Dr. Larry Wienkers.

REFERENCES

Abbott RW, Brudny-Kloeppel M. 2009. Conference Report: European Bioanalysis Forum. *Bioanalysis* 1:273–276.

Anderson S, Luffer-Atlas D, Knadler MP. 2009. Predicting circulating human metabolites: How good are we? *Chem Res Toxicol* 22(2):243–256.

Atrakchi A. 2009. Interpretation and considerations on the safety evaluation of human drug metabolites. *Chem Res Toxicol* 22(7):1217–1220.

Baillie TA. 2009. Approaches to the assessment of stable and chemically reactive drug metabolites in early clinical trials. *Chem Res Toxicol* 22(2):263–266.

Baillie TA, Cayen MN, Fouda H, Gerson RJ, Green JD, Grossman SJ, Klunk LJ, LeBlanc B, Perkins DG, Shipley LA. 2002. Drug metabolites in safety testing. *Toxicol Appl Pharmacol* 182(3):188–196.

Baillie TA, Cayen MN, Fouda H, Gerson RJ, Green JD, Grossman SJ, Klunk LJ, LeBlanc B, Perkins DG, Shipley LA. 2003. Reply. *Toxicol Appl Pharmacol* 190: 93–94.

CDER. 2005. Draft guidance for industry: Safety testing of drug metabolites. (Announcement, draft guidance was superseded by the final guidance in February 2008.) http://edocket.access.gpo.gov/2005/pdf/05-11169.pdf.

CDER. 2008. Guidance for industry: Safety testing of drug metabolites. http://www.fda.gov/downloads/Drugs/GuidanceComplianceRegulatoryInformation/Guidances/ucm079266.pdf.

Dain JG, Collins JM, Robinson WT. 1994. A regulatory and industrial perspective of the use of carbon-14 and tritium isotopes in human ADME studies. *Pharm Res* 11(6):925–928.

Dalvie D, Obach RS, Kang P, Prakash C, Loi CM, Hurst S, Nedderman A, Goulet L, Smith E, Bu HZ, Smith DA. 2009. Assessment of three human in vitro systems in the generation of major human excretory and circulating metabolites. *Chem Res Toxicol* 22(2):357–368.

Davis-Bruno KL, Atrakchi A. 2006. A regulatory perspective on issues and approaches in characterizing human metabolites. *Chem Res Toxicol* 19(12):1561–1563.

Espina R, Yu L, Wang J, Tong Z, Vashishtha S, Talaat R, Scatina J, Mutlib A. 2009. Nuclear magnetic resonance spectroscopy as a quantitative tool to determine the concentrations of biologically produced metabolites: Implications in metabolites in safety testing. *Chem Res Toxicol* 22(2):299–310.

Guengerich FP. 2006. Safety assessment of stable drug metabolites. *Chem Res Toxicol* 19(12):1559–1560.

Guengerich FP. 2009. Introduction: Human metabolites in safety testing (MIST) issue. *Chem Res Toxicol* 22(2):237–238.

Hamilton RA, Garnett WR, Kline BJ. 1981. Determination of mean valproic acid serum level by assay of a single pooled sample. *Clin Pharmacol Ther* 29:408–413.

Hanzlik RP, Fang J, Koen YM. 2009. Filling and mining the reactive metabolite target protein database. *Chem-Biol Interact* 179:38–44.

Hastings KL, El-Hage J, Jacobs A, Leighton J, Morse D, Osterberg RE. 2003. Letter to the Editor. *Toxicol Appl Pharmacol* 190:91–92.

Hop CE, Wang Z, Chen Q, Kwei G. 1998. Plasma-pooling methods to increase throughput for in vivo pharmacokinetic screening. *J Pharm Sci* 87(7):901–903.

Huang S-M. 2007. Making a decision on drug metabolism data—Go or no go (Human-specific metabolites—Regulatory perspectives). AAPS-PPDM section round table, San Diego, CA. http://www.fda.gov/cder/drug/drugInteractions/presents/2007-12-11_Huang_20_min_final_BW.pdf. CDER website, AAPS-PPDM Section Roundtable.

Humphreys WG, Unger SE. 2006. Safety assessment of drug metabolites: Characterization of chemically stable metabolites. *Chem Res Toxicol* 19(12):1564–1569.

ICH. 2008. Draft consensus guideline nonclinical evaluation for anticancer pharmaceuticals S9 current Step 2 version dated November 13, 2008. http://www.ich.org/LOB/media/MEDIA4917.pdf.

ICH. 2009. ICH guidance on nonclinical safety studies for the conduct of human clinical trials and marketing authorization for pharmaceuticals M3 (Revision 2) current Step 4 version dated June 11, 2009. http://www.ich.org/LOB/media/MEDIA4744.pdf.

Isoherranen N, Hachad H, Yeung CK, Levy RH. 2009. Qualitative analysis of the role of metabolites in inhibitory drug-drug interactions: Literature evaluation based on the metabolism and transport drug interaction database. *Chem Res Toxicol* 22(2):294–298.

Lappin GSL. 2008. Biomedical accelerator mass spectrometry: Recent applications in metabolism and pharmacokinetics. *Expert Opin Drug Metab Toxicol* 4(8):1021–1033.

Lauer B, Tuschl G, Kling M, Mueller SO. 2009. Species-specific toxicity of diclofenac and troglitazone in primary human and rat hepatocytes. *Chem-Biol Interact* 179:17–24.

Leclercq L, Cuyckens F, Mannens GS, de Vries R, Timmerman P, Evans DC. 2009. Which human metabolites have we MIST? Retrospective analysis, practical aspects, and perspectives for metabolite identification and quantification in pharmaceutical development. *Chem Res Toxicol* 22(2):280–293.

Li AP. 2009a. Overview: Evaluation of metabolism-based drug toxicity in drug development. *Chem-Biol Interact* 179:1–3.

Li AP. 2009b. Metabolism comparative cytotoxicity assay (MCCA) and cytotoxic metabolic pathway identification assay (CMPIA) with cryopreserved human hepatocytes for the evaluation of metabolism-based cytotoxicity in vitro: Proof-of-concept study with aflatoxin B1. *Chem-Biol Interact* 179:4–8.

Luffer-Atlas D. 2008. Unique/major human metabolites: Why, how, and when to test for safety in animals. *Drug Metab Rev* 40(3):447–463.

Ma S, Zhu M. 2009. Recent advances in applications of liquid chromatography-tandem mass spectrometry to the analysis of reactive drug metabolites. *Chem-Biol Interact* 179:25–37.

Naito S, Furuta S, Yoshida T, Kitada M, Fueki O, Unno T, Ohno Y, Onodera H, Kawamura N, Kurokawa M, Sagami F, Shinoda K, Nakazawa T, Yamazaki T. 2007. Current opinion: Safety evaluation of drug metabolites in development of pharmaceuticals. *J Toxicol Sci* 32(4):329–341.

Pang KS. 2009. Safety testing of metabolites: Expectations and outcomes. *Chem-Biol Interact* 179:45–59.

Park BK, Pirmohamed M, Kitteringham NR. 1998. Role of drug disposition in drug hypersensitivity: A chemical, molecular, and clinical perspective. *Chem Res Toxicol* 11(9):969–988.

Powley MW, Frederick CB, Sistare FD, DeGeorge JJ. 2009. Safety assessment of drug metabolites: Implications of regulatory guidance and potential application of genetically engineered mouse models that express human P450s. *Chem Res Toxicol* 22(2):257–262.

Prueksaritanont T, Lin JH, Baillie TA. 2006. Complicating factors in safety testing of drug metabolites: Kinetic differences between generated and preformed metabolites. *Toxicol Appl Pharmacol* 217(2):143–152.

Roffey SJ, Obach RS, Gedge JI, Smith DA. 2007. What is the objective of the mass balance study? A retrospective analysis of data in animal and human excretion studies employing radiolabeled drugs. *Drug Metab Rev* 39(1):17–43.

Smith DA, Obach RS. 2005. Seeing through the mist: Abundance versus percentage. Commentary on metabolites in safety testing. *Drug Metab Disp* 33(10):1409–1417.

Smith DA, Obach RS. 2006. Metabolites and safety: What are the concerns, and how should we address them? *Chem Res Toxicol* 19(12):1570–1579.

Smith DA, Obach RS. 2009. Metabolites in safety testing (MIST): Considerations of mechanisms of toxicity with dose, abundance, and duration of treatment. *Chem Res Toxicol* 22(2):267–279.

Smith DA, Obach RS, Williams DP, Park BK. 2009. Clearing the MIST (metabolites in safety testing) of time: The impact of duration of administration on drug metabolite toxicity. *Chem-Biol Interact* 179:60–67.

Vickers AE. 2009. Tissue slices for the evaluation of metabolism-based toxicity with the example of diclofenac. *Chem-Biol Interact* 179:9–16.

Vishwanathan K, Babalola K, Wang J, Espina R, Yu L, Adedoyin A, Talaat R, Mutlib A, Scatina J. 2009. Obtaining exposures of metabolites in preclinical species through plasma pooling and quantitative NMR: Addressing Metabolites in Safety Testing (MIST) Guidance without using radiolabeled compounds and chemically synthesized metabolite standards. *Chem Res Toxicol* 22(2):311–322.

Viswanathan CT, Bansal S, Booth B, DeStefano AJ, Rose MJ, Sailstad J, Shah VP , Skelly JP, Swann PG, Weiner R. 2007. Workshop/conference report—Quantitative bioanalytical methods validation and implementation: Best practices for chromatographic and ligand binding assays. *The AAPS Journal* 9(1):E30–E43.

Xceleron. 2009. Xceleron launches solution to FDA MIST. http://www.xceleron.com/metadot/index.pl?id=3105&isa=DBRow&op=show&dbview_id=2596.

Yu C, Chen CL, Gorycki FL, Neiss TG. 2007. A rapid method for quantitatively estimating metabolites in human plasma in the absence of synthetic standards using a combination of liquid chromatography/mass spectrometry and radiometric detection. *Rapid Commun Mass Spectrom* 21(4):497–502.

INDEX

Biotransformation and Metabolite Elucidation of Xenobiotics, Edited by Ala F. Nassar

Printed in the United States
By Bookmasters